Pretreatment of Textile Substrates

Handbook of Textile Processors Series

Pretreatment of Textile Substrates

Mathews Kolanjikombil

Senior Textile Processing Technologist

WOODHEAD PUBLISHING INDIA PVT LTD

New Delhi

Published by Woodhead Publishing India Pvt. Ltd.
Woodhead Publishing India Pvt. Ltd.,
303, Vardaan House, 7/28, Ansari Road,
Daryaganj, New Delhi - 110002, India
www.woodheadpublishingindia.com

First published 2019, Woodhead Publishing India Pvt. Ltd.
© Woodhead Publishing India Pvt. Ltd., 2019
Reprint, 2020

Woodhead Publishing India Pvt. Ltd. ISBN: 978-93-85059-42-1
Woodhead Publishing India Pvt. Ltd. e-ISBN: 978-93-85059-99-5

Typeset by Allen Smalley, Chennai

Printed and bound in India by Replika Press Pvt. Ltd.

Dedicated to my beloved son Vinay
and daughter-in-law Preethi

Contents

Preface

This is the second book in the Textile Processors Handbook series. This book deals with the area of the preparation of all the substrates. I tried to explain all the preparation processes of natural, synthetic and their blends. The preparation of the general blends in the market is given, but there can be some other blends which may not have been explained in this book. I have tried to give the guideline recipes for all the processes. These recipes is only for guideline, I would advise to take trials and fix the recipes as per the conditions of work. Any important precautions to be taken is mostly given as notes under the process details.

Bio-preparation has been of more importance lately. I have included a separate chapter giving the details of bio-preparation mainly for natural fibres or blends of natural fibres.

Hope this book will help the processors to have a better insight into preparation of various fibres and their blends. Yarn preparation will be given in a separate book on yarn dyeing.

Bangalore
15.10.2018
Mathews Kolanjikombil

PART I

Pretreatment of cellulosics – singeing & desizing

In textiles, pre-treatment is a processes applied on textile material (viz. fibres, yarn, woven and knit fabrics) to prepare the material for dyeing, printing or finishing. As the grey material will have lot of impurities, these impurities hampering the subsequent processes has to be removed and made absorbent to ease the further processes like bleaching, dyeing, finishing, etc. For all practical purposes pre-treatments are carried out in continuation of dyeing or printing and their equipment is part of the wet processing plant.

The main object of pre-treatment is to impart a uniform and high degree of absorptivity for aqueous liquors with the minimum possible damage to the material. The cotton fabric, for example, after the pre-treatment should become free of all natural impurities like pectin, wax, protein and husks and the sizing chemicals comprising of adhesives and softeners. Besides high and uniform absorptivity, the textile materials should have adequate degree of whiteness so as not to hamper colour and brilliance of the applied colours. Normally achievement of whiteness of about 80% remission (As. C 100% reflectance from barium sulphate) and a D.P. of 1,600–2,000 are aimed at for the cotton goods.

Given below an example of preparation processing (cotton) flow chart (There can be variations).

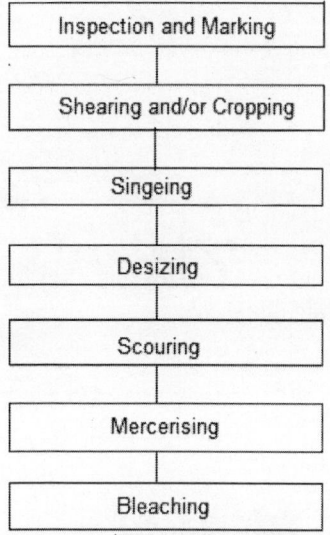

Figure 1.1. Flow Chart

Shearing and singeing are always carried out in open-width but desizing, scouring and bleaching of woven fabric may be done either in rope or in open-width forms by batch, semi-continuous or continuous systems. In the rope form, woven and knit fabrics are scoured and bleached in kiers (autoclaves), winches and jets for batch and in J box or a similar storage system for continuous working which are not so common these days. In open-width processing, jiggers, pad-batch and pad- roll machines are used for batch system and J-box, U-box, normal steamer, pressure steamer, roller bed conveyor steamer or perforated belt steamer are employed for the continuous working. Choice of the equipment depends mainly on the volume of cloth to be processed but the present clay trend is to install the open-width continuous pre-treatment machines.

1.1   Cotton

1.1.1   Composition of cotton

The composition of cotton varies as per the variety of cotton, geographical region, cultivation methods, maturity, harvesting time, etc. However, on an average the composition of the fibre may be considered as follows:

Table 1.1. Composition of cotton fibre

Percentage of	Whole fibre	Primary wall
Cellulose	88-96%	52%
Pectin	0.7–1.2	12
Oils and waxes	0.4–1.0	7
Protein	1.1–1.0	12
Minerals	0.7–1.6	3
Other organic compounds	0.5–1.0	14

The main component is cellulose, but the content of cellulose varies in different the parts of the fibre. Oils and waxes in cotton fibre contain saturated and unsaturated fatty acids, alcohols, etc.

Table 1.2. Composition of cotton wax

Cotton wax	Percentage
Saturated fatty acids	24
Unsaturated fatty acids	1

Cotton wax	Percentage
Alcohols	52
Sterols	10
Hydrocarbons	7
Inert matter	6

The mineral matter contains salts ant it widely varies as per the area where cotton is grown. This necessitates different processes during the preparation. For example, when the cotton contains more of iron process should contain demineralising to remove iron to make the cotton cleaner and whiter. If the cotton contains more of magnesium and calcium chelating agents has to be used to avoid these ions interfering in the dyeing processes. A typical composition of mineral matter in cotton may be:

Table 1.3. Composition of mineral matter in cotton

Mineral matter	Percentage
Sodium carbonate	44.8
Potassium chloride	9.9
Potassium sulphate	9.3
Calcium sulphate	10.6
Magnesium sulphate	4.8
Ferric oxide	3
Aluminium oxide	5

In case of other cellulosic fibres the percentage of impurities is even higher and hence need extensive treatment to remove the impurities.

Table 1.4. Composition of cellulosic fibres

Substances	Cotton	Linen	Hemp	Ramie	Jute
Cellulose (%)	82.7	62	67	68.8	64.4
Hemicellulose (%)	5	17.2	16.1	13.1	12
Pectin (%)	0.7	2.3	0.8	1.9	0.3
Lignin (%)	0	2.5	3.3	0.6	11.8
Soluble Substances (%)	1	3.9	2.1	5.3	1.2
Fats (%)	0.6	1.7	0.7	0.3	0.3

During the scouring process fats get saponified and rendered soluble and the minerals get complexed, by complexing agents. The alkali available in the scouring bath transforms the pectins, and proteins which are readily soluble. Fats are usually in the form of esters such as the following fatty acids and alcohols: Palmitic, Oleic, Tearyl (about 40% on the weight if the waxes) as well as the other lauric, myristic, cetylic etc. types and cholesterol esters, all in the mixed form of mono- di-, and triglycerides.

Depending on the geographic origin the mineral compositions can vary considerably.

Table 1.5. Residual Ash composition of cotton of different geographical region

Cotton from different region	Residual ash %
Dharwar	4.16
Dollerah	6.22
Sea Islands	1.25
Peru	1.68
Bengal	3.98
Broach	3.14
Domrawottee	2.52
Egypt	1.73
Pernambuco	1.6
USA	1.52

Residual ash content gives an idea of mineral contamination in the natural fibre. It is always better to find the ppm (parts per million) content of calcium, magnesium and iron. While calcium and magnesium interference in dyeing process increases the water hardness and the iron can give problems in bleaching like pinholes. The success of a scouring recipe depends on reducing the levels of this mineral contamination.

Table 1.6. Mineral composition of various cotton fibres of different origin

Cotton from different region	Calcium (ppm)	Magnesium (ppm)	Iron (ppm)	Aluminium (ppm)	Manganese (ppm)	Copper (ppm)
Cotton – Turkey – Izmir	910	890	22	15	3	
Cotton – Turkey – Hatay	725	640	24	17	3	

Cotton from different region	Calcium (ppm)	Magnesium (ppm)	Iron (ppm)	Aluminium (ppm)	Manganese (ppm)	Copper (ppm)
Cotton – Turkey – Urfa	6290	1190	63	45	31	
Cotton – Turkey – Tarsus	985	620	29	23	3	
Cotton – China	514	502	68			
Cotton – Russia	1320	567	112			
Cotton – Pakistan	550	480	65			
Cotton – India – Sendhwa	1000	600	125	45	5.9	
Cotton – India – Bailhongal	1030	845	115	64	5.6	
Cotton – India – Pandurna	980	790	475	220	9.9	
Cotton – India – Jetpur	580	585	84	65	3.9	
Cotton – Brazil – Assai – Piranha	3147	1156	680		30	6
Cotton – Brazil – Sao Paulo	845	555	46		11	6
Cotton – Brazil – Paranah	2700	1100	250		30	6
Cotton – Peru	700	440			>1	>1
Cotton – Brazil (GO)	1024	625	238			
Cotton – Brazil (MS)	2425	1085	208			
Cotton – USA – Texas	810	365	75		>1	>1
Cotton – USA – California	600	540	40		>1	>1
Cotton – Egypt – Makko	640	452	11		>1	>1
Linen Normandy	3412	328	92			

Even in the same region there can be different mineral content in the fibre from different areas. Given below the ash content variation in the cotton fibre from Maharashtra and Punjab (India).

Table 1.7. Composition of the Ash Content of Indian Cotton

	Maharashtra	Punjab
	%	%
Ash content of dry fibre	**3.99**	**1.85**
Composition of the ash		
SiO_2	15.56	14.4
Al_2O_3	10.8	12.87
CaO	9.75	10.65
MgO	1.87	4.36
K_2O	27.32	26.03
Na_2O	4.51	8.4
Fe_2O_3	5.89	1.92
SO_3	1.96	2.52
P_2O_5	3.26	4.46
CO_2	12.19	8.03
Cl	6.55	3.84
Others	0.34	2.52

1.1.2   Other impurities added to cotton during weaving or knitting

During manufacturing processes like spinning, warping, weaving, knitting other additives can be added on to the fibre which also has to be removed from the fibre during preparation processes. They are

- Sizes
- Mineral oil from knitting machines
- Waxes
- Rust
- Dirt
- Microbial/fungus spots
- Spinning oils
- Knitting oils.

The process of pre-treatment is more complicated in case of woven fabric both due to the low grade of cotton yarn used and also due to sizes which are added on to the fabric.

Normal pre-treatment procedure in any process will depend on the machinery availability, capacity of machines, the quality of fabrics, and other procedural constraints.

Table 1.8. Detectable waxes and alcohols in cotton

Chemical formula	Systematic name	Trivial name	Melting point °C
$C_{15}H_{31}COOH$	Hexadecanoic acid	Palmitic acid	64
$C_{17}H_{35}COOH$	Octadecanoic acid	Stearic acid	69
$C_{19}H_{39}COOH$	Eicosanoic acid	Arachinic acid	76
$C_{21}H_{43}COOH$	Docosanoic acid	Behenic acid	81
$C_{23}H_{47}COOH$	Tetracosanoic acid	Lignoceric acid	81
$C_{25}H_{51}COOH$	Hexacosanoic acid	Cerotic acid	88
$C_{27}H_{55}COOH$	Octacosanoic acid	Montanic acid	91–93
$C_{29}H_{59}COOH$	Triacontanoic acid	Mellisic acid	92
$C_{31}H_{63}COOH$	Dotriacontanoic acid	Locca acid	
$C_{33}H_{67}COOH$	Tetratriacontanoic acid	Ghedda acid	
$C_{17}H_{35}COOH$	Octadecanoic-9-acid	Oleic acid	13
$C_{19}H_{39}COOH$	Eicosanoic-9-acid	Gadoleic acid	
$C_{24}H_{49}OH$	Tetracosanol	Lignoceryl alcohol	75–77
$C_{26}H_{53}O$	Hexacosanol	Ceryl alcohol	79–81
$C_{28}H_{57}O$	Octacosanol	Montanyl alcohol	83
$C_{30}H61OH$	Triacontanol	Gossypyl alcohol	
$C_{32}H_{65}OH$	Dotriacontanol		
$C_{34}H_{69}OH$	Tetratriacontanol		92
$C_{30}H_{60}(OH)_2$	Tricontandiol	Coceryl alcohol	
$C_3H_5(OH)_3$	Propantriol	Glycerol	18

From the above table, one can see waxes having melting point as low as 64 and as high as 93 is available and alcohols of 18–92. The higher melting point means we have to have higher temperature of treatment as boiling water.

Table 1.9. Composition and removal properties of wax

Component	Content (%)
Wax ester	22
Phytosterols	12–14
Polyterpenes	1–4
Hydrocarbons	7–8
Free wax alcohols	42–46
Saponifiable	36–50
Non-saponifiable	50–63
Inert	0–3

Saponifiable material needs caustic soda and non-saponifiable matter need detergents and emulsifiers to effectively remove impurities.

Given below the certain processing routes followed-up to bleaching in case of woven goods. Selection of process route depends on many things like cotton used, impurities in the cotton, yarn used (carded, combed, twist, count, etc.), size used machinery availability, whiteness requirements, shade to be dyed, customers' requirements, etc.

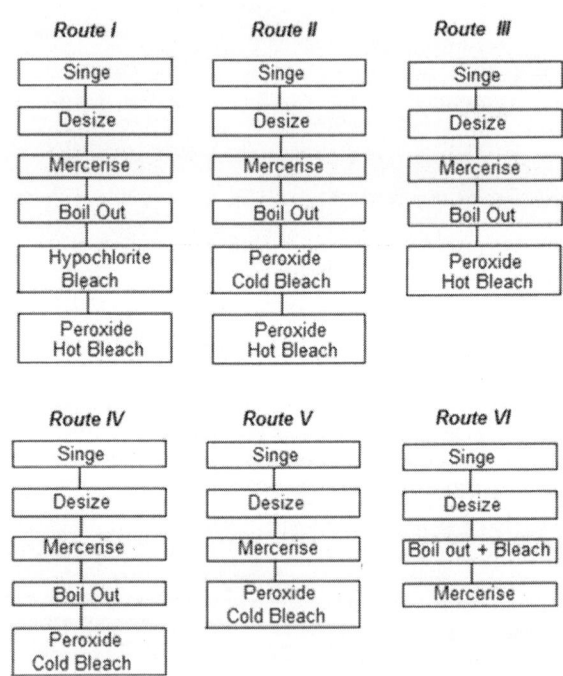

Figure 1.2. Different process routes for preparation

1.2   Singeing

When a yarn is spun from staple fibres it is natural that the ends of the fibre (may or may not) will be projected outside the body of the yarn. In a multi-filament yarn there may be broken out or cut out ends which can be projected outside the yarn even though the number of ends may be very less. Even after weaving/knitting these fibre ends will be shown on the surface of the fabric, which may not give a good appearance to the fabric/yarn. In case of a brushed or napped fabric it may not be a problem. In case of a normal fabric it may hamper the reflection of the shade and will affect the brightness of the fabric. Hence it is necessary to remove these fibres, which is done by either singeing or cropping/shearing.

1. The surface of yarn or fabric appears fuzzy or hairy due to the presence of projected fibres or yarns and gets soiled easily. The smooth lustrous appearance of yarn (e.g., sewing and embroidery threads) and fabric is achieved by singeing.
2. The protruding fibres obstruct subsequent dyeing and printing processes in an uncertain manner. Singeing is almost essential with goods intended for printing.
3. The goods to be mercerised are singed to develop maximum lustre in the fabric.
4. For polyester/cellulosic blended fabrics, singeing is the best method for controlling pilling. Sometimes double singeing is done to minimise pilling.

Singeing is the process of burning-off of the protruding fibres by a suitable method. The natural fabrics are usually singed in grey form as there can be burned ashes and some sort of yellowing on the fabric which are cleared in the subsequent processes. On the other hand, synthetic fibres and their blends are to be singed after dyeing, otherwise molten amorphous beads take up more dyes making dark spots. Polyester melts at higher temperature and requires a powerful flame. There are various methods of singeing:

1. Plate singeing machine
2. Rotary cylinder or roller singeing machine
3. Gas singeing machine.

The singeing is carried out in yarn or fabric form.

1.2.1   Plate singeing

This method of singeing is almost obsolete. In plate singeing the fabric is passed over a curved heated red hot (750°C) copper plate (2.5–5 cm thickness) at high speeds so that the protruding fibres are instantly burned-off. The heating of the plates is done by a suitable burning arrangement of gas mixed with air. The plates are heated to bright redness and the cloth passes over and in contact with these plates at a speed ranging from 150 to 250 yards per minute. The passage of the cloth can be arranged in such a manner that one or both sides of the fabric may pass over and in contact with the heated plate(s), in order to accomplish singeing of one or both sides of the fabric in a single passage. In order to avoid local cooling of a certain part of the plate(s) by constant passage of cloth over it, an automatic traversing mechanism is fitted to the machine. This mechanism brings the cloth into contact with a constantly changing part of the plate(s), not only to avoid local cooling but also local wearing of the plate(s). This method of singeing gives harshness to the fabric and any disturbance gives uneven singeing and hence this method is seldom used now.

1.2.2   Roller singeing

The next development in a machine design to get a more perfect singeing was roller singeing. In this method the singeing surface and fabric is moving to get better even singeing. Here the singeing heated surface is of a rotating copper or cast iron cylinder running very high speed with special traversing arrangements for the fabric to prevent local cooling and wear and tear of the metallic parts and constantly a fresh surface of the roller comes in contact with the cloth.

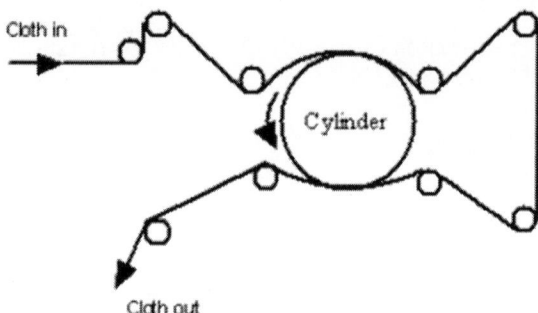

Figure 1.3. Schematic digram of Roller singeing machine

The rotary cylinder has internal firing and revolves slowly. The direction of rotation of the cylinder is opposite to the direction of the fabric so that the protruding fibres or nap of the fabric is raised. This type of machine is particularly suitable for the singeing of velvets and other pile fabrics. If the singeing of both sides of the fabric is required, then two cylinders are employed, one for each side of the fabric. The singeing may be better but the harshness of the fabric was yet a problem.

1.2.3   Gas singeing

From the problems facing with the plate and roller singeing machines it was understood that the fabric should not be touching a hot surface during singeing. Hence the gas singeing machines were designed, where a width wise flame was created and the fabric is passed through the flame where by the fibres are burned-off to get a clean surface. The fabric in open-width is first passed (with the help of a guide roller) through a pair of hot metallic drying cylinders for drying as well as for warming up. The fabric is then passed through a number of weft straighteners or curved rollers to remove weft wise creases from the fabric. The fabric is then subjected to cleaning by means of brushes, which rotate in a direction opposite to the fabric motion. The brushes are placed inside a chamber and an exhaust fan removes the fluff and dust collected by the brushes. Then the fabric passes over gas burners with ceramic nozzles where singeing is carried out. Two or more gas burners are used and the passage of the fabric is so arranged with the help of a large number of guide rollers so that both sides of the fabric can be singed in a single passage.

The same set or two different sets of burners may be used for singeing both sides of the fabric by threading the fabric suitably. It is also possible to adjust the flame height and consequently the heat intensity, by altering the pressure of the gas or air–gas mixture, and is expressed in terms of the height of the manometer in centimetres.

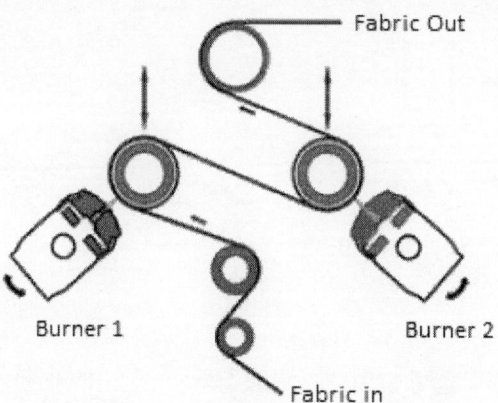

Figure 1.4. Schematic digram of Gas Singeing Machine

After the burning-off operation there are chances that the fabric may carry some sparks/flame on the fabric (especially selvedges of the fabric in case of a split fabric) which can be a danger and burn-off the fabric in due course hence, it has to be completely put off. This is done by passing the fabric through water and a squeezing mangle called quench box. In practice, generally, the quench box is suitably used for impregnating the desizing chemicals, which is the next process and the fabric is batched. In case of any undue happening and the fabric is burned-off there will be provision for the cut-off ends of the fabric with live flame to fall into a pool of water so that the flame is put-off.

The efficiency of singeing depends on the following:

1. Speed of the fabric
2. Singeing position
3. Distance of the flame from the fabric
4. Flame intensity
5. Moisture on the fabric
6. Brushing and raising of the protruding yarn.

With multiple trials the optimum parameters for a particular fabric has to be reached to get the best result. Any faulty singeing can show and confuse for different reasons. For example, an uneven singeing width way can be confused for a dyeing problem like selvedge to selvedge shade variation due to dyeing. A very good singeing is a must for printing goods.

Singeing burner can be adjusted normally adjusted in three different ways as shown in the figure below:

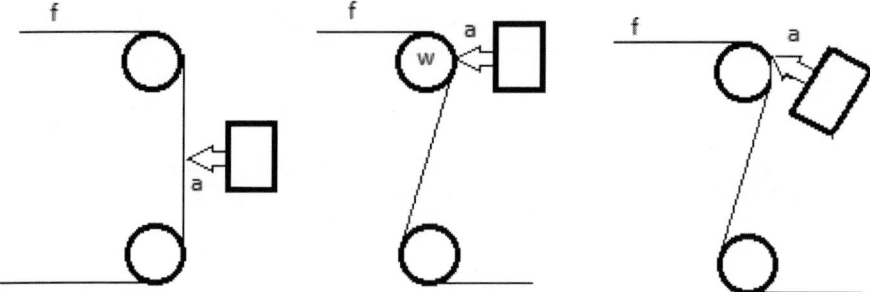

Figure 1.5. Different positions of singeing. f = Fabric, a = Flame, w = water cooled roll

1.2.3.1   Singeing on to the fabric

The burner is arranged vertical to the fabric. The flame impinges on to the fabric and there is possibility of flame entering the interstices of the fabric and, therefore, imparts good singeing. This method is very suitable for fabrics made from natural and man-made cellulosic fibres, and for heavy industrial and technical fabrics.

1.2.3.2   Tangential singeing

The flame impinges on the fabric at an angle. Consequently, only protruding fibres are burnt and the fabric surface is not directly affected. This is ideally suited for lightweight and sensitive fabrics. It is also suitable for synthetic filament fabrics which require mild singeing just to remove tiny broken filament-ends only. Fibres are burnt and the fabric surface is not directly affected. This is ideally suited for lightweight and sensitive fabrics. It is also suitable for synthetic filament fabrics which require mild singeing just to remove tiny broken filament-ends only.

1.2.3.3   Singeing on to the water-cooled roll

Generally recommended for synthetic, blended and open-structured fabrics, in this method the flame is impinges on the fabric when the fabric is moving over a roll. As the flame is against the roll there are chances of roll getting heated and give harshness to the fabric as in the case of roller singeing. Hence the rolls are kept cooled by cold water passing through the hollow roll. An elastic steam-air cushion, formed inside the fabric, resists deeper penetration of the flame and singeing occurs only at the surface.

1.2.4   Important gas singeing parameters

Following are the important gas singeing parameters:
1. Flame intensity
2. Fabric speed
3. Singeing position
4. Distance between flame burner and fabric
5. Flame width.

1.2.5   Essential conditions for good gas singeing

Following are three essential conditions for good singeing:
1. A flame with high mechanical and thermal energy to quickly burn thermoplastic protruding fibres (e.g., polyester) without any molten beads formation
2. A homogeneous flame with uniform mechanical and thermal energy to result in uniform singeing
3. An optimal flame/fabric contact time to neither result in incomplete not over-singeing.

1.2.6   Important care to be taken while singeing

(a) Make sure that the flame is more bluish (less yellowish) to give the maximum temperature.

(b) Control and maintain the recommended flame length and angle of contact, depending on the fabric construction, thickness, weight, heat sensitivity, etc.

(c) Regulate the fabric speed according to the fabric construction/thickness/weight etc.

(d) Make sure that all the burner nozzles are free from choking. Choking of nozzles may result in the appearance of haziness, patchy appearance or faint lines, which become apparent after dyeing.

(e) Make sure that the machine is threaded through the machine correctly. Rubbing marks may show up if the fabric is threaded wrongly over defective stationery bars and/or if the fabric rubs against the burners. (As the guide rollers are very close to the burners, any problems due to loose brackets supporting the rollers will guide the fabric to touch the burner block/nozzle). Invariably such defects only show-up after dyeing or finishing, at which stage it is difficult to attribute the cause to the singeing machine operation.

(f) Make sure that the threading is as per specification drawing so that both sides of the fabric are singed. Singeing on only one side of the fabric may result in face to back shade variation after dyeing.

(g) Make sure that the width of the flame is set to cover just a little more than the fabric width. This will ensure conservation of energy.

(h) Make sure that the exhaust blowers over the burners are in proper operation. If not, it can lead to redeposition of the burnt out fibres on the fabric causing black specks.

(i) Ensure appropriate quenching into water/desize bath after singeing. Otherwise, the entrapped smouldering particles may lead to fabric getting burnt (holes).

(j) Guide rolls next to the flames or the guide rollers on which flame is directed in case of heat-sensitive fabrics should be cooled, generally by cold water circulating through the guide rollers. Otherwise they could become red hot and scorch the fabric.

(k) Interlinking of stop button/flame switch-off mechanism/ quenching system should be effective to avoid burning of the fabric and any incident of fire.

Table 1.10. Recommended settings (Parameters) of singeing for different fabrics

Fabric	GSM	Method of singeing	Fabric/flame distance (mm)	Machine speed (m/min)
Cotton	80–140	Onto the fabric	8	90–110
Cotton/viscose	160–220	Onto the fabric	7	80–90
Cotton/polyester	70–100	Water cooled roller	7	100
P/V suitings	Above 250	Onto the fabric	6	80–90
Lt. Wt. cotton/blend	50–120	Tangential		100
Woollen		Water cooled roller	10	45–70
Corduroy or similar		Water cooled roller	8–10	100
Velour/velvet		Roller singeing	at 700–800°C	

1.2.7   Bio-wash

The singeing effect can be made by a bio-wash process especially for knitted good which is explained in this book elsewhere.

1.2.8   Demineralisation

Until relatively recently incoming dye house water has been considered to be the single most likely source of metal contamination. This is no longer true. Today, there is the distinct possibility of metal contamination in the cotton itself being greater than that found in the hardest of hard waters. Nowadays, it is simply not sufficient to use softened water. The need for effective, and appropriate, sequestration to achieve this goal consistently, and reproducibly, is becoming ever more important. The metal impurities that can most seriously impair "Right First Time" production are:

- Alkaline Earth metals
- Heavy metals

Broadly speaking, the potential dangers are:

- Physical and chemical damage to the cotton during preparation
- Reduced depth of shade (Loss of economy of dye recipe)
- Dulling of shade (Less attractive to customer).

- Unlevel dyeing (Unacceptable to customer)
- Quasi Unlevelness (Complaint from garment converters)
- Poor shade reproducibility (Lower productivity)
- Reduced fastness (Customer complaint).

1.2.9   Potential sources of metallic impurities

- Dye house water
- Greige/prepared cotton mineral matter
- Common salt/Glauber's salt
- Chemical auxiliaries, if manufactured or diluted (or both) in hard water stored / transported in MS containers
- Reactive dyes
- Free metal present in metal complex dyes
- Dye paste
- Finished in hard water before drying
- Tray drying in MS trays, storage in MS drums.

1.3   Hardness in cotton

The total level of impurities in cotton can range from 4% to 12% of total weight of fibre. Actual levels of metallic impurities will inevitably vary. These will vary from one country to another, from one region to another within a given country, from one year to another within a given region (of a given country) or from one harvest to another within a given year (of a given region within a given country).

For the dyer, the elements that pose the greatest threat are alkaline earth and heavy metal contaminants such as calcium, magnesium, manganese, iron and copper. Depending on its origin, raw cotton can exhibit widely different contents of alkaline earth and heavy metal ions. The metal content of cotton grown in different regions is given earlier in this chapter (Refer Composition of Cotton).

1.3.1   Effect of metal ions in preparation

If calcium and magnesium ions are not sequestered, there is the strong possibility of their combining with natural "soaps" which have been generated during the alkaline scouring process to form waxy substances. These have been referred to as "Lime soap deposits". These can deposit not only on the substrate itself, but also on the surfaces of machinery. It is not only the current production batch which is at risk, but also subsequent ones, as these lime soap deposits can dislodge at random and deposit on subsequent batches also.

Uneven deposits of waxes on the cotton carried forward into dyeing can cause problems such as:

• Unevenly prepared surface presented for dyeing
• Rejection of a production batch at the garment stage, even for a production batch which passed initial QC inspection
• Poorly penetrated "streaky" dyeing
• Some dyestuff "swimming" on the surface of the fibre (particularly reactive red dyes) "Cloudy" dyeing.

In addition to the re-deposition of waxy substances, there is a further risk of white deposits of calcium and magnesium hydroxides and carbonates on to the cotton goods which apart from appearance leads to poor absorbency, handle and frictional properties leading to abraded fabric.

These comments apply equally to alkaline scouring and scour/bleaching. There are further dangers which relate specifically to the peroxide bleaching of cotton. There is a need to promote, and ensure, a controlled rate of release of the active "peroxy" radical. This being the transient species promotes the removal of seed husks and natural colour. The presence of iron, copper or manganese (even in trace amounts) can cause catalytic and uncontrolled decomposition of hydrogen peroxide. This can result in:

• Inefficient use of (expensive)peroxide
• Reduced degree of whiteness*
• Loss of tensile strength of the fibre "Pinhole" damage.

As little as 1 ppm of heavy metal can catalyse such decomposition. Metals, however, cannot be adequately removed by conventional alkaline processes since in an alkaline medium, sequestering agents cannot quantitatively separate the minerals of a complex structure containing heavy metals. Moreover, in the alkaline pH region, cellulose swells rapidly and strongly, thus impairing the transport of crystalline minerals from the core to the periphery of the fibre. Demineralisation with organic or inorganic acid is more effective as compared to the alkaline treatment process. However regardless of the efficacy of an acid treatment the use of organic or inorganic acids for the demineralisation of cellulosic fibres involves a number of disadvantages such as corrosion of machine parts, difficulties in handling, and risk of fibre damage with strong inorganic acids. While organic acids give lower demineralisation.

1.3.2   Desizing

Desizing is the removal of sizing products to ensure the success of scour boiling and bleaching and production of faultless dyeing. For an effective desizing operation it is necessary to find out sizing products used.

- Size is generally used to improve the weavability of the yarn.
- In order to avoid abrasion of the weaving instruments (i.e., reed, shuttle) on the warp threads, the warp threats are sized. To keep them smooth softeners and preservatives (to avoid mildew attack at long and damp stocking) are additionally added. It often happens that the weaver makes "after-waxing" directly at the loom in case of bad running properties.
- A distinction is made between water-soluble and water-insoluble sizing agents, mixtures are also applied quite often.
- Desizing can be executed in a discontinuous, half-continuous or continuous process.

Sizing products are generally natural polymers (e.g., Starch) as is or modified to improve certain of its properties with respect to the yarn to be sized (e.g., Carboxymethyl starch (CMS)) as well as synthetic polymers and co-polymers, (e.g., Polyvinyl Alcohol (PVA), Polyacrylate (PAC)) etc. Any one or a mixture of two or more sizing products are used along with waxes, fats and lubricating agents.

1.3.3   Main characteristics of sizes

Sizing compounds and lubricants are applied to yarns before fabric formation to protect the integrity of the yarns. While increasingly faster weaving processes demand more enduring sizes, such as acrylic-based compounds, natural sizes that can be decomposed by enzymes are still on the market.

Sizing materials has to fulfill certain requirements in sizing and subsequent process steps.

1.3.4   Sizing requirements

- Reduction of friction
- Good film formation
- No skin formation
- Water solubility (cold or hot water, hard water)
- Low-costs.

1.3.5   Needs of subsequent processes

- High weaving efficiency
- Stability to electrolytes and alkaline solutions
- Readily removable (desizing)
- Compatibility with other process steps (e.g., singeing/thermo fixation)
- Non-foaming (in the bath or on rollers)
- Ecological compatibility
- Identifiable on the fibre.

The preparation of threads for weaving involves covering the warp threads with a size with specific properties (see table below). A fibre is easier to *weave* if its porosity is reduced: its resistance to rubbing is significantly improved. Sizing therefore improves the weavability of warp threads by reducing thread–thread and thread–metal friction. The efficiency of a size is measured by the weaving yield. For high quality desizing it is essential to know the chemical and physical properties of the sizes in order to apply a process which will ensure their total elimination.

The sizing agents are categorized into two types based on their solubility which is most important for a processor that are as follows:

Water insoluble sizes – Starch (potato, maize, wheat, rice, etc.) and modified starches.

Water-soluble sizes – CMC, PVA, Polyacrylates, Polyester, etc. Water-soluble sizes (PAC, PVA, CMC, polyester) first swell in water and are then removed with the help of surfactants and dispersing agents at high temperature and turbulence.

Following products are common in size for spun yarn for dry weaving

- Starches
- Cellulose derivatives sodium salt of CMC (CarboxyMethyl Cellulose)
- Polyvinyl alcohol
- Acrylic sizes
- Alginates
- And others like gelatine, chalk powder, mutton tallow, waxes, gum, softeners, etc.

It goes without saying that the fabric can contain a single size or a mixture of different sizes depending on the type and structure of the textile substrate.

1.3.6   Starches

Even though synthetic sizing agents are available in the market with many advantages but traditional sizing agents due to its cheapness and free availability is still being used. Such compounds comprise starch and starch derivatives, as well as soluble cellulose derivatives, with waxes often admixed. Cheapest sizing material available in the market (India) was starch which was used almost 100%, along with some gums, gelatine as binder, but this size is used to give harshness hence some softeners, mutton tallow, waxes were introduced to give some lubrication softness to the yarn. However with evolution of modern high speed looms, requirements also changed. Starch rich recipes give less flexibility, abrasion resistance is poor, and hence synthetic products came in to picture, like PVA, acrylic sizes. These products give added advantages of better binding, better abrasion resistance, etc., so that the droppings are less and loom efficiency is better, also the total add-on on the yarn can also be reduced. However the disadvantage is that these products are expensive and cost of size increases. To strike the right balance is important. Thus the optimum recipe is made to overcome the problems at the reasonable cost.

Starch is a polysaccharide and consists of
14–27 % Amylose (water soluble)
73–86 % Amylopectin (water insoluble)

Table 1.11. Components of different starch source

Starch origin	Starch %	Amylose % in starch	Amylopectin % in starch	Decomposition possibilities
Potato	20	23	77	Easy
Manioc/tapioca	25	18	82	Very difficult
Sago	27	26	74	
Wheat/maize	60	25	75	Difficult
Rice	75	19	81	Difficult

Amylose molecule is in the form of helix with six glucose units per turn (Figure below). The low molecular weight of amylose is water soluble straight chain polysaccharides of glucose whereas amylopectin (70–80%) being water insoluble is difficult to remove from cotton due to its higher molecular weight and branches chain.

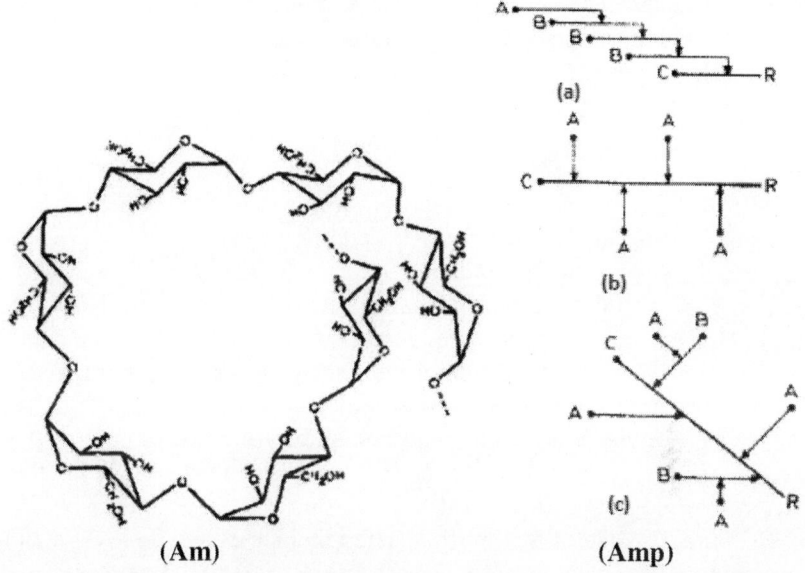

(Am)   (Amp)

Figure 1.6. (Am) – Helical structure of amylose (Amp) – Schematic representation of amylopectin. R–Reducing end, A – Non-reducing end of singly connected chain, B – Non-reducing end of multiply connected chain, C–Non-reducing end of chain which carries a reducing end.

Table 1.12. Characteristics of amylose and amylopectin

	Amylose	Amylopectin
Molecular weight	10000–50,000	50,000–100,000
Solubility	Soluble (colloidal disperse)	Insoluble (can only swell)
Enzymatic decomposition	100%	up to 60%
Chemical structure		

Due to the low water-solubility of starch, it must be decomposed into fragments which dissolve more easily. This can be carried out both with enzymes, mostly *α*-amylases, or oxidising agents, e.g., sodium persulphate or hydrogen peroxide. Depending on the origin of the greige material, different starch size types may have been applied. Their enzymatic decomposition may be very easy.

Starch is the most difficult size to remove. It does not readily dissolve in water and must be broken down chemically into water-soluble compounds by either enzymes, oxidizing agents, or acids. Enzymes breakdown starch into water-soluble sugars and dextrines, oxidizing agents oxidize starch into compounds that are soluble in alkaline solution, while acids hydrolyse starch into water-soluble compounds. Enzymes are specific in their action in that they do not attack cotton, while oxidizing agents and acids can degrade cotton in addition to starch. Starch is therefore usually desized with enzymes. Starch is actually composed of two components, a straight chain polysaccharide of glucose and a branched chain polysaccharide of glucose. Amylose, the straight chain component, is relatively low in molecular weight, water-soluble, and makes up 20–30% of starch. Amylopectin, the branched chain component, is relatively high in molecular weight, water insoluble, and makes up 70–80%of starch. It is therefore the amylopectin component that is difficult to remove in desizing.

Enzymes capable of hydrolysing starch include α- and β-amylase, amyloglucosidase (glucoamylase) and isoamylase. Both α- and β-amylases attack the α-1,4-linkage, but are unable to break the 1,6-branched linkages. They predominantly produce maltose and dextrins as end products. Glucoamylase liberates glucose from non-reducing ends at α-1,4 and 1,6-linkages and generates glucose. Isoamylase is a debranching enzyme, producing mainly maltose.

The desizing recipe must be selected in the corresponding way. This means that easily degradable starch sizes can be removed very well through an enzymatic treatment, whereas some size types which are very hard to remove may be removed more successfully with oxidising agents. An oxidative desizing can also be combined with cold bleach by adding sodium persulphate. We recommend an α-amylase for the enzymatic decomposition of starch since it is highly efficient throughout the whole temperature range.

1.3.7   Modified starches

Other starch related products used in sizing are starch derivatives. To improve the disadvantages of starch as such, changes can be done on the starch molecule. Thus the hydroxyl groups in starch can be functionalised to form acetates, ethers or esters to various degrees of substitution to change the gelation and swelling behaviour. Among others, hydroxyethyl-, methyl- and carboxymethyl-starches are important as thickeners. Corresponding water-soluble cellulose derivatives can also be produced for application in thickener formulations.

1.3.8   Cellulose derivatives

Cellulose is another natural polymer belonging to the polysaccharides group. The chemical formula for cellulose is $(C_6H_{10}O_5)n$. It has a chemical structure very similar to amylose. Instead of α-D 1,4 glycosidic bonds, cellulose has only β-D 1,4 glycosidic bonds as shown below.

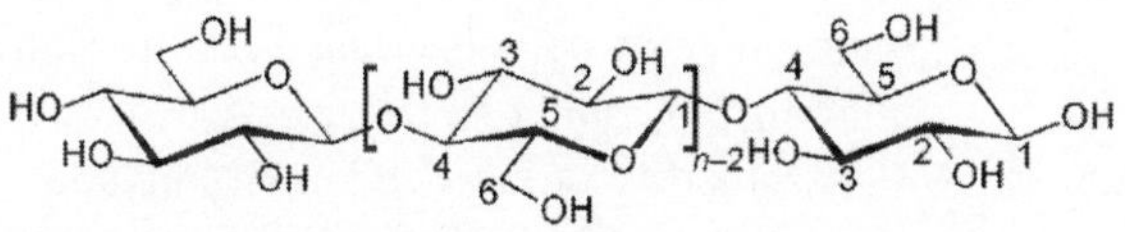

Figure 1.7. Chemical structure of cellulose

Because of these β-D 1,4 glycosidic bonds, the molecular chain of cellulose can extend quite linearly, making it a good fibre-forming polymer. In order to use cellulose for sizing purposes, it should be modified to shorten the molecular structure. Of the two, most used cellulose derivatives are carboxymethyl cellulose (CMC) and hydroxyethyl cellulose (HEC). Carboxymethyl cellulose (CMC) is manufactured from alkali cellulose and sodium chloroacetate. The hydrogen atoms of hydroxyl groups on C2 and C6 are partially substituted with $-CH_2COONa$ or $-CH_2COOH$ depending on reaction conditions. The degree of substitution (DS) is usually between 0.2 and 1.5 (0.2–1.5 carboxymethyl groups ($-CH_2COOH$) per anhydroglucose unit). CMC with DS 1.2 or below is water-soluble. The final product always contains sodium salt. A foaming test can distinguish sodium CMC from other cellulose ethers, alginates and natural gums. Sodium CMC solution, after vigorous agitation, would not produce any foam layer. Uranyl nitrate can be used to detect the existence of CMC. A 4% uranyl nitrate is used to precipitate CMC between pH 3.5 and 4.4. A 0.5% methylene blue methanol solution may also be used to detect CMC on the fabric. After rinsing in distilled water and drying, the methylene blue treated sample may show a blue/purple colour which confirms the existence of CMC or acrylic sizes. A separate extraction of the fabric with toluene can exclude the acrylic sizes from the test.

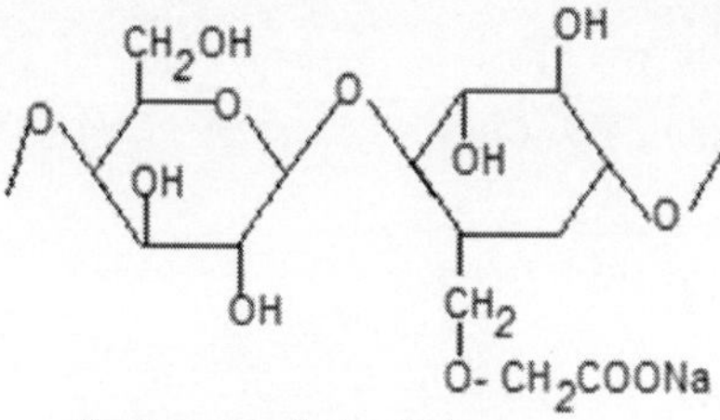

Carboxymethylcellulose

Figure 1.8.

When the hydrogen atom of the hydroxyl group on C6 of cellulose is partially substituted with a hydroxyethyl ($-CH_2CH_2OH$) group in a reaction with ethylene oxide under alkaline condition, hydroxyethyl cellulose (HEC) is produced. So far there are no known testing methods for HEC detection. However, if one wants to distinguish CMC from HEC, an ion tolerance test can be conducted. CMC is anionic and can be precipitated from an aqueous solution with a cationic surfactant. Since HEC is non-ionic, its aqueous solution is compatible with cationic surfactants. Based on the same ionic tolerance principle, a high salt concentration can precipitate CMC, not HEC.

All grades of CMC are readily solubilized in water and can be easily desized with a warm water wash at 65–70°C. CMC increases viscosity of the liquor after removal in the bath, which further stains the fabric. These can be easily removed by enzymatic treatment or by solvent based desizing agents.

1.3.9   Alginates

Alginates are linear co-polymers of randomly arranged β-D 1,4 mannuronic acid (M) and α-L 1,4 guluronic acid (G) blocks as shown below. It's chemical structure is similar to that of cellulose except that it has a carboxylic group on the C5 position instead of a methyl group in the case of cellulose. Alginates have good water solubility. Two- and higher valent metal ions, strong acids and bases can precipitate alginates out of its aqueous solutions. In order to distinguish alginates from other thickening agents, precipitation methods can be tried. About 2.5% of $CaCl_2$ can cause a 0.5% sodium alginate solution to precipitate. Aqueous solutions of gum arabic, sodium carboxymethyl cellulose, carrageenan, gelatin, gum ghatti, karaya gum, carob bean gum, methyl cellulose and tragacanth gum would not be affected. Saturated $(NH_4)_2SO_4$ would not precipitate 0.5% sodium alginate.

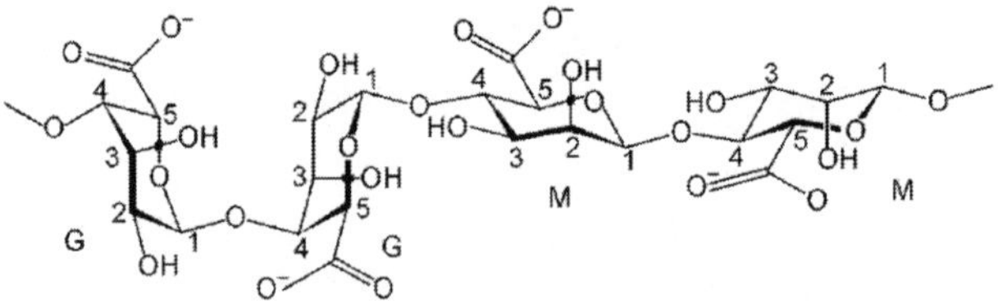

Figure 1.9. *Chemical structure of alginate*

1.3.10   Polyvinyl alcohols

Synthetic polymer sizes such as polyvinyl alcohol (PVA) and carboxymethyl cellulose (CMC) are very popular because in most cases they are very easy to remove compared to starch. Care must be taken in desizing these sizes, because these are available in many grades with varying solubility proper- ties. Polyvinyl alcohol (PVA) is the hydrolysis product of polyvinyl acetate. Depending on the hydrolysis conditions, there are fully hydrolysed PVA and partially hydrolysed PVA, as shown below: PVA, for instance, is available

$$- CH_2 - CH - CH_2 - CH - CH_2 - CH -$$
$$\qquad\quad OH \qquad\quad OH \qquad\quad OH$$

Chemical structure of PVA (Fully hydrolysed)

$$- CH_2 - CH - CH_2 - CH - CH_2 - CH -$$
$$\qquad\quad OH \qquad\quad COCH_3 \quad OH$$

Chemical structure of PVA (Partially hydrolysed)

in grades ranging from fully hydrolysed (FH Grade) to partially acetylated (PA Grade). The FH grade is more difficult to solubilise and thus requires more time. Fully hydrolysed PVA usually has a degree of hydrolysis (DH) of 98–99.8% and can dissolve in water only at >80°C. The solubility of partially hydrolysed PVA with a DH between 85% and 90% is dependent upon its molecular weight. Partially hydrolysed PVA with the higher molecular weight requires a high temperature to dissolve and higher temperature for complete removal. The grade of PVA on sized fabrics can be determined by spotting with an iodine/boric acid solution. Fabric sized with the partially acetylated grade will turn deep blue purple when spotted with the solution while no colouration will result when fabric sized with the fully hydrolysed grade is spotted. In general, PVA is desired with a hot water wash at 80–85°C at a neutral pH. Conditions that can make PVA very difficult to remove in desizing include over drying of the fabric and alkaline pH during hot water desizing. The fabric should not be heat set prior to desizing. PVA gels under alkaline conditions so the wash water should be neutral. PVA often combined with starch size, hence can be removed by enzymatic or oxidative desizing.

Specific detection of PVA on fabrics can be achieved using potassium dichromate ($K_2Cr_2O_7$). Two solutions are used. Solution A consists of 11.88 g $K_2Cr_2O_7$ and 25 ml concentrated H_2SO_4 in 50 ml distilled water. Solution B contains 30 g NaOH in 70 ml distilled water. After solutions A and B are applied to a white fabric sample sequentially, the brown colour developed indicates the existence of PVA. A yellow–green colour can be triggered by unsized goods, potato starch, styrene maleic anhydride co-polymer, alginates, guar, gelatin or CMC.

1.3.11   Polyacrylates

Acrylic is a generic term for a large group of homopolymers and co-polymers derived from acrylic acid, shown below. Since the hydrogen atoms of the carboxylic group and the vinyl group can be substituted by many different chemical groups, a huge variety of polyacrylic acid and polyacrylates is currently available for many different applications. Most of them are used as an emulsion.

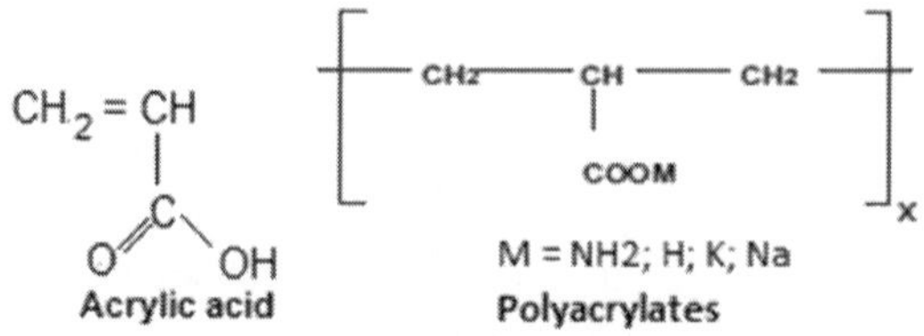

Figure 1.10.

There are several polyacrylates (PAC) prepared from acrylic acids, methacrylic acids or acrylic esters. The most commonly used acrylate salts are ammonium, sodium, potassium. Polyacrylates are strong electrolytes and are therefore readily soluble in a neutral or alkaline medium. The analysis of acrylics is almost impossible without using sophisticated instruments.

1.3.12   Polyester (PES)

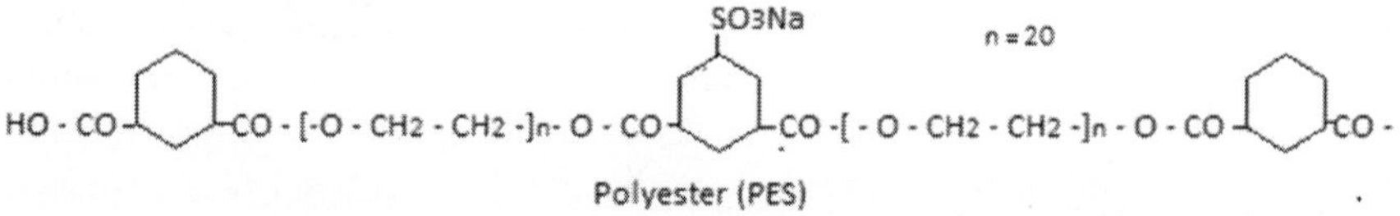

Figure 1.11.

Polyester (PES) is obtained by the polycondensation of polyglycols with aliphatic or aromatic acids containing hydrophilic groups such as sulphate, phosphate, carboxylate, etc. Their main disadvantage is their sensitivity to alkali and electrolytes.

The percentage of size used in the yarn depends on what you are sizing (cotton, blends, PES, etc.) as the requirement of size add-on varies from yarn quality, on which loom, yarn will be used (like, normal shuttle loom, rapier, Sulzer, Air Jet, etc., count of the yarn (finer the count more is the add-on), density of the yarn (ends per inch-epi) – more the dense more is the size add-on.

Thus the recipe is made taking into consideration of all above points and so also if there are other associated problems like droppings, hairiness of the yarn, etc. However in general now the trend is to use optimised starch recipe for normal looms with average speed of less than 200 ppm (approximate recipe is based on 85–90% starch + 10–15% synthetic). As the speed of the loom, fine counts like 60's, 80's, 100's with high epi, synthetic component use is more (approximate recipe is based on 50–80% starch and 20–50% of synthetic component).

1.4   Size possibilities in different fibres

It can be noted that cotton is for example, sized both with natural and synthetic sizes or combinations thereof, whereas pure PA or PES woven are generally only sized with synthetic sizes.

Table 1.13. Size components of different fibre yarns

	Starch and derivative	CMC	PVA	Polyacrylate	Polyester	Co-polymers
Cotton	x	x	x	x		
Viscose	x	x	x	x		
Acetate/triacetate			x			
Polyamide				x		x
Polyester				x	x	x
Polyacrylonitrile	x	x	x	x	x	
Wool		x	x	x		
Silk						x

x = Suitable

Other additives

Waxes are a group of organic compounds consisting mainly of heat-sensitive hydrocarbons which are insoluble in water but soluble in most organic solvents and, most of them are free from glyceride. Sources of waxes can be animal, vegetable, mineral, synthetic and petroleum. The waxes used for fibre finishing are mostly petroleum-based, for example, paraffin wax. They are inexpensive and easily available. However, their performance is not as good as synthetic compounds. If used alone, waxes are very difficult to remove in the textile processes that follow their application and use, causing dyeing and finishing problems. Tests for waxes are mainly for their physical properties, such as melting point, flash point, colour, density, odour and so on. Waxes can be detected by following test. In a small crucible, take one drop of a saturated ethanol solution of hydroxamic acid in a cruscible and mix with one drop of an ethanol solution of the sample and one drop of an ethanol solution of KOH. The mixture is heated to bubbling (evaporation) and then allowed to cool and is acidified with HCl. A drop of $FeCl_3$ is added to the acidified sample. A violet colour confirms the existence of waxes and fatty oils.

Other additives in a size mixture are small amounts of oils, greases, silicones as lubricants and defoamers and anti-microbials.

1.5   Common sizing materials and their detection

In the literature, numerous analysis methods are given for detecting sizes. In practice, quite simple staining and precipitation reactions can be used. However, these staining and precipitation reactions require a certain experience and training of an analyst since they are not always definite. Therefore, several detection reactions must be carried out to achieve a possibly reliable result. The size is basically analysed through a combination of detection reactions directly on the fabric surface and in an aqueous extract.

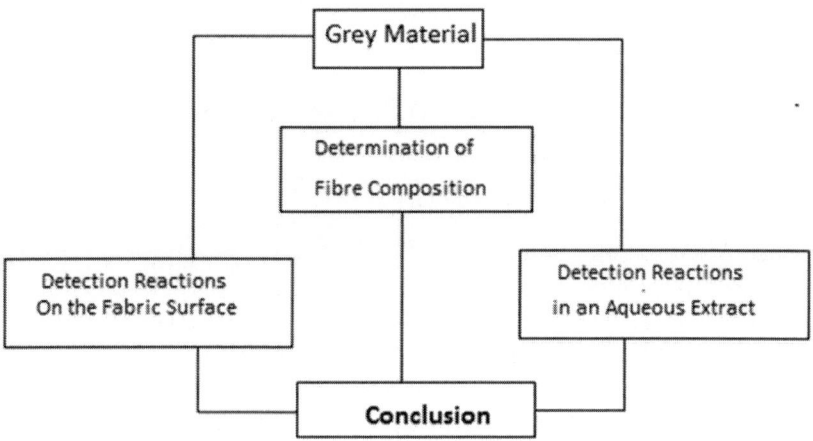

Figure 1.12. Precedure for size Detection

1.5.1   Detection of size on the fabric surface

The fastest method to detect sizes on fabric surfaces is staining of the size with the corresponding colour reactants. However, these tests are only possible if the fibre material itself is not stained by the reactant. In most cases, staining tests only react to certain components or final groups within the size molecule. Some sizes which may seem different have equal or at least similar components so that they can only be distinguished through a combination of different detections. Basic dyestuffs e.g., react with anionic groups in polyacrylate, polyester and polyvinyl acetate sizes. These size types which are mainly applied on PES and PA fibres can only be distinguished through different detection methods.

The detection of sizes directly on the fabric surface is carried out by dripping on different reactants. The detections on the fabric surface described ought to generally serve as preliminary tests for the size detection in the aqueous extract.

Prepare the following solutions:

1. *Basic dyestuff Benzacryl red GRL 100%*

    Size types: polyvinyl acetate, polyacrylate, polyester, polyvinyl alcohol (PVA).

    Basic dyestuffs e.g., Benzacryl red GRL 100% stain synthetic sizes with anionic groups (e.g., polyester, polyacrylate, polyvinyl acetate). The detection reactions with a basic dyestuff ought to be only applied on synthetic fibres since the natural fibre share of certain fibre mixtures (natural/synthetic fibres) may be stained, too. Polyamide fibres are also slightly stained with the basic dyestuff. The colour depth is simultaneously an indication for the add-on. The deeper the colour, the higher the size add-on. However, the different sizes cannot be distinguished through the staining reaction with the dyestuff alone. Therefore, various detection reactions are required to be able to clearly identify a size type.

    Test procedure

    Implementation of the staining with basic dyestuff: About 10 cm² of a fabric sample are wetted with the 0.5% dyestuff solution (for about 5 s), let dwell in the air for 30 s and subsequently rinsed while being stirred in a beaker glass filled with water (liquor ratio 1:1000) for 15 s. Afterwards, the sample is dried in the air and the staining is evaluated.

Polyacrylate   Polyester   Polyvinyl acetate

Figure 1.13. Staining of different fibres

2. *Differentiation between polyester and polyacrylate*

Polyester and polyacrylate sizes are stained with the basic dyestuff to an equal colour depth. Polyester and polyacrylate can be distinguished with the different dissolution behaviour of the dyestuff size salt (staining with the basic dyestuff) in methanol or dichloromethane. The stained polyacrylate size can be completely extracted with cold or 40°C warm methanol by means of two to three treatments, whereas the stained polyester size dissolves only partially. The dissolving behaviour with dichloromethane is just the other way around.

Test procedure

As described above, the fabric samples are stained with the basic dyestuff (staining is carried out with Benzacryl red GRL 100%). Some fibres from the warp yarn are then treated with solvent and the solubility of the size dyestuff salt is evaluated.

Table 1.14. Detection solubility of different sizing agents

Polyacrylate	Soluble in 40°C methanol. Almost insoluble in dichloromethane
Polyester	Soluble in dichloromethane. Insoluble in 40°C methanol

3. *Differentiation between polyvinyl acetate and PYA (delimitation from polyester and acrylate)*

Polyvinyl acetate is stained with basic dyestuffs like polyester and polyacrylate. For delimiting polyvinyl acetate from polyester or polyacrylate a detection with iodine/potassium iodide ought to be carried out. PYA has also a colour reaction with iodine/potassium iodide but can be distinguished from polyvinyl acetate.

After the treatment of the fabric sample with iodine/potassium iodide and a subsequent air drying the fabric sized with polyvinyl acetate remains brown, whereas the fabric sized with PVA turns colourless. This delimitation can only be made if the combination PVA/polyvinyl acetate is not used.

Test procedure

About 10 cm² of a fabric sample are wetted with iodine! Potassium iodide solution (for about 5 s), let dwell in the air for 30 s and then rinsed while being stirred in a beaker glass filled with water (liquor ratio 1:1000) for 15 s. The sample is then dried in the air.

Table 1.15. Staining of PVA and PVAc

Polyvinyl alcohol (PVA)	Brown staining disappears
Polyvinyl acetate	Brown staining remains

4. *Acrylic acid*

Some of the acrylic acid sizes are only slightly stained with the basic dyestuff or not at all and pretend thus a lower size add-on. If the greige material is only slightly stained with the dyestuff, a new sample is treated with 0.1 mol/l caustic soda and the staining with the basic dyestuff is carried out without intermediate rinsing. The sample treated this way is stained more deeply if it is sized with a size based on acrylic acid.

Test procedure

Some acrylic acid sizes are only stained slightly pink with the acetic dyestuff solution (carried out like above). In such a case, proceed as follows:

The fabric sample is wetted for 2 s with a 0.1 mol/l NaOH solution and treated with the dyestuff solution. Continue the procedure as described above (staining carried out with Benzacryl red GRL 100%).

Before Treatment   After Treatment
Figure 1.14. Staining with Acrylic acid

5. *Iodine/potassium iodide*

1.3 g iodine + 2.4 g potassium iodide in 1 l of water. (*Solution 1*)

Starch is a macromolecular polysaccharide.

Starch on the fabric is treated with hot water and this solution drops are added to this will give a bluish violet colouration which identifies starch.

Depending on the add-on starch and starch derivatives are stained to an intense blue to violet with iodine/potassium iodide. High concentrations of PYA could disturb the proof for starch.

Only if starch is absent, does PVA become visible through a brown staining. However, this PVA detection is not always definite, so the detections described below ought to be always carried out.

Test procedure

Iodine/potassium iodide solution is dripped onto the fabric sample.

If starch is absent, the PYA size is stained brownish but this detection is not definite.

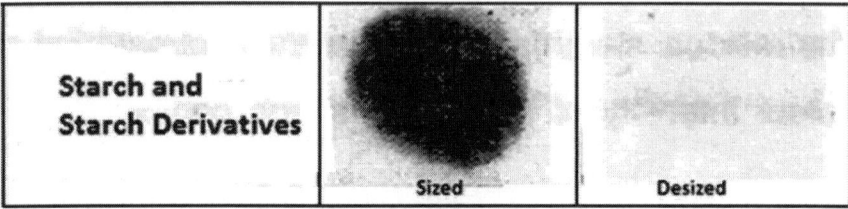

Figure 1.15. Staining with Iodine

6. *Mixed indicator 5 (mixture of methyl red and methylene blue)*

Size types: polyvinyl alcohol (PVA), polyacrylate, polyester, polyvinyl acetate.

Synthetic sizes have acid groups with different strengths and can be neutralised with strong or weak bases. With this indicator, highly acid polyester groups produce a blue stain with an intensely red edge, i.e., the indicator mixture is separated.

An acrylate size does not separate the indicator mixture or if so, only to a slight extent, which can be seen by a brown stain. Polyacrylate or acrylic acid sizes with a stronger acid reaction may have an orange edge which, however, is not as strongly delimited as with polyester sizes. Green stains are produced with polyvinyl acetate sizes. A differentiation between polyvinyl acetate and PVA is not possible with this indicator. In most cases a differentiation cannot be made with size mixtures. The presence of preparations with a different ionic character may impair the specific detection. The mixed indicator 5 ought to be generally only applied on purely synthetic fabrics and should be seen as a pre-trial.

Test procedure

The mixed indicator 5 is dripped onto the fabric sample and the colour reaction is evaluated.

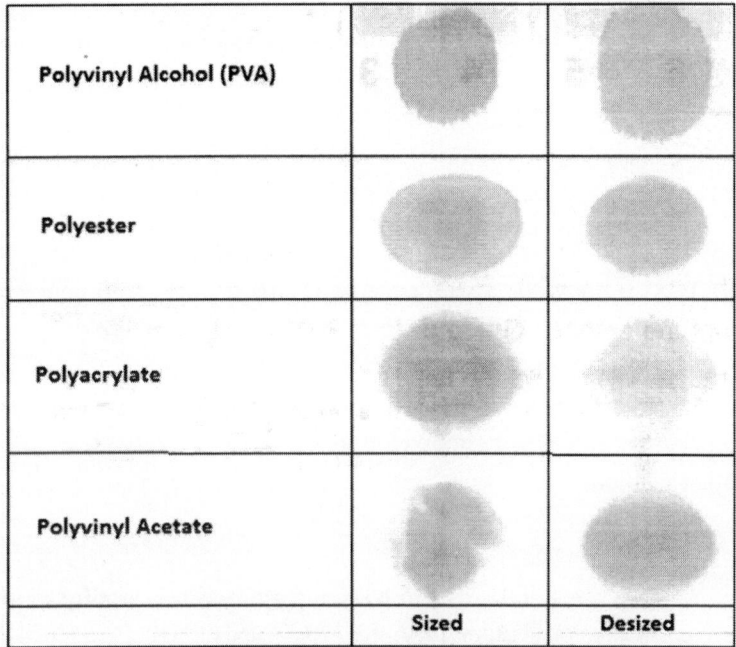

Figure 1.16. Staining with mixed indicator

7. *Chromic acid/caustic soda*

Size types: polyvinyl alcohol (PVA).

PVA size is stained brown with chromic acid/caustic soda. This detection is normally not impaired through the presence of other sizes.

Test procedure

Chromic acid is dripped onto the fabric sample. After 1–2 s two drops of NaOH 50% are dripped onto the dripped area and rubbed with a glass rod.

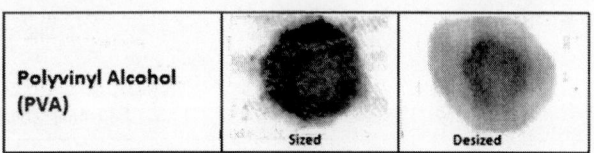

Figure 1.17. Staining with Chromic acid

8. 0.13 g iodine + 2.6 g potassium iodide + 4 g boric acid (*Solution 2*).

PVA or polyvinyl acetate size is stained dark brown to grey–blue with iodine/boric acid.

Test procedure

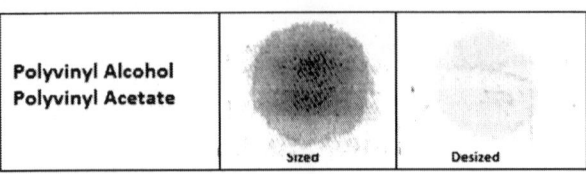

Figure 1.18. Staining with Iodine and KI

9. 11.88 g potassium dichromate in 50 ml water + 25 ml concentrated sulphuric acid (*Solution 3*).

10. 30 g sodium hydroxide in 70 ml water (*Solution 4*).

11. Uranyl nitrate solution 4% (Aqueous solution) (*Solution 5*) CMC s are identifiable with uranyl nitrate or with acridine orange.

Other methods

    a. CarboxyMethyl Cellulose (CMC)

    b. Polyacrylate (PAC)

    PAC s are strong electrolytes and are therefore readily soluble in neutral or alkaline medium. PAC s are directly identifiable with ninhydrin.

    c. Polyester

    Their main disadvantage is their sensitivity to alkali and electrolytes. It is identifiable with Astrazon Red F3BL.

As described in the tables below, various reactants are dripped onto the fabric sample. Then, the colour reactions are evaluated. If the fibre material is known, some detection reactions will not be absolutely required since some size types are hardly applied on certain fibre types (e.g., starch size on a pure polyester fabric).

The following table gives a summary of the sizes and the corresponding detections. A detailed description of the detection reactions and the implementation is given on the following pages.

Table 1.16.

Size type	Reactants	Staining properties
Starch and starch derivatives	Iodine/Potassium iodide	Blue/violet
	Iodine/ Potassium iodide	Brown
	Chromic acid/NaOH	Brown
Polyester (PES)	Mixed indicator 5	Blue with orange edge

Size type	Reactants	Staining properties
Polyvinyl alcohol (PVA)	Iodine/Boric acid	Blue/grey
	Mixed indicator 5	Brown to blue/grey
	Benzacryl Red GRL 100%	Red
Polyvinyl acetate (PVAc)	Mixed indicator 5	Green to brown
	Iodine/boric acid	Blue/grey
	Benzacryl Red GRL 100%	Slightly red
Polyacrylate (PAC)	Mixed indicator 5	Blue with orange edge
	Benzacryl Red GRL 100%	Red
Acrylic acid	Mixed indicator 5	Blue with orange edge
	Benzacryl Red GRL 100%	Red
Polyester (PES)	Mixed indicator 5	Blue with orange edge

1.5.2   Size detections in the aqueous extract

Besides some preliminary samples which are directly carried out on the fabric, the more exact detection of the size on the fabric is carried out in the aqueous extract of the fabric sample. Sizes on the sample are sometimes completely dissolved in the aqueous extract. The reactions with different reactants help distinguish the individual sizes. An important fact is to always take several reactions for a definite detection of a certain size type.

Fibre materials are sometimes not sized with synthetic or natural sizes. In this case, sizing greases are often present on the fibre which can be detected with the Draggendorff reactant.

In general, the individual sizes are preferably detected from the hot water extract. The detection of carboxymethyl cellulose (CMC) is an exception. In the case of cellulose as substrate, the detection of CMC from the hot water extract is always positive. Therefore, it is necessary to use a cold water extract for the CMC detection.

Some sizes, e.g., acrylic acid, are almost insoluble in hot water, so the wrong conclusions may be made. Acrylic acid, however, can be extracted from the fabric in diluted caustic soda, so an alkaline extract ought to be preferably carried out, too. After a subsequent neutralisation the acrylic acid is present as polyacrylate and can be detected as such.

1.5.3   Methods of extraction

Hot water extract

Mix as much fabric as possible with water in a beaker glass and heat it up. After about 15 min filter out and reduce it by boiling. The extract can be used for the individual detections.

Cold water extract (especially for the CMC detection)

In the hot water extract the detection of carboxymethyl cellulose (CMC) is always positive. Therefore, a cold water extract must always be used, too if CMC is assumed to be present. Dip as much fabric as possible into cold water. Filter out after about 15 min and reduce it by boiling. This extract is then available for detecting CMC.

Alkaline extract, e.g., acrylic acid

An alkaline extract is carried out if the proof for polyacrylate was negative in the normal hot water extract and if acrylic acid is assumed to be present.

Mix as much fabric as possible with a solution of 4 g/l NaOH 100% in a beaker glass and heat it up. Filter out after about 15 min and reduce it by boiling. Afterwards, neutralise the alkaline extract with hydrochloric acid exactly to pH 7 (use pH meter) and use it for the individual detection reactions.

1.5.4   Test and observations

After the production of the extracts the individual samples are taken out and mixed with the different reactants as described in the overview.

The following table gives an overview of the individual reactions of the sizes with the detection reactants.

Table 1.17. Reactions with hot water extracts

	Iodine/ Potassium iodide	Borax test	CuSO$_4$	HCl	Molisch reaction
Starch	Blue staining				
CMC			White deposits soluble in HAc		Red/violet ring
PVA	Brown staining if starch is absent	Blue staining if starch is absent			

Polyacrylate	White deposits partially soluble in acetic acid	White deposits soluble in concentrated NH_3
Acrylic acid	White deposits soluble in acetic acid	No deposits
Polyester	White deposits insoluble in acetic acid	White deposits insoluble in concentrated NH_3
Polyvinyl acetate	White deposits soluble in acetic acid	White deposits insoluble in concentrated NH_3

1.5.4.1   Detection of starch, starch ether and polyvinyl alcohol (PVA)

The hot water extract is mixed with iodine/potassium iodide. A deep blue staining indicates starch or starch ether. If starch is absent, the sample will be stained brown with the presence of PVA. For checking PVA the starch must first be enzymatically decomposed before the actual PVA detection can be made.

1.5.4.2   Enzymatic starch decomposition

Treat part of the extract with 2.0 g/l desizing enzyme for 10 min at 98°C.

Differentiation between natural starch and starch ether

Natural starch and starch ether can e.g., be differentiated by means of the different speed of the enzymatic decomposition. The aqueous extract is treated with desizing enzyme as described above and mixed with 1–2 drops of iodine/potassium iodide after about 10 min. If a blue staining occurs with the same intensity as before the enzymatic treatment, it could be a starch ether. Natural starch shows a red–violet 10 brownish staining. A blank trial with natural starch is absolutely recommended. Afterwards, the PVA detection with borax can be carried out.

PVA detection with borax

The extract is mixed with diluted HCl + 0.05 mol/l (= 0.1 N) iodine/potassium iodide and a pinch of borax is added. A blue staining of the borax indicates the presence of PVA.

Detection of polyester, polyacrylate, acrylic acid and CMC

Alkaline extract with acrylic acid size (see above)
Some special acrylates (e.g., acrylic acid) can often not be detected in a simple aqueous extract. Therefore, the detections ought to be also carried out in an alkaline extract.

Detections with copper sulphate solution (CuSO₄)
The hot water extract is mixed with 10% $CuSO_4$ solution. If polyester or polyacrylate or acrylic acid or CMC or a mixture are present, white deposits will be produced. The deposits of acrylic acid or CMC completely dissolve in acetic acid, whereas the deposits of polyacrylate or polyester dissolve only partially.

If CMC is assumed to be present a Molisch reaction in the cold water extract ought to be absolutely carried out.

Detections with hydrochloric acid (HCl)
The hot water extract is mixed with 10% hydrochloric acid. If polyester or polyacrylate or a mixture are present, white deposits or turbidity will be produced. The deposits of polyacrylate completely dissolve in ammonia solution, whereas the deposits of polyester and polyvinyl acetate are insoluble.

Polyvinyl acetate can be detected as PVA after the saponification with caustic soda. The aqueous extract is scoured with caustic soda, neutralised and then the borax test is carried out on PVA.

Detection of CMC in the cold water extract (Molisch reaction)

In the cold water extract, too, CMC produces white deposits with copper sulphate solution. The deposits dissolve in acetic acid. Therefore, the Molisch reaction ought to be carried out for clearly identifying CMC. About 1 ml of the cold aqueous extract is mixed with 1 ml of an about 10% solution of a naphthol in chloroform (always prepare freshly). Carefully undercoat with 1–2 ml concentrated sulphuric acid by slowly adding the latter.

A red–violet ring at the marginal area of both liquids indicates CMC.

Detection of starch, starch ether and PVA

Add iodine/KI to hot water extract (see above) – Stained brown – PVA confirmed.

Borax test – Stained blue – PVA confirmed.

Add iodine/KI to hot water extract – Stained blue – Starch + PVA, starch ether, starch.

Enzymatic starch decomposition if starch detection is positive.

Borax test – Stained blue – (Test positive) – Starch + PVA.

Borax test – Borax remains unstained – (Test negative) – Starch.

Detection of polyester, polyacrylate, acrylic acid and CMC

Add drops of 10% $CuSO_4$ solution to hot water extract – White deposits – Polyester, Polyacrylate, Acrylic acid, CMC.

Add acetic acid to the above precipitate	– Dissolved – CMC, Acrylic acid present
	– Undissolved – Polyacrylate (partially dissolved) or Polyester present
Add 10% HCl to hot water extract	– White deposits – Polyacrylate, polyester, Polyvinyl acetate present
Add ammonia to the above solution	– Dissolved – Polyacrylate
	– Undissolved – PES or PVAc

Detection of CMC in the cold water extract (Molisch reaction)

Cold water extract – Molisch reaction – Red–violet ring – CMC
– Add 10% $CuSO_4$ – White deposits – CMC
– Add acetic acid to above – Soluble – CMC

1.5.4.3   Preparation of chemicals for the above tests

Iodine solution for the qualitative starch detection

For producing a 0.05 mol/l (= 0.1 N) iodine/potassium iodide solution, preferably ready for use solutions (e.g., Fixanol or Titrisol) are used. However, if the solution is to be produced freshly, the following procedure ought to be carried out:

12.8–13.0 g iodine (I_2) are put in a litre flask where 20–25 g iodate-free potassium iodide (KI) were prepared in about 50 ml distilled water. The closed flask is shaken until the iodine has completely dissolved, and then the flask is filled up to the mark with distilled water.

The iodine solution ought to be stored in a brown bottle, closed with a ground-in stopper, in the refrigerator. Due to the volatility of the iodine the bottle ought to be only briefly opened for taking out product.

Iodine solution for determining the residual size amount according to the TEGEWA violet scale

For producing the 0.005 mol/l iodine solution preferably ready for use solutions (e.g., Fixanol or Titrisol) are used with a concentration of c(12) = 0.05 mol/l (= 0.1 N). From this solution 50 ml are taken, diluted with distilled water filled up to 1l.

If the solution is to be produced freshly, the following procedure ought to be carried out:

10 g potassium iodide are dissolved in 100 ml water. Add 0.65 g iodine and shake until it is completely dissolved. Afterwards, fill up to 800 ml with distilled water and fill then up to 1l with ethanol.

Chromic acid

11.8 g potassium dichromate ($K_2Cr_2O_7$) are dissolved in 50 ml hot water while being stirred. After cooling down, 25 ml sulphuric acid concentrated are carefully stirred into it. When sulphuric acid is added to the orange-coloured potassium dichromate solution, chromium trioxide partially detaches in the form of red needles which are, however, not annoying. The cooled down-solution ought to be stored in a brown glass bottle.

Iodine/boric acid

20 ml iodine solution with a concentration of c (I_2) = 0.05 mol/l (= 0.1 N) (e.g., Fixanol or Titrisol) are mixed with 80 ml 1,4 dioxan. Then, 1 g boric acid is dissolved in it while being heated up. The solution is then cooled down to 25°C and filled up to 100 ml with 1,4 dioxan. Add 7 ml water and mix the whole solution thoroughly. The finished solution ought to be stored in a brown glass bottle.

Basic dyestuff

An 0.5% solution of Benzacryl red GRL 100% or a 0.5% acetic solution of Astrazon red F3BL is produced.

α naphthol (Molisch reaction)

A 10% solution of α-naphthol is produced in chloroform. The solution ought to be prepared freshly before every use.

Other chemicals

- Distilled water
- Copper sulphate solution 10%
- Hydrochloric acid solution 10%
- Borax
- Boric acid
- Sulphuric acid conc.
- Caustic soda 0.1 mol/l
- Mixed indicator 5 from Merck

- Ammonia solution 25%
- Desizing enzyme–any α Amylase.

1.6   Cotton desizing process

A desizing recipe can be optimised for a single known size (ideal case) or for a mixture of unknown sizes (frequent case). The golden rule in case of doubt as to the nature of the size is to adapt a desizing recipe capable of eliminating the most difficult size; if necessary, enzymatic desizing for bio-conversion of the insoluble starch into water-soluble glucose.

Different methods of desizing are:

1. Enzymatic desizing
2. Oxidative desizing
3. Acid steeping
4. Rot steeping (use of bacteria)
5. Desizing with hot caustic soda treatment
6. Hot washing with detergents.

Most widely practiced is the first method.

1.6.1   Enzymatic desizing

Table 1.18. Elimination of sizing material by different methods

	Enzymatic elimination		Washing	Sensitivity to pH	
	α-Amylase	Cellulases	>80°C	Alkaline	Acid
Starch	*				
CMS	*		(*)		
CMC		*	*		
PVA			*	*	
PAC			*		*
PES			*	*	

The earlier methods of the desizing involved the treatment of the textile materials using severe and harsh oxidative chemicals like hypochlorites, peroxides and chlorites or treating the fabric with very dilute solutions of mineral acids for a very long time. Both the methods, often, resulted in the damaging the substrate when the attempts were made to remove the sizing ingredients to a higher level. It has been recommended that at least 85% or more of the size should be removed for further processing.

To ensure a high whiteness and/or good dyeability, the size and other applied or natural impurities must be removed in the desizing stage. Removal of hydrophobic parts of size, of the lubricant, is often problematic. They are not removed during desizing but are expected to be solubilised or emulsified during alkaline scouring. However, many times, the removal is not complete.

A large amount of woven cotton is sized with a mixture of starch based polymers and tallow based lubricants. Less than 10% of tallow is saponified in a traditional desizing or scouring process. Besides, use of these agents and the resulting products of process also have their own impact on the effluent generated in the processing. Since starch is the major component, once it is degraded, the size film breaks and the size components are washed away. Drawbacks with acids and oxidising agents are not specific to starch.

In order to overcome many operational, and economic problems associated with the preparatory process, the modern approach to integrate the chemical pre-treatment process to combined desize/scour/bleach operation has, also, been developed in the past and is also based on the non-enzymatic methods. Notwithstanding these developments, the pollution and effluent related problems associated with these processes are very high.

For continuous desizing along with continuous scouring treatment, the slow reacting enzymes was posing a problem. Thermophilic α-amylase enzymes are capable of desizing starch based on size formulation to an adequate degree in as little as 1 min, in a steamer. These are available and used commercially.

There was another problem of removal of other additives in size when lower temperature active enzymes are used. Compared to these, enzyme concentration is not that much significant. High temperature enzymes are apparently more efficient than medium temperature enzymes. Amylases degrade only starch and do not degrade the other additives and impurities present. If the fabric contains only PVA size, enzyme desizing would have minimal effect on its removal and a considerable amount of PVA would pass into the alkali scour.

Tamarind kernal powder in place of starch can be used in the size paste formulation, which is prepared by roasting dehulled seeds of tamarinds. Low viscosity tamarind kernel powder (LTKP) is removed by cellulolytic degradation of tamarind powder. LTKP sized fabric can be desized by cellulase steeping and cellulase padding methods.

Fabrics are initially treated with hot water before soaking it in cellulase-buffer bath at certain temperature. The variables in the cellulase-steeping process such as pH, temperature, time of incubation, enzyme concentration, etc. affect the efficiency of desizing. Padding with cellulose enzymes can be carried out at 50°C, the incubation at 50°C, and finally washed with hot and cold water. The desizing efficiency is increased up to pH 5.6, thereafter it decreases and this confirms the earlier observation of effect of pH on process efficiency. The cellulase action seems to be active in the range of 4.5–5.6 where the removal of LKTP is more than 90%, which levels off after 3 h. Buffered systems yield similar results in lesser duration and steeping method produces better results than padding results. But starch is still the predominant size mainly for economic reasons.

Earlier, malt based enzymes were used, which were inconsistent in their power and unstable at higher temperatures. This made them unsuitable for quicker process. But, with bacterial amylases, the desizing process can be accelerated considerably because of the higher working temperature of the enzymes and their better stability with the size ingredients. Starch and cellulose are chemically closely related but the amylases, because of their specificity, can only attack the linkage in amylase and amylopectin. The enzymes breakdown starch to dextrins and sugars that are more soluble in water and can be washed out of the cloth easier. Liquefaction is the main action responsible for solubility. Since the enzymes are biochemical catalysts, their effect is highly dependent on the chemical and physical conditions to which they are exposed. Enzymes are complex and have high molecular weights. Today enzymes are produced by biotechnological processes in great amount of constant quality, and are therefore applicable to large-scale processes. Advances in the field of genetic engineering allow enzyme manufacturers to design specific enzymes for specific processes (with regard to temperature stability or an optimum pH, for example).

For continuous desizing operations, the use of thermo stable enzymes is highly recommended. The wetting of starch after swelling by hot water is crucial for penetration of the enzyme inside the starch molecules. So the choice of wetting agent is very important. Best choice is non-ionic type. The process of desizing the cotton textiles treated with the starch based sizes, using the enzymes can be divided into three stages.

- Impregnation
- Dwelling
- After wash.

There can be an additional step of impregnation prior to impregnation to remove the water soluble size ingredients. But given the very high hydrophobic nature of the substrate after the sizing, the efficiency of such step remains untouched. Incubation period enables to understand whether hydrolysis or oxidation in the substrate is carried out completely.

Enzymes exercise a selective action over the substrate and convert starch into soluble state. Amylolatic enzymes are capable of breaking starch sizes completely without attacking the cellulose. Malt amylases fall into two categories and are named α (alpha) and β (beta) species which are found to be present in the ratio of 1.5–1.6. α, β amylases cause the rupture of starch macromolecules at 1–4 carbon linkages. α- amylase is of greater importance in desizing, as it rapidly splits up starch molecules into low molecular weight dextrin. β-amylase attacks straight chains, cleaves the units and produces maltose, so that molecular chain of starch is shortened gradually. Malt enzymes have slower action as compared with pancreatic enzymes. When an (α-amylase is applied to a starch solution, it is found that viscosity of the solution decreases rapidly, but for β-amylase the viscosity drops slowly. Thus it is clear that the proportion of (α and β-amylases in a desizing mixture determines the period (time) of effective desizing. It is also clear that the molecular structure of cotton is unaffected by amylases.

Bacterial amylases are more stable than other amylases. α-amylase brings about a random scission of amylase and amylopectin components of starch. Amylase is obtained from pancreas, bacteria and fungi. Differences exist in the action of amylase on starch solutions and films. Starch exerts a protective effect so that the stability of the absorbed enzymes may be appreciable even at boiling point of water. Bacterial amylase exerts highest thermal stability and activity to achieve the most effective size removal.

The stability of the enzyme decreases with the increase in the temperature. Presence of enzyme activators, and sodium chloride, increases the enzyme activity (bacterial amylase) up to 4 g/l, pH 6.5, 60°C; bacterial amylase – 2 g/l results in the efficiency almost 100% for a dwell time of 24 h.

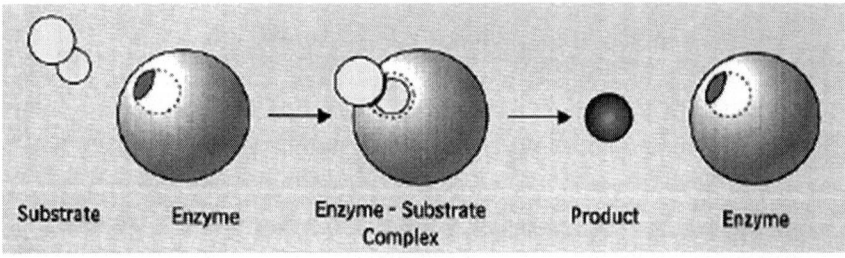

Substrate   Enzyme   Enzyme - Substrate   Product   Enzyme
Complex

Figure 1.19.

In a typical enzyme desizing, the grey fabric is first treated with 0.5 g/l non-ionic wetting agent at 90°C which is known as pre-wash. In the enzymatic scouring operation, lipase and protease enzymes are used. Stock solution of 18% weight by volume solution of sodium chloride is used to reduce the activity to an appropriate level. But at high temperatures, certain metal ions are detrimental to the enzymes, including Cu, Cd, Zn, Hg and oxidising.

1.6.2   Batch wise processes

Desizing is more important for cotton fabrics, but in case of synthetic fabric also we have to remove the additions done during spinning which has to be removed before further processes. These are also termed as desizing.

Desizing is carried out in order to get level dyeing by removing sizing agents added to the fibre in the weaving process. The desizing method to be applied should meet the type of sizing agents used in the process.

Starches are removed with enzymatic type desizing agent or oxidizing type desizing agent of ether persulphate or sodium bromite.

PVA and CMC can be removed in boiling bath, oxidative type desizing agents are applied for productive desizing them acrylic ester based sizing agents are removed by conversion into water soluble sodium acrylate by treatment with hot bath containing alkali such as sodium carbonate or caustic soda. In either case, a surfactant is always added to the bath in order to promote the penetration of the liquor and to prevent the redeposition of the sizing agent. Uniform desizing is important and, for this purpose, it is desireable to desize the material in open-width form on a jig or continuous machine. Knitted materials on the other hand are usually desized on a winch or a liquor flow dyeing machine.

Table 1.19. Best pH range for best removability of sizes

Sizes	pH range of best removability
Polyacrylates/polymethacrylates	10–12
PES–Polymerisates	7.5–9
Modified starches	7.5–12
Polyvnylic alcohols, CMC	6–8

Most of the commercial desizing agents are designed to remove starch, PVA, CMC, that cannot be removed by alkali or hot water treatment. Starch is removed either by aqueous solubilisation through the action of amylase, starch decomposing enzymes, or by decomposition through the action of oxidizing agents.

Desizing agents of the enzymatic type have their own optimum temperature and pH values for the treatment. In general suitable conditions are 40 – 60°C in a neutral bath. The oxidative desizing agents are used in alkaline bath and, in particular, can be applied to short term desizing. As for bromites such as sodium bromite ($NaBrO_2$), it should be borne in mind that the desizing agent of this type has a corrosive action on metal.

Table 1.20. Composition of sizing agents used for cotton and its blends

Fabric →	Cotton		Rayon		PE/cotton blends	PE/rayon blends
Component	%	%	%	%	%	%
PVA	4	1.4		2.8	5.5	5.8
Starch	2	6	7	1	2.5	2.2
CMC					0.2	
Acrylic ester	0.5		0.62	0.5	0.3	0.2
Oil	0.5	0.5	0.5	0.6	0.6	0.7

1.6.3   Guideline recipes

Table 1.21. Batch wise process based on enzymes

Chemicals	Unit	Size used in the fabric			
		Starch/CMC	Polyester	Acrylate	PVA
Wetting agent	g/l	1–3	1–3	1–3	1–3
Seq. agent	ml/l				2-Jan
Solvent	ml/l		1–2	0–2	
Enzyme	g/l	1–3			
Common salt	g/l	0–2			
Soda ash/NaOH	g/l		0-4		
NaOH 100%	g/l			1-4	
Defoamer	g/l	0.1–0.3	0.1–0.3	0.1–0.3	0.1–0.3

Apply at 85–95°C for 30–60 min at a liquor ration of 20:1–5:1 and wash-off hot.

Processes and recipes for enzymatic desizing with amylases
Machines employed
Discontinuous
Jig, Winch, Jet, Overflow, etc.

Continuous
Pad-batch, Cold and Warm

Pad-steam: e.g., Of machinery used	– Combi-steamer, U Box, L Box, J Box, etc. Open-width washing machine: Enzyme shock, Open width Steamer

Time required for desizing depends on the method of application, kind of fabric (tightness of weaving, weight, etc.) and the temperature and concentration of the enzyme. The incubation time in various methods may be as follows:

Table 1.22. Batch processes - Conditions for desizing

Machine	Time required	Temperature °C
Jigs	2–4 ends	60–100
Winch	30 min	90–100
Cold pad–batch	6–24 h	25–40
Hot pad–Batch	3–8 h	60–70
Pad steam/J-box	15–120 s	90–110

Guideline recipes for Jig process

Figure 1.20. Jigger

Table 1.23. Jig Process – Warm and Hot process

Substrate →	Cellulosic and their blends with synthetics	
Machine	Jig	
MLR	2:1–5:1	
Guide recipes	Warm	Hot
Water hardness	10	10
Suitable enzyme ml/l	2–3	

Suitable enzyme ml/l		2–4
Common salt g/l	3	3
Wetting agent g/l	1–2	1–2
Liquor pH	6–6.5	6–6.5

Process: Enter cold, raise temperature to 70°C for warm process, 90–100°C for hot process, Treat for 1–2 h.

Guideline recipes for winch process

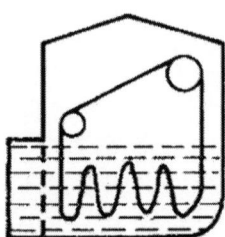

Figure 1.21. Winch

Table 1.24. Winch Process – Warm or Hot

Substrate →	Cellulosic and their blends with synthetics	
Machine	Winch	
MLR	8:1–20:1	
Guide recipes	Warm	Hot
Water hardness	10	10
Suitable enzyme ml/l	2–3	
Suitable enzyme ml/l		2–4
Common salt g/l	1	1
Wetting agent g/l	0.5–1	0.5–1
Liquor pH	6–6.5	6–6.5

Process: Enter cold, raise temperature to 70°C for warm process, 90–100°C for hot process, Treat for 1 h.

Guideline recipes for Jet, overflow, etc.

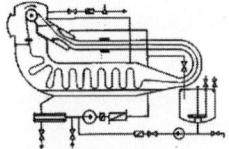

Figure 1.22.

Table 1.25. Jet Process – Warm or Hot

Substrate →	Cellulosic and their blends with synthetics	
Machine	Jet, Overflow etc.	
MLR	6:1–10:1	
Guide recipes	Hot	Hot
Water hardness	10	10
Suitable enzyme (warm) ml/l		
Suitable enzyme (Hot) ml/l	2–4	2–4
Common salt g/l	1	1
Wetting agent g/l	0.5–1	1–2 (low foaming)
Liquor pH	6–6.5	6–6.5

Process: Enter cold, raise temperature to 70°C for Warm processand 90–100°C for hot process, treat for 1 h. Drain hot, Rinse hot.

Guide line recipe for desizing of wovens

Recipe

Quantity	Unit	Bath additions
2.0–5.0	ml/kg	Suitable wetting cum scouring agent
1.0–3.0	ml/kg	Sequestering agent
1.0–2.0	ml/kg	Suitable emulsifier

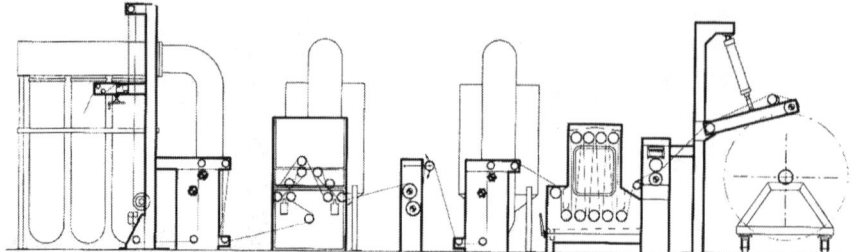

Figure 1.23. Pad batch enzymatic cracking

Process: Impregnate at 60°C, pick-up 100% and batch for 6–24 h.

1.6.4   Semi-continuous and continuous processes

In semi-continuous process a pad roll machine may be used. The impregnated material is steamed through a small steaming unit whereby the fabric temperature is increased and batched in a reaction chamber where the closed chamber is connected to steam. After the chamber is filled with

Figure 1.24. *Pad-roll unit*

material of its capacity and removed and moved to a separate area and allowed to rotate inside the chamber for the required time while steam connection is given to the enclosed chamber to maintain the temperature. After the reaction time is over the material is washed.

In continuous process an impregnator and a steaming chamber is used. The steaming can be done in a regular steamer or J box. A usual steamer for desizing consist of section with tight strand fabric guiding and roller bed.

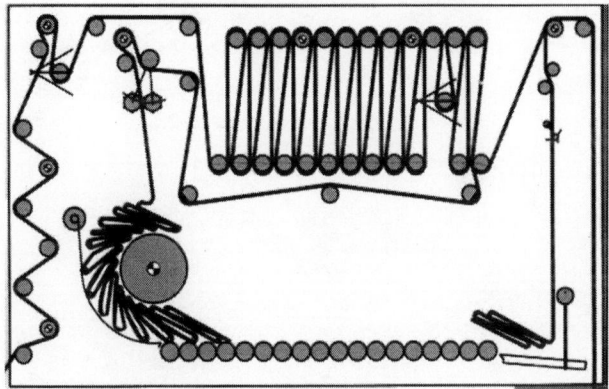

Figure 1.25. Schematic diagram of a steamer

In tight strand section the fabric which has been impregnated with the bleaching liquor is heated-up as per the requirement and on roller bed we maintain sufficient dwelling time for reaction purpose. The dwelling time can be adjusted as per the requirement by controlling the cloth content inside the steamer.

1.6.5   Pad batch or pad-steam methods

1.6.5.1   Pad batch desizing

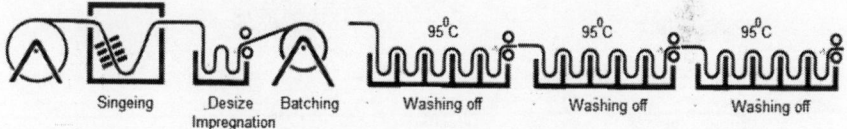

Figure 1.26. Pad batch desizing

Table 1.26. Guideline recipe for pad batch desizing

Chemicals	Unit	Size used in the fabric			
		Starch/CMC	Polyester	Acrylate	PVA
Wetting agent	g/l	2–8	2–8	2–8	2–8
Sequestering agent	ml/l				2–4
Solvent	ml/l		2–4	0–3	
Suitable enzyme	g/l	3–6			
Common salt	g/l	0–5			

Soda ash/NaOH	g/l		0–10		
NaOH 100%	g/l			2–8	
Defoamer	g/l	0.2–0.5	0.2–0.5	0.2–0.5	
Impregnation temperature	°C	50–60	20–30	20–30	20–30
pH (as per enzyme used)		5.5	5.5	5.5	5.5
Liquor pick-up	%	80–100	80–100	80–100	80–100
Reaction time–Pad batch	h	4–16	4–16	4–16	4–16
Reaction time–Pad steam @°C	min	1–3 @98–100°C	1–3 @98–100°C	1–3 @98–100°C	1–3 @98–100°C

Wash -ff as hot as possible.

General temperature used of different processes, when enzymes are selected as per temperature and condition of the machine/method used.

Table 1.27. Temperature of treatment for different desizing process

Machine/method	Temperature, °C
Pad-steam	98–100
Injecta module	90–95
U-box, J-box	85–90
Enzymatic shock desizing	80
Jigger, Winch	75–80
Warm pad-batch	50–70
Cold pad-batch	25–30

Alternative recipes for enzymatic desizing on Jig, Pad-steam, Pad-roll, Pad-batch for cotton, polyester/cotton, polyester/viscose

Enzymatic desizing of starch sizes

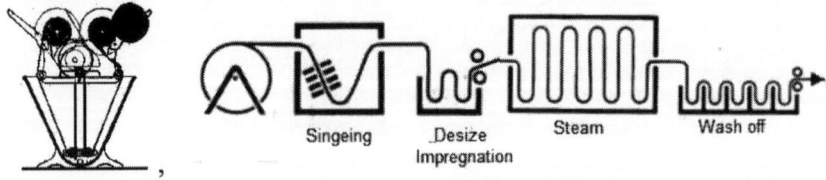

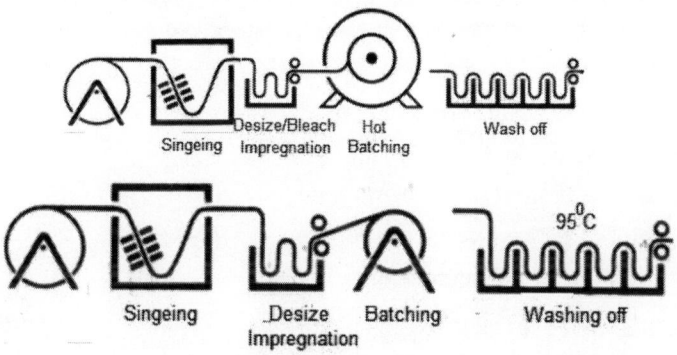

Figure 1.27.

Table 1.28. Guideline recipes

Chemicals	Jig	Pad-steam	Pad-roll	Pad-batch
Suitable enzyme	2–4 g/l%	2–4 g/l%	2–4 g/l	2–4 g/l
Common salt	3 g/l			
Non-ionic wetting agent	1–2 g/l	2–3 g/l	2–3 g/l	2–3 g/l
Acetic acid to pH	6.2–6.5	6.2–6.5	6.2–6.5	6.2–6.5
Padding temperature °C		70	70	70
Steaming temperature °C		103–105	90	R.\T
Time	1 h at 70°C	60–80 s	60–90 s	3–5 h

Table 1.29. Other recipes

Quantity	Unit	Bath additions
5–15	gpl	Oxidative type desizing agent (e.g., Persulphate)
5–10	gpl	Caustic soda
1.0–3.0	gpl	Scouring agent (anionic)

Temperature–90–95°C, time–60 min, liquor ratio–1:5–1:10.

Notes:

i. Caustic soda is added to permit the persulphae type desizing agent to exhibit its full oxidative decomposing effect. NaOH also acts on cotton as a scouring agent

ii. As the surfactant used together with the persulphate type desizing agent, the anionic type is preferred to the non-ionic type from the viewpoint of the stability of the desizing agent.

Advantages of enzymatic desizing

1. No fibre damage
2. No use of aggressive chemicals
3. A lot of process possibilities
4. High biological degradability.

Disadvantages of enzymatic desizing

1. No additional cleaning and cracking effect
2. Low-effects on certain starches (e.g., Tapioca starch)
3. Effects can be reduced by certain size additives and other impurities.

1.7   Acid desizing

1.7.1   Water soluble starches

The synthetic sizes used in sizing are easily removed by a hot alkaline scour/wash. It is advisable to impregnate the goods with water and allowed to stand for 2–3 h in order to allow the size to swell and which is subsequently washed-off in an open-width washer for example a 6 compartment open soaper as follows:

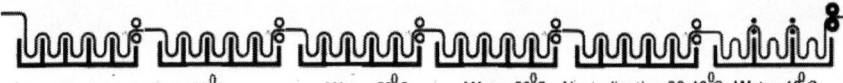

Persulphate Bath 70°C   Water 60°C   Water 60°C   Neutralisation 30-40°C   Water 40°C

Figure 1.28. Water soluble starch desizing process

Table 1.30. Process details for water soluble starch desizing

Compartment	Recipe
1st compartment	1–2 g/l Trisodium phosphate
	2 g/l Suitable detergent
	70°C Temperature
2nd compartment	1–2 g/l Trisodium phosphate/soda ash
	2 g/l Suitable detergent
	60–70°C Temperature

3rd compartment	Water at 60°C
4th compartment	Water at 60°C
5th compartment	3-4 ml/l Acetic acid or
	1-2 ml/l Hydrochloric acid
	0–40°C Temperature
6th compartment	Water at 40°C

N.B. A five compartment soaper or 4 compartment soaper also can be suitably manipulated to be used as a washing-off unit.

1.7.1.1   Acid treatment recipes

For removing cationic auxiliary agents from polyester/cotton polyester/viscose blends and polyester.

(a) Discontinuous method (MLR 5:1–10:1)

Formic acid 85%   2 ml/l

Detergent   1–2 ml/l

Treat for 20 min at 70°C

(b) Continuous method

In a five compartment soaper

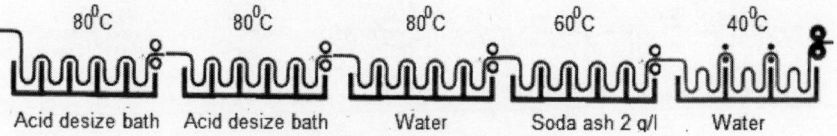

Figure 1.29. Acid desizing process

Table 1.31.

Compartment	Recipe
1st compartment	10 ml/l Formic acid 85%
	5 ml/l Suitable detergent
	Temperature 80°C
2nd compartment	Same as above, 80°C
3rd compartment	Water at 80°C
4th compartment	Soda ash 2 g/l at 60°C
5th compartment	Water at 40°C

1.7.1.2   Enzyme and acid desizing recipes

Table 1.32. Guideline recipes

Fabric	Enzyme desizing				Acid desizing			
	Poplin	Long cloth	Cambric	Mulls (Voil)	Poplin	Long cloth	Cambric	Mulls (Voil)
Metre/Kg	8 m/kg	10 m/kg	12 m/kg	20 m/kg	8 m/kg	10 m/kg	12 m/kg	20 m/kg
Enzyme g/l	1	2	2.5	1.5				
HCl g/l					2	3	4	2.5
Salt g/l	2	4	4	3				
Wetting agent	0.2	0.3	0.4	0.2				
Temperature °C	50	50	55	60	RT	RT	RT	RT
Time h	12	8	8	10	6	6	6	6
pH	6.5–7.0	6.5–7.0	6.5–7.0	6.5–7.0	4	4	4	4
Desizing efficiency %	70	78	85	82	75	80	90	85
Ash content %	0.28	0.25	0.31	0.27	0.05	0.06	0.07	0.07

1.7.2   Continuous processes

1.7.2.1   Pad-J box desizing

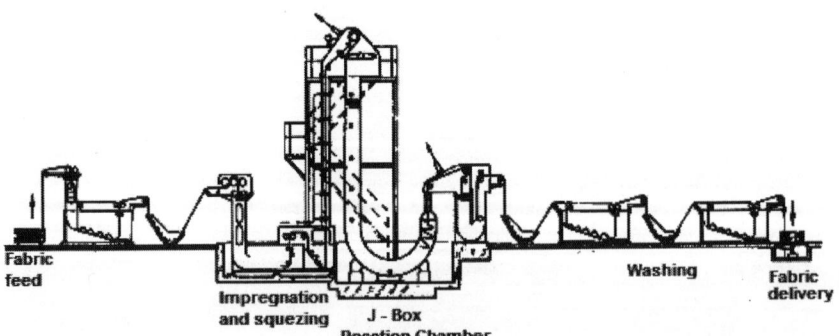

Figure 1.30.

Table 1.33. Guideline recipe

Quantity	Unit	Chemicals
15	g/l	Desizing agent
7	g/l	Caustic soda
3	g/l	Anionic scouring agent

Pick up 100–110%
Steaming 95–100°C, Time–20 min.

1.8   Oxidative desizing

An alternative desizing procedure is oxidative desizing with peroxides. The oxidizing agent is hydrogen peroxide or in some cases a combination of hydrogen peroxide and potassium persulphate, and the process is carried under highly alkaline condition.

Starch size can be removed by oxidative method using persulphates, which becomes handy in some process routes. Oxidative desizing is comparatively unimportant as the oxidative agent used can attack the cotton and damage them as the chemical structure of cellulose and starch are similar. Hence the maximum persulphate usage in cold pad batch process may be limited to 0.5% and then and not more than 0.2–0.35% (in relation to the material weight) in the steaming process. Application takes place in an alkaline medium. Without the alkali, complete desizing is impossible. The oxidizing agent is hydrogen peroxide or in some cases a combination of hydrogen peroxide and potassium persulphate, and the process is carried under neutral and highly alkaline conditions. It is always better to give an intensive alkaline after-treatment to achieve complete desizing.

Persulphate breaks down the carboxyl and fixed groups of polysaccharide into low molecular and water soluble size fragments:

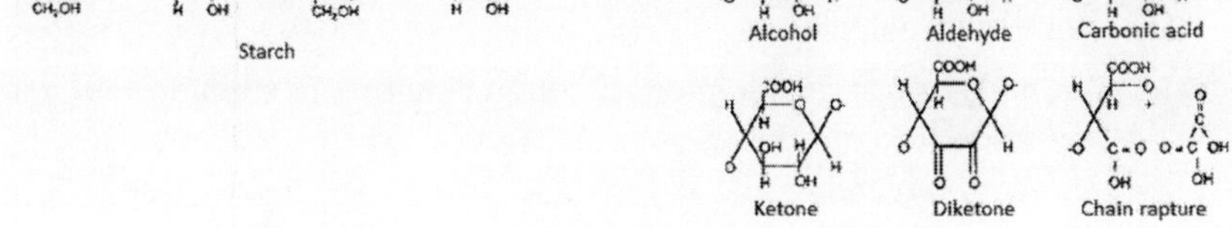

Figure 1.31. Reaction of starch with persulphate

As said earlier, the risk here is that the persulphate can attack cellulose which is of similar structure to that of the starch. The danger of oxidative damage can be minimised by using stabilised persulphate in place of pure persulphate (example Lufibrol O – BASF).

The most important aspects of oxidising agents are that they can be applicable to wide range of fabrics, the size content of which is often not known. Common oxidising agent other than persulphates is hydrogen peroxide, persulphate + peroxide, sodium bromite, etc. Other oxidising agents which are not commonly used but has been found effective in the removal of starch are peroxymonosulphuric acid (H_2SO_5), peroxydisulphates [$Na_2S_2O_8$, $K_2S_2O_8$ and (NH_4)$_2S_2O_8$], acid hydrogen permonosulphates ($KHSO_5$ and NH_4HSO_5) and potassium peroxydiphosphate ($K_4P_2O_8$) have been shown to be effective for desizing.

Sodium chlorite, even though a commonly used oxidising (bleaching) agent in textile, should not be used as a desizing agent, as it "fix" starch rather than removing them.

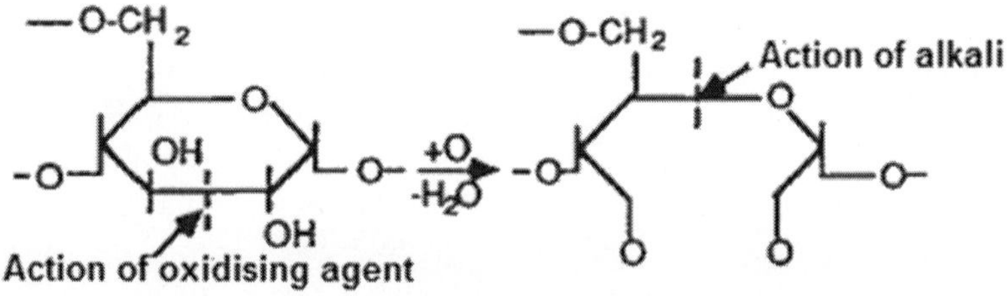

Figure 1.32. Oxdative degradation of Starch and action of alkali

Advantages of oxidative desizing include

1. It degrades and removes starch, PVA, and CMC so the type of size on the fabric or size blends do not affect the procedure, effective for tapioca starches and no loss in effectiveness due to enzyme poisons.

2. Some scouring and bleaching actions are also obtained, and

3. In some cases the scouring process can actually be combined with desizing.

Disadvantages include:

1. The high-risk of degradation of cotton resulting in strength loss, oxycellulose formation, and possible strength loss of polyester in polyester/cotton blend.

2. With problematic sizes it is advisable to carry out combined treatment, i.e., enzymatic desizing followed by an alkaline stage desizing with persulphate.

3. Normally sodium, potassium, or ammonium persulphate can be used. Sodium persulphate is most widely used because of its good solubility and because it has no disturbing odour as is the case of ammonium persulphate.

Desizing with persulphate

Table 1.34. Guideline Recipe

Quantity	Unit	Chemicals
2	ml/l	Sequestering agent
3–5	ml/l	Non-ionic washing agent
1–2	ml/l	Wetting/deaeating agent
3–5	g/l	Sodium persulphate

Pick-up   = 80–100%
Temperature = Cold
Dwelling   = 4–12 h
Desizing with persulphate (One stage desizing/scouring)
Machinery Used : Pad–Roll, Pad-steam, Combi-steamer,
U box, L box, J box, etc.

Alternate recipes

All natural cellulosic fibres and their blends with synthetics. One stage desizing and scouring.

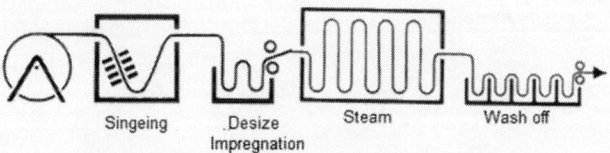

Singeing   .Desize   Steam   Wash off
            Impregnation

Figure 1.33. One stage desizing and scouring

Table 1.35. Guideline recipe

Quantity	Unit	Chemicals
3–6	g/l	Sodium persulphate
8–10	g/l	Caustic soda
5–10	g/l	Wetting agent

Machine   – Pad-steam

pH $-$ 10–10.5
Time $-$ 1–3 min
Temperature $-$ 95–100°C

Guideline recipe for pad-roll, pad-steam

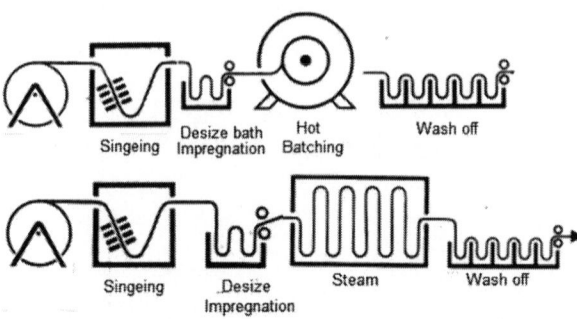

Singeing   Desize bath   Hot   Wash off
           Impregnation Batching

Singeing   Desize   Steam   Wash off
           Impregnation

Figure 1.34. Pad-roll, pad-steam process

Procedure: Saturator temperature 30°C, Liquor pick up to 80–100%.

Table 1.36. Guideline recipe

Substrate	Cellulosic and their blends with synthetics	
Machine	**Pad-steam all systems**	
Type of treatment	**Med. to long steaming times**	
Guide recipes		
Caustic soda solid (g)	40–80	40–80
Seq. agent for heavy metals ml/l	8–15	
Suitable scouring agent ml/l		4–8
Sodium persulphate	2–4	2–4
Working method:		
(a) Saturator impregnation	6–6.5	6–6.5
Pick-up %	80–100	80–100
Temperature °C	30 max.	30 max.
(b) Steaming and batching		
Temperature NT °C	90–100	90–100
Time Min	60–120	60–120

Rinse as hot as possible.

Sodium persulphate has to be dissolved in water below 30°C and added to the bath.

NT – Normal temperature and HT – High temperature.

Alternate recipes

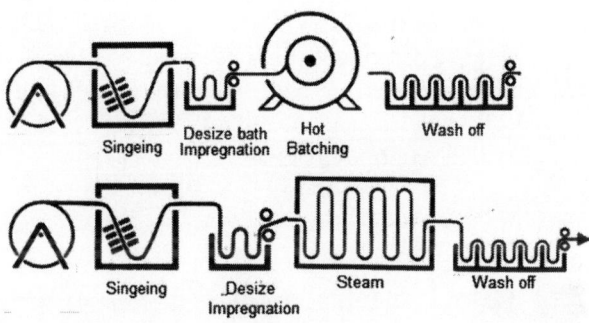

Figure 1.35.

Table 1.37. Guide recipes

Substrate →	Cellulosic and their blends with synthetics	
Machine	**Pad roll, pad- steam**	
Type of treatment	**Med. to long steaming times**	
Guide recipes		
Caustic soda solid (g)	20–40	20–40
Seq. agent for heavy metal ml/l	5–2	
Suitable scouring agent (P/S) ml/l		2–5
Sodium persulphate		1–2
Working method		
(a) Saturator impregnation	6–6.5	6–6.5
Pick-up %	80–100	80–100
Temperature °C	30 max.	30 max.
(b) Steaming		
Temperature NT °C	100–106	100–106
HT °C	130–140	130–140
Time NT min	2–20	2–20
HT min		2–20

1.8.1   Desizing with hydrogen peroxide (oxidative shock desizing)

This method is suitable for PE/cotton blends and in some cases of cotton containing sizing agents of modified starches and mixed synthetic sizes.

Fully continuous, multi-stage ranges are used for pre-treatment. It is applied instead of enzymatic desizing, e.g., in open-width on a pad-steam range with short steaming times. The pre-bleached material generally exhibits a fairly good degree of whiteness and cleanliness which facilitates the subsequent wet treatments like scouring or scouring and bleaching.

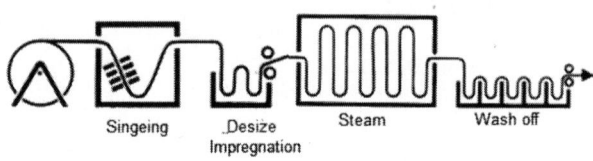

Singeing   Desize   Steam   Wash off
Impregnation

Figure 1.36. Desizing Process with H_2O_2

Table 1.38. Guide Recipes

Substrate →	Natural cellulosics and their blends with synthetics	
Machine	Pad-steam normal temperature	
Type of treatment	Oxidative shock desizing–Short steaming time	
Method	**Containing silicate**	**Silicate free**
Water hardness °Bh	6.25–12.5	6.25–12.5
Scouring agent	3–6	3–6
Suitable stabiliser ml/l	4–6	
Suitable stabiliser ml/l		5–8
Sodium silicate 38 °Bé ml/l	5–10	
Caustic soda solid g/l	10–40	10–40
Hydrogen peroxide 35% ml/l	30–60	30–60
(a) Saturator impregnation	6–6.5	6–6.5
Pick-up %	80–100	80–100
Temperature °C	20–30	20–30

(b) Steaming

Temperature	100–102	100–102
Time Min	40–120	40–120

Rinse as hot as possible.

Discontinuous scouring and washing (removal of water soluble starches)
Cellulosic fibres and their blends with synthetic fibres

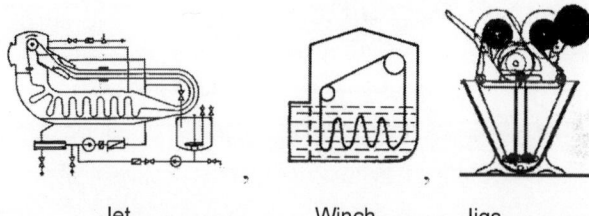

Jet   Winch   Jigs

Figure 1.37. Batchwise scouring and washing

Table 1.39.

Substrate →	Cellulosic fibres and their blends with synthetics		
Form	Loose fibres, tow, yarn, woven and knit goods		
Machines/liquor ratio	Circulating machine–Jet, overflow. MLR 6:1–15:1	Winch MLR 8:1–20:1	Jiggers MLR 2:1–5:1
Wetting and scouring agent		1–2	2–4
Low forming scouring agent ml/l	0.75–1.5	3–6	
Lubricating agent	1–2	0.5–1.0	
Soda ash	0.5–2.0	0.5–1.0	2–4
Working method			
If necessary rinse first with water for 10 min at 40–45°C			
In fresh liquor			
Starting temperature °C	40	30–40	30–40
Treatment temperature °C	60–90	60–90	60–90
Treatment time min	20–30	20–30	45–60

Rinse hot and cold, neutralise with acetic or formic acid.

1.8.2   Bromite desizing

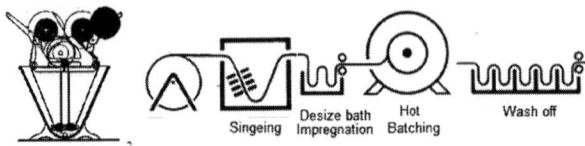

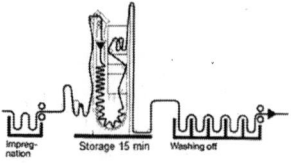

Figure 1.38. Bromite desizing process

Table 1.40. Guide recipes

	Jigs	Pad-roll
		J-box
Available bromine	1 g/l	1.5–2 g/l
Borax	5 g/l	
Wetting/scouring agent	1 ml/l	2 ml/l
Caustic soda 36°Bé	2 ml/l	
Soda ash for pH		10
Time/temperature	15–20 min	15 min cold
Time h	Wash-off	Impregnate with NaOH convey through pad-roll, J-box, rinse hot, cold

A quick reference table for various oxidative bleach methods

Table 1.41. Guide Recipes

Oxidising agent	Process	Recipe	pH	Time min.	Temperature °C
Hydrogen peroxide	Pad-steam	1–2 Vol. H_2O_2 7–15 g/l NaOH	8–9	1–5	90

Oxidising agent	Process	Recipe	pH	Time min.	Temperature °C
Sodium bromite	Cold pad batch	1–3 g/l active Br$_2$ 20–30 g/l NaOH 5–10 g/l wetting agent	7.5–8.5	15	20–40
Persulphate	Pad-steam	3–6 g/l sodium persulphate 8–10 g/l caustic soda 5–10 g/l wetting agent	10–10.5	1–3	95–100
Persulphate + H$_2$O$_2$	Cold pad batch	40 g/l H$_2$O$_2$ (25%) 10 g/l sodium persulphate 10 ml/l sodium silicate 10 ml/l caustic soda 5 g/l stabiliser 5 g/l wetting agent	10–10.5	6–20	20–40

1.8.3   Desizing of garments

Efficient desizing of garments (i.e., of Denim Jeans, Denim jackets and other cellulosic garments) gives better washing and preshrinking effects.

Process stage 1
Wet out garments with 0.5–1 g/l wetting agent in drum washing machines or pits.

Process stage 2
Desize with 1–5 g/l enzyme (e.g., Bactosol AMI liq-Clariant)
    2–4 g/l Common salt
    At 60–90°C, pH 6–6.5 for 60–75 min on drum washing machines or soaked in pits for overnight.

Process stage 3
Thorough washing, preferably hot wash.

1. **Rot steeping**

    In this method, grey cotton fabric is steeped in water in suitable box at a temperature of about 30–40°C or pad the material with warm water and squeezed to 100% expression and left it in open space or box for 24 h at 35–40°C or overnight at 60°C. During the storage micro-organisms naturally present in the water develop excreting enzymes which attack the starch. The swollen and hydrolysed starch is thus partially converted into soluble state which is then removed from the fabric by normal washing with water. The main problems in this method are low-efficiency due to longer treatment time and degradation of cellulose due to cross-infections of mildew if the fermentation process is not properly controlled, requirement of large floor space for storing the goods, etc.

2. **Acid steeping**

   Not very well-practiced now, in this method the material is impregnated with 2.5 g/l sulphuric or hydrochloric acid solution at room temperature (25–30°C) followed by storage for about 6–8 h. Faster desizing (within 1–2 h) is possible using stronger solution of acid (e.g., 10 g/l). The process called "souring" is used for low-quality cotton fabric. Hydrolysis of starch by mineral acid is an exothermic process and the temperature may even rise up to 50°C. During storage the fabric should not be allowed to dry, otherwise the concentration of the acid in the fabric will increase with subsequent damage of the fabric. By the acid steeping method, almost all the starch present in the fabric may be converted to soluble form and can be washed-off.

   Even though the desizing efficiency is much better than the rot steeping there are many disadvantages for this method. The main disadvantage is the damage of the fabric due to any localised drying during storage which can appear as patches in dyeing or subsequent processes. The main disadvantage of acid steeping is that there is always a chance of tendering of the fabric by acid hydrolysis unless proper care is taken. The loss in weight after acid desizing is always more than that after rot steeping, because the former process dissolves nearly all the mineral impurities.

3. **Solvent desizing**

   Solvent can easily remove synthetic sizing agents and with proper selection of the solvent any sizing agent can be removed. Since most of the solvent is able to remove fats, waxes, spinning oils, etc., usually desizing and scouring can be done in one operation.

   In a typical process the fabric is padded through the solvent formulation, steamed to remove the solvent and to activate the enzyme. A wash-off with water completes the process. These processes are suitable for polyester/cotton, polyester/rayon and other blends as well as for cotton.

1.8.4   Iodine test for starch–TEGEWA violet scale

Thanks to the blue scale it is easy to identify and quantify residues of starch on the fabric. The iodine reactant (Minilab) produces a more or less intense bluish violet colouration depending on the residual quantity of starchy sizes.

A quantitative approximation is associated: rating 9 represents a quantity of starch less than 0.04%, rating 1 more than 2.5% starch. Rating 5, less than 0.2%, is acceptable for dyeing. PVA can also be identified by a touch of boric acid in an iodine medium. Colouration is dark blue.

1.8.4.1   The procedure

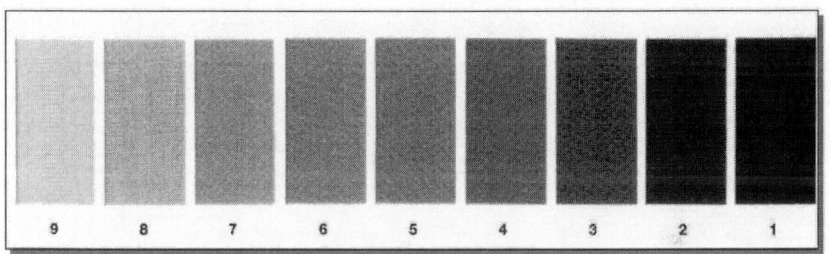

Figure 1.39. TEGEWA-Violet scale

Target: Assessment of the degradation degree of starch size.

Procedure: Put fabric (about 4 × 4 cm) sample in iodine solution (Concentration c = 0.005 mol/l) for about 1 min, short washing out with cold water, dap with starch free filter paper and compare immediately with TEGEWA violet scale.

1.8.4.2   Assessment

Table 1.42.

Violet scale	Warp thread		% Starch
	% Starch	Mean	fabric
1	>2.6		
2	2.6–1.6	2	1
3	1.6–0.90	1.2	0.6
4	0.90–0.60	0.7	0.35
5	0.60–0.30	0.4	0.2
6	0.30–0.20	0.26	0.13
7	0.20–0.14	0.17	0.09
8	0.14–0.10	0.12	0.06
9	c.a. 0.08	0.08	0.04

Grade 9 on the violet scale (= TEGEWA scale) indicates the complete desizing of the sample. Grade 6 can be generally considered to be still sufficient for reactive dyeings.

Other tests

Extraction with petrol ether or dichloromethane
The residual fat content provides information about the removal of naturally and artificially applied fats, waxes and preparations. The waxes which have been applied additionally, often parts of sizes, are especially difficult to remove. The effects of insufficient removal are bad hydrophilicity and uneven dying results.

Table 1.43. Solvent extraction results

Fabric	Petrol ether extraction
	3 h extraction in Soxhelet
Grey fabric	0.8–1.02%
Good scouring/bleaching	<0.4%
Moderate scouring/bleaching	0.40–0.60%

Table 1.44. Modified wicking-test (measurement of the capillary rise)

Time (sec/min)	Goodness of hydrophilicity
–3	Extremely high
3–5	Very good
5–8	Good, acceptable
8–	Process to be checked

Table 1.45. Desirable metal content

	Calcium/Magnesium ppm	Iron ppm
Grey	600–2500	10–100
Good pre-treatment	<300	<10

Table 1.46. Desireable Ash content

	Ash content 2–3 h at approx. 800°C
Grey	0.80–1.20%
Good pre-treatment	<0.20%
Moderate pre-treatment	0.20–0.40%

PART II

Pretreatment of cellulosics - mercerising and scouring

2.1 Mercerising and causticising

Mercer observed that treatment of cotton with NaOH increases the dye exhaustion and fixation and Horace observed that treatment with NaOH by preventing shrinkage the fabric becomes more lustrous. This treatment is developed as Mercerisation. He discovered the actual mercerising process and applied for its patent in 1890.

The treatment is carried out using caustic soda (28–30°Bé or 48–52°Tw) or 270–330 g/l, which determines the contraction and swelling of the fibres; they become translucent and increase their tensile strength, but reduce their flexural and torsion strength. The bean-like section of the fibre becomes first elliptic and then circular, allowing a better reflection of light with a consequent increase of lustre. The treatment time is 30–40 s can be carried either under control or without control commonly referred as

- Tension mercerisation (treatment under tension/control)
- Slack mercerisation (treatment without tension/control)

The former method (tension mercerisation) is most widely practiced mercerisation process.

2.1.1 The advantages of mercerisation

2.1.1.1 Improving lustre

During mercerisation under tension (when mercerised without tension following changes may not take place) the ribbon like structure swells, untwists and form a cylindrical cross-section like a man-made fibre, thus reflecting more light. The lustre depends on the quality of raw material. Long-staple fibre acquires a greater lustre than short staple. Twisted double yarn becomes more lustrous than non-twisted loose yarn. To retain an optimum level of lustre, it is necessary not to release tension before caustic soda concentration is lowered to about 60 g/l. For this purpose, the first rinsing must be carried out under tension.

2.1.1.2 Obtaining a silky look

Along with the lustre the fibre attains a silky look.

2.1.1.3   Improving the tensile strength

Mercerisation tends to increase the tensile strength of cotton fibres. There are many theories regarding the reason for the increase in strength–the changes in spiral angle and orientation, strengthening of weak points along the fibre the weak points include internal strains (spiral reversals), the places of low cross-sectional areas and the places of high distortions. The other important factors may be crystallite length and degree of crystallinity as well as the removal of cellulose of very low degree of polymerisation. With decrease in temperature, the increase in tenacity is more pronounced.

2.1.1.4   Retaining a greater proportion of tensile strength after easy care finishing

The decrease in strength due to the resin finishes is reduced due to mercerisation.

2.1.1.5   Obtaining improved dimensional stability

When mercerised under tension the fabric stability increases like heat setting process of man-made fibres. After setting the width in mercerising the fabric, it retains the same width after washing test.

2.1.1.6   Improving elasticity and obtaining stretch property

When mercerisation is carried out without tension, the value of elongation at break is three times the value of the unmercerised material, thus conferring elastic properties to cotton fibres. This elasticity is the basis of the preparation of stretch materials.

2.1.1.7   Increasing uniformity of dyeing

The uniformity, purity and brilliancy of the shades are also higher due to the change of optical properties of the fibre by mercerisation.

2.1.1.8   Improving reactivity

Mercerised goods are more reactive to chemical agents, which manifest themselves in the acceleration of hydrolysis under the action of acids and damage by oxidising agents due to decrease in degree of crystallinity.

2.1.1.9   Improving % dye exhaustion

Exhaustion of dyes on mercerised materials are faster than those on unmercerised materials. This is probably due to swelling and opening of the structure with consequent higher penetration of the dyes.

2.1.1.10   Improving % dye fixation

Due to decrease in the degree of crystallinity, the resulting fibres have greater absorptive capacity for dyes.

2.1.1.11   Improved colour yield

Mercerised material requires less dye to obtain the same colour strength than

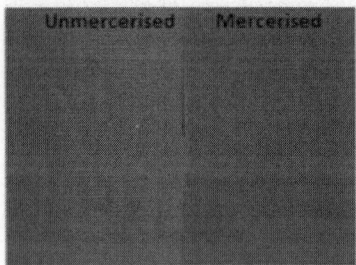

Figure 2.1. Improved colour yield by mercerisation

unmercerised material. The dye consumption is reduced by about 10–15% in light shades and about 25–30% in deep shades.

2.1.1.12   Increase in Young's modulus

The elongation at break decreases and the Young's modulus increases significantly on mercerisation under tension.

2.1.1.13   Improving moisture absorption

Improving easy care finishing properties.

Improvement in CRA

Ability to cover and removal of immature (dead) fibres
Improvement in cover factor
The characteristic hollow channel inside the cotton fibre or lumen almost completely disappears.

Lengthwise shrinkage

Considerable lengthways shrinkage occurs when fibres, yarns and fabrics are mercerised without tension. The fibre as well as yarn diameter, increases with corresponding decrease in metric count.

Improved resistance to whether

Mercerised fibre is more resistant to light and weather effects than non-mercerised fibres.

The effect of mercerisation is different in different materials

Yarn	Woven fabric	Knitted fabric
Improvement of tensile strength	Improvement of colour yield	Equalising of mesh structure
Lustre enhancement	Compensating of yarn quality	
Smooth surface woven fabric	Improvement of the dimensional stability	Improvement of the dimensional stability
Improvement of colour yield	Lustre enhancement	Realisation of silky structure and touch
	Covering of dead cotton	

2.1.2   Mercerisation methods

The wide range of treatment methods can approximately be broken down into the following divisions. Parentheses denote established terminology.

1. Classification according to the form of the product
    a) Yarn mercerisation

    Batch: Hank mercerisation

    Cheese mercerisation

    Continuous: Single end mercerisation

    • Tow mercerisation

    • Warp mercerisation

    b) Knit mercerisation

    Open mercerisation

    Closed mercerisation (round mercerisation, tubular knit mercerisation)

    c) Cloth mercerisation

    Chainless mercerisation (roller mercerisation)

    Chain mercerisation (stenter mercerisation)

    Batch-up mercerisation

2. Classification according to the mercerising conditions
    a) Water content

    Dry mercerisation

    Wet mercerisation

    b) Tension

       Fixed-length mercerisation

       Tension mercerisation

       Tensionless mercerisation

    c) Alkaline concentration

       Low-concentration alkaline mercerisation

       High-concentration alkaline mercerisation

       Two-step mercerisation

    d) Temperature

       Ambient-temperature mercerisation

       High-temperature mercerisation

       Low-temperature mercerisation

3. Classification according to timing

    a) Grey mercerisation

    b) Pre-dyeing mercerisation

    c) Post-dyeing mercerisation

4. Classification according to the number of treatments

    a) Single mercerisation

    b) Double mercerisation

5. Classification according to the type of alkali used

    a) Caustic soda mercerisation

    b) Ammonia mercerisation

6. Other

    a) Alkali pad-dry method

    b) Alkali pad-steam method.

Mercerisation is an exothermic reaction and heat is liberated, swelling, crystallisation (Cellulosic chain gets rearranged and becomes more parallel).

- Swelling of the fibre takes place.
- Cellulosic chains gets rearranged and become parallel.
- Convolutions disappear.
- Horizontal cross-sectional view shows that it becomes circular.
- Shrinkage takes place.
- Inter chain forces are weakened (due to the presence of solvated dipole hydrates).

- Change of crystal lattice from cellulose I–II.
- Modification of crystalline network (The fibrils do not lose their identity but change in dimension and orientation.).
- The most important effect of mercerisation on the fine structure of cellulose fibre consists of changing the crystal lattice from cellulose I–II.

2.1.2.1   Parameters of mercerisation

- Concentration of NaOH

After swelling reaches its greatest point, NaOH thoroughly penetrates the interior of the micelle, and a reaction between the alkali and the micelle occurs, completing the generation of alkali cellulose I. From 18–22°Bé, the range displaying the greatest degree of swelling, to around 24°Bé, the cotton hair first contracts momentarily, and then swells again, and at 24°Bé or above a second swelling peak was observed. These observations cannot be disregarded as baseless occurrences.

According to the behaviour observed for woven fabric and fibres of raw cotton exposed to alkali, woven fabric only displays around half the degree of swelling displayed by fibres if both are treated with the same concentration of alkaline solution, but no such difference exists between their respective degrees of alkali absorption.

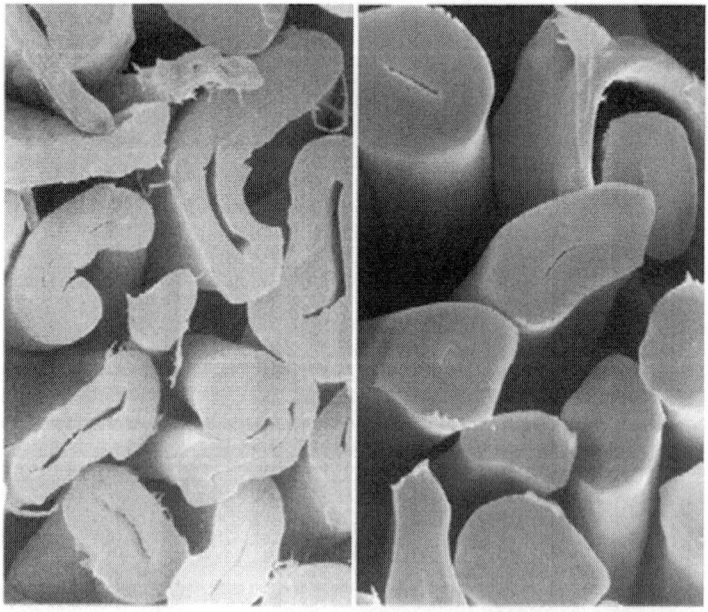

Unmercerised cotton fibre   Mercerised cotton fibre

Figure 2.2. Electron microscopic view of cross-section of cotton fibre

Investigations by Bechter reveal that with an impregnation concentration of 28–30°Bé, or 250–300 g NaOH 100% per litre (215–240 g NaOH 100% per kg) there is no significant increase of iodine sorption or dyestuff uptake. This means optimal mercerisation is accomplished with a concentration of 220–240 g NaOH 100% per kg cloth on the fabric.

The concentration at which yarn showed the greatest degree of swelling was determined to be 20% NaOH (26.5°Bé). Caustic concentration above 240 g/l don't improve the mercerising effects.

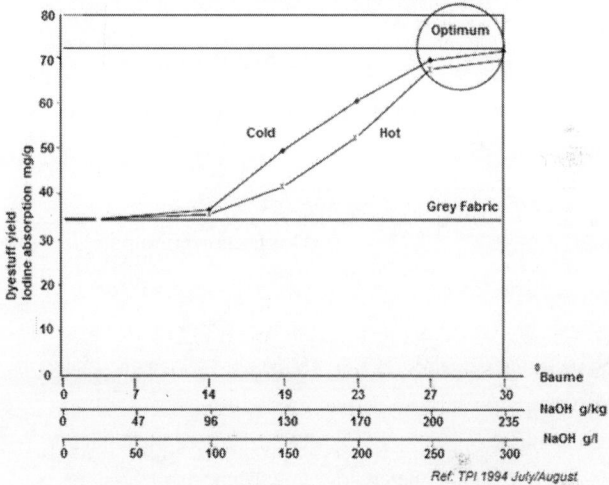

Figure 2.3. Effect of concentration of NaOH

Cotton hair under physical restraint, when converted into a yarn or a woven fabric restrains the hair's freedom due to twisting and to crossing of twisted yarn, when treated with an alkali, displays different behaviour because the shrinking and swelling displayed originally cannot occur due to the constraining forces. Researches have proved that even for the same type of yarn, the behaviour of single, double and triple yarns display differences, as does the shrinkage of woven fabric and hairs.

• Time of treatment

Time of treatment is an important factor in the mercerisation process. As can be seen in the table given below the time of mercerisation varies as per temperature. Once the mercerisation process is over at a given temperature and concentration further treatment will not have any effect on mercerisation. The speed of bulk mercerisation machine and impregnation time is all designed in such a way as to give the optimum time for mercerisation.

Table 2.1. Conditions of Mercerisation

Caustic concentration	270–330 g/l NaOH 100% or 28–32°Bé
Caustic temperature (cold)	15–20°C
Caustic temperature (hot)	60–90°C
Tension	Against shrinkage
Stabilising	Up to about 50 g/l NaOH 100%
Wetting agent	For quick and even wetting
Pre-treatment	Raw, desized, boiled-off, bleached

2.1.3　The practical mercerisation process

Mainly there are two type of mercerisation–hot mercerisation and cold mercerisation.

Table 2.2. Features of hot and cold mercerisation

Cold mercerisation	Hot mercerisation
Faster caustic soda penetration, speedier swelling	Lower investment cost through shorter impregnation sections
High swelling uniformity, no ring mercerising	Maximum squeezing action at in feed squeeze
Better dye penetration	Shorter impregnation times
In wet on wet impregnation a cooling device is necessary to compensate exothermic reaction	In wet impregnation there is "automatically" a mixing temperature of 55°C
Smoother cloth appearance	Dyestuff affinity decreases at mercerising >60°C
Softer handle	No cooling needed for mercerising

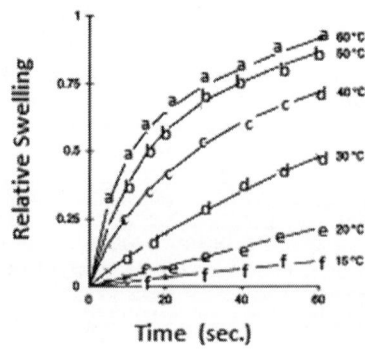

Figure 2.4. Relative Swelling of Cotton vs. Temp.

Advantages of hot mercerisation

- Higher degree of lustre can be anticipated (if stretching takes place while cooling)
- Higher tensile strength can be anticipated (if stretching takes place while cooling)
- Improved dye exhaustion
- Higher degree of penetration
- Uniformity
- Higher productivity
- Desired effect is achieved rapidly
- Contact time between fabric and NaOH is reduced by 50%
- When the fabric is impregnated in hot NaOH the fabric becomes highly plastic and less elastic and can be stretched to greater extent
- Improved colour yield/dye-uptake
- Improvement in easy-care finishing property.

Disadvantages of hot mercerising

Due to rapid and extensive swelling of cotton during their impregnation with cold NaOH, the structure is compacted at the surface of the mercerised goods.

- This makes further penetration of NaOH difficult and almost impossible.
- The result is lack of uniformity.
- Mercerisation at elevated temperatures to avoid lower swelling offer some advantages, which is known as hot mercerisation.
- Lower swelling.
- Conversion of crystalline structure from cellulose I–II is retarded.
- Cross-sectional view reveals a skin-core appearance.
- Higher energy consumption.
- Costlier.

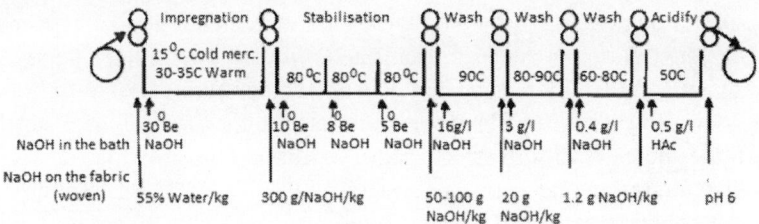

Figure 2.5. Processing sequence of a classic mercerising process

A reasonable approach is to use hot caustic for better penetration and allow the impregnated material to cool to an adequate temperature in order to have good swelling known as two-step process.

Step 1: *Impregnation*–Sodium hydroxide solution has been shown to be most effective at a concentration of 28–32°Bé, i.e., if the caustic has a content of 270–330 g/l NaOH 100%. The key factor is that the caustic content of the fabric is 220–240 g NaOH 100% per kg of fabric, and that the caustic film has a concentration of 28°Bé.

Saturation of the cotton material with NaOH, preferably under relaxed conditions, at a temperature between 50°C and boiling temperature of NaOH. The immersion time ranges from 4–60 s.

Dwell time during mercerisation

The following impregnation times are required in order to achieve the best possible mercerising effects:

- Approx. 50–60 s for cold mercerising.
- Approx. 25–30 s for hot mercerising.

Step 2: *Stabilisation*–Controlled hot stretching following the saturation and cooling of the stretched material to less than 25°C. Tension controlled washing to a NaOH concentration of 60 gpl (stabilisation), followed by final washing under normal conditions. In order to prevent shrinkage of the fabric in the washing zone, the caustic concentration needs to be lowered before the fabric reaches the washing compartments. The caustic content on the fabric should not exceed the following limits:

50 g NaOH 100%/kg of fabric for light to medium weight articles.

70 g NaOH 100%/kg of fabric for heavy articles.

Impregnation temperature

There are advantages and disadvantages for both hot and cold mercerising. Below two graphs gives an idea of the effect of cold and hot mercerising. Also given a comparison of hot and cold mercerising. Maximum swelling of the cotton occurs at a temperature of 12–15°C. Due to rapid and extensive swelling of cotton during their impregnation with cold NaOH, the structure is compacted at the surface of the mercerised goods. This makes further penetration of NaOH difficult and almost impossible. The result is lack of uniformity. Increasing the caustic temperature to 50–60°C results in less swelling, but the fibres are penetrated right through, as a result of which the mercerising effects are more uniform. This prevents subsequent ring discolouration and improves resistance to fraying. Mercerisation at elevated temperatures to avoid lower swelling offer some advantages, which is known as hot mercerisation.

Table 2.3. Temperature of different mercerisings

Mercerisation process	Temperature
Cold mercerising	15–20°C
Warm mercerising	30–40°C
Hot mercerising	60–65°C

At lower temperatures the viscosity of caustic soda increases the handling of the solution and washing-off is more difficult. Hence warm or hot mercerising is practiced more commonly.

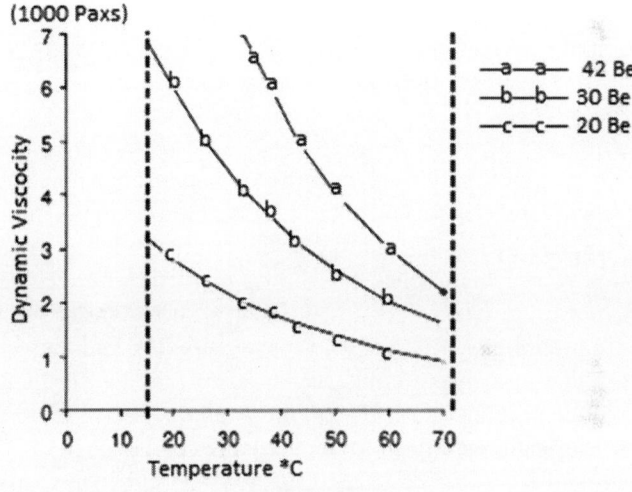

Figure 2.6.

In another possible method of hot mercerisation consists of wet-on-wet impregnation of the fabric immediately after washing at 95°C and a subsequent hot squeezing with high speed steam injection. The heated fabric is impregnated with NaOH solution at 30°C in the first step and at 20°C in the second step. Thereafter the stabilisation sequence is normally carried out. Impregnation with hot alkali of 250–300 gpl concentration leads to the formation of only soda cellulose I. The transformation of soda cellulose I–V takes place only during cooling step. While hot mercerisation appears to be advantages, it is important to remember that cotton fibres impregnated with concentrated alkali solution are subject to alkali degradation and this will be higher in the presence of air and temperature.

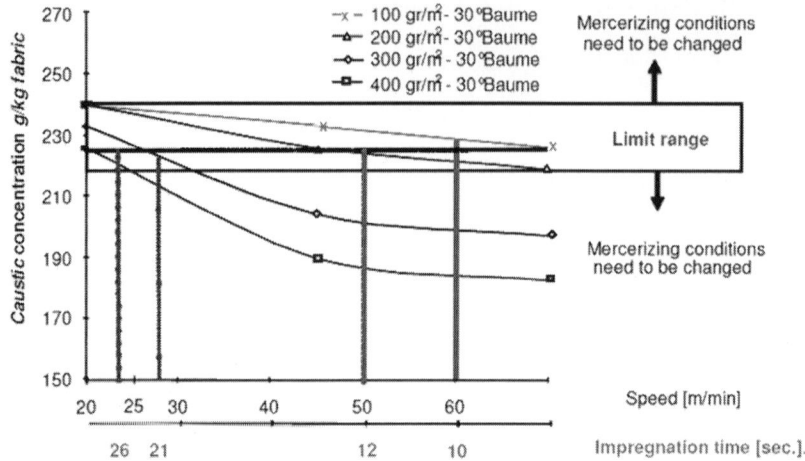

Figure 2.7. Optimum conditions of mercerisation

Tension applied
There were many investigations regarding the relationship between shrinkage and the effects of temperature and alkaline concentration in the mercerisation of cotton yarn, and the results show that the concentration of 20°Bé represents a line beyond which behaviour during mercerisation changes. It was seen that temperature has its greatest influence on the shrinkage of cotton yarn at concentrations of 3 N NaOH, (approx. 16°Bé), and above 4 N (approx. 20°Bé), that influence diminishes. On measuring the shrinkage of cotton yarn at concentrations of 30°Bé for temperatures from 0 to 40°C, shown that there is very small variations for temperatures in the range of 10–30°C at a concentration of 30°Bé.

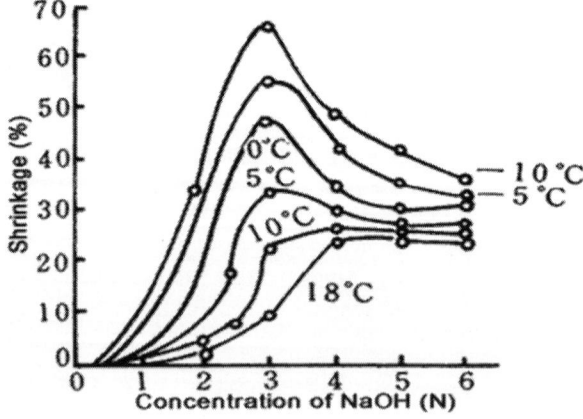

Figure 2.8. The shrinkage of cotton yarn in NaOH

• **Drying:** Fibres in their wet state at the conclusion of mercerisation have a very high degree of swelling, and have large internal air gaps. But if the fibres are dried, these gaps will contract as the water evaporates. The final fixed sizes of the gaps inside the fibres are altered by different temperature and tension conditions during the drying. In comparison to untreated fibre, absorption of dyestuff is twice as high after mercerisation, and 2.4 times as high after tensionless mercerisation. Furthermore, the absorption of dyestuff is reduced by one third after natural drying in air, and by nearly one half after drying at 110°C in comparison with non-dried fibre, which suggests the contraction of the air gaps inside the fibre.

The size of the air gaps in the cellulose's micelles, that is, the fibre's internal volumetric capacity, is altered by the tension and temperature conditions during the rinsing and drying at the end of the mercerisation process which, in addition to altering the hygroscopic characteristics and producing variations in the absorption and reactivity of the dye, also significantly influence the hand of the fabric.

The heat generation due to the reaction between the cellulose and the alkali, like heat generation in other typical chemical reactions, is not simple, and varies according to the concentration of the alkali and other factors. From a concentration of around 120 g/l NaOH, or about 15–16°Bé, the reaction heat increases sharply, and from around 240 g/l, or 26°Bé, the reaction heat increases in more or less constant proportion with increases in concentration. While different researchers recorded different results for this heat generation, it is in the range of 2.6–4 cal/g of cotton. In contrast to this, with an increase in the concentration of the alkali solution of 1 M in the range from 15–16°Bé to 19–20°Bé for which the swelling was greatest, the heat generation was 8.93 cal/g, and from a concentration of around 22°Bé, at which the production of alkali cellulose I was more or less complete, the rate of increase of generated heat was 3.40 cal/g cotton/M NaOH. Above 42°Bé, this became 3.0 Cal/g cotton/M NaOH. While the time required for the completion of this heat generation for concentrations below about 34–35°Bé was less than 5 min, above 42°Bé it was 10 min, and above 22M NaOH, 100 min was required, the reaction needing a very long period of time to complete. At concentrations below 22°Bé, significant differences in behaviour with regards to heat generation are displayed between fibres that have been mercerised once before and fibres that have not been mercerised at all. Items that have been mercerised once before display the absorption of alkali and generation of heat even in alkali solutions of low concentration.

Points to be taken care while mercerising

- It is important to bear in mind that tension plays a key role which brings out not only lustre, but also increases the tenacity and Young's modulus, however, the elongation at break decreases.
- As the concentration of NaOH increases from 150 to 350 gpl similar results as in the case of tension applied can be seen. The lustre of cotton fibres is decided by the ratio between the long and short axes of the cross-section of the cellulose air, and it improves as the cross-section becomes more circular.
- Fibre shrinkage in width and length direction is inversely proportional to the tension applied.
- It is important to note that efficiency of mercerisation is not dependent on swelling of cotton in alkali since swelling value of cotton is the least above 150 gpl concentration of NaOH.

The axial ratio of the cross-section of single cotton fibres and their lustre

While mercerisation can greatly improve lustre, it cannot make up for deficiencies in the lustre of the raw cotton itself, and in order to produce products of superior lustre, primary considerations relate to the choice of raw cotton, the twisting and manufacture of the cotton yarn and the structure of the fabric. While changes in the mercerisation process do influence the improvement of lustre to a certain extent, any effect that surpasses the more basic variations cannot be expected.

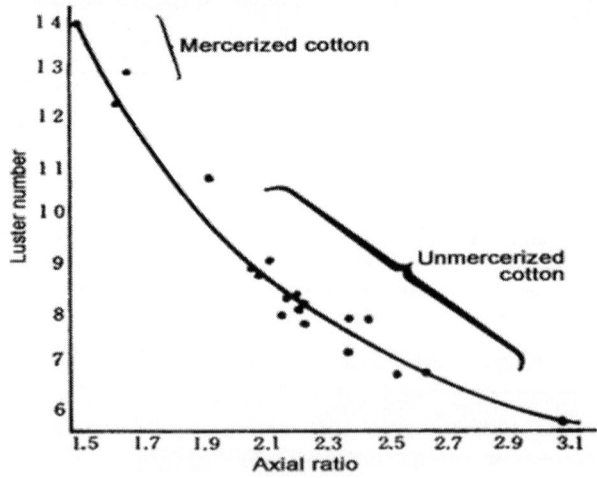

Figure 2.9.

Mercerisation can be carried out at

- At grey stage (with the addition of alkali stable, low-foaming, efficient wetting agent)
- After desizing
- After scouring
- After scouring and bleaching
- Every stage of mercerising has its own advantages and dis-advantages. It is suggested to carry out mercerising preferably after scouring to get best of the best of the process.

Guideline conditions of mercerisation

- Application of NaOH solution around 55–60°TW (31–35%) at a temperature of 15–18°C.
- A dwell period of 55 s on an average, to permit diffusion of alkali into the fibre.
- Warp tension during alkali treatment and stretching the weft (width) of the fabric during washing to prevent shrinkage.
- Finally washing-off of the traces of alkali from the treated fibres.
- Investigations reveal that with an impregnation concentration of 28–30°Bé or 250–300 gpl of 100% NaOH or 215–240 gpl (grams of 100% NaOH per kg), there is no significant increase of iodine absorption or dyestuff uptake.
- This means optimum mercerisation is accomplished with a concentration of 220–240 g of 100% NaOH per Kg of cloth on the fabric.

2.1.4   Changes in the properties of cellulose due to mercerisation

Mercerisation reduces the axial ratio and increase the light scattering within the fibre (transparency) and thus increases the lustre. Here it is important to bear in mind that short fibres on the surface decreases the lustre which can be avoided by singeing. It is also important to note that increase in lustre can be seen from 180 gpl concentrated NaOH to a maximum of 260–300 gpl and then decreases slightly. Swelling and shrinkage are more when there is no tension in the fibre, but the alteration in cross-section caused by swelling is more when mercerisation is carried out under tension.

Under optimum conditions each cotton fibre may contract nearly 9% in length and swell nearly 150%. Swelling of the cotton fibre has also has a dis-advantage.

The fibre becomes more compact in its swollen condition. The compacting further diminishes the further penetration of caustic soda into the fibre. In other words penetration slows down and the effect of mercerisation in the fibre's core is lower than on its surface giving rise to non-uniformity of mercerising. Mercerisation also affects the size of crystallites and orientation of crystalline region and the extent of orientation depends on the tension during mercerisation. Mercerisation, both slack and with tension, increases the strength uniformity along the fibre length. The increase in TS is due to the elimination of weakest points in the fibre. Mercerised fibre with tension shows greater gain in strength than that of without tension. Mercerisation increases the cohesion between individual cotton hairs and this closer embedding of the hairs in the yarn not only increases the strength but makes it more uniform in strength and less in diameter. In practice, the improvements in strength are noticed mostly on mercerised yarn only, with fabric the major effect is only on the surface, along with dye fixation.

Formation of different celluloses on structural modification

1. At 12–19% (by weight) NaOH (concentrated)   Soda cellulose I formed

2. At 20–45% (by weight) NaOH (concentrated)   Soda cellulose II formed

3. By drying soda cellulose I   Soda cellulose III formed

4. At 20–25% (by weight) NaOH (concentrated) and above 30°C   Soda cellulose II and III formed

5. At 40–45% concentrated NaOH between –10 and 20°C   Soda cellulose V formed.

Cellulose I content and total crystallinity index decreases as the temperature of mercerisation increases. Cellulose II follows an opposite course due to the better penetration of NaOH hydrates–cellulose III, IV and X can also be obtained on treatment of cotton with ammonia at –35°C, hot glycerine and phosphoric acid.

2.1.5   Mercerising machines

Generally a mercerising machine consists of four sections which will have different tasks according to the process technique.

- Mercerising section: Impregnation with caustic

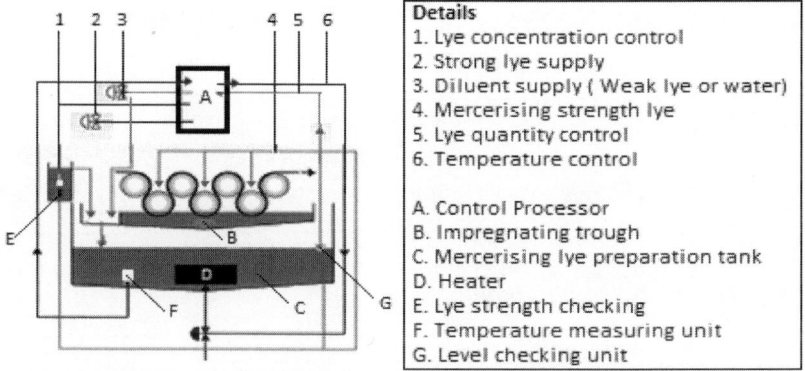

Details
1. Lye concentration control
2. Strong lye supply
3. Diluent supply (Weak lye or water)
4. Mercerising strength lye
5. Lye quantity control
6. Temperature control
A. Control Processor
B. Impregnating trough
C. Mercerising lye preparation tank
D. Heater
E. Lye strength checking
F. Temperature measuring unit
G. Level checking unit

Figure 2.10. Schematic diagram of impregnation zone of Dimensa (Benninger)

- Intermediate squeezing aggregate: Dividing of mercerising section against stabilising section. Here itself the required time factor is taken care to allow the caustic to penetrate and mercerising to take place.

Caustic content for impregnation

Sodium hydroxide solution has been shown to be most effective at a concentration of 28–32°Bé, i.e., if the caustic has a content of 270–330 g/l NaOH 100%.

The key factor is that the caustic content of the fabric is 220–240 g NaOH 100% per kg of fabric, and that the caustic film has a concentration of 28°Bé.

Impregnating temperature

Maximum swelling of the cotton occurs at a temperature of 12–15°C. However, heavy swelling in the selvedge areas prevents the caustic from penetrating any further into the cotton. Increasing the caustic temperature to 50–60° C results in less swelling, but the fibres are penetrated right through, as a result of which the mercerising effects are more uniform. This prevents subsequent ring discolouration and improves resistance to fraying.

Standard temperatures

Table 2.4.

Method of mercerising	Temperature
Cold mercerising	15–20°C
Warm mercerising	30–40°C
Hot mercerising	60–65°C

Dwell time during mercerisation

The following impregnation times are required in order to achieve the best possible mercerizing effects:

- Approx. 50–60 s for cold mercerising.

- Approx. 25–30 s for hot mercerising.

- Stabilising section: Water treatment and thinning down concentration of the dyelye. At this zone structural modifications will take place.

Caustic content for stabilisation

In order to prevent shrinkage of the fabric in the washing zone, the caustic concentration needs to be lowered before the fabric reaches the washing compartments. The caustic content on the fabric should not exceed the following limits:

50 g NaOH 100%/kg of fabric for light to medium weight articles.
70 g NaOH 100%/kg of fabric for heavy articles.

- Intermediate squeezing aggregate: Dividing of stabilising section against washing section.

- Washing section: Washing-off the alkali and neutralisation.

After having achieved the predetermined dye lye concentration in the stabilising compartment, the residual caustic soda on the fabric is washed out, at the highest possible temperature, in the subsequent washing compartments. The higher the temperature, the better the washing effect.

Washing temperature: 90–95°C.

To ensure that the fabric is neutral, the second to last bath is used to neutralise it using acetic acid (CH_3COOH) 80%. The pH-values are determined by the customer and are strongly dependent on the processes which follow.

Generally speaking, a pH-value between 6.5 and 7.5 can be left on the fabric.

The pH-value is normally higher on the fabric than the pH-value in the compartment.

In order to achieve a pH-value of e.g., pH 6.5 on the fabric, the pH in the washing compartment must be set to approx. pH 5.5. The pH-value of the acid dosage must be lowered down when processing heavy fabrics.

The fabric is rinsed a second time, after neutralisation, in order to remove the acetate which forms when neutralising with acetic acid.

Caustic soda circulation

For the efficient use of caustic soda and control the cost of mercerisation, the caustic soda has to be circulated meticulously. There are three strength of caustic soda liquor in a merceriser caustic soda circulation.

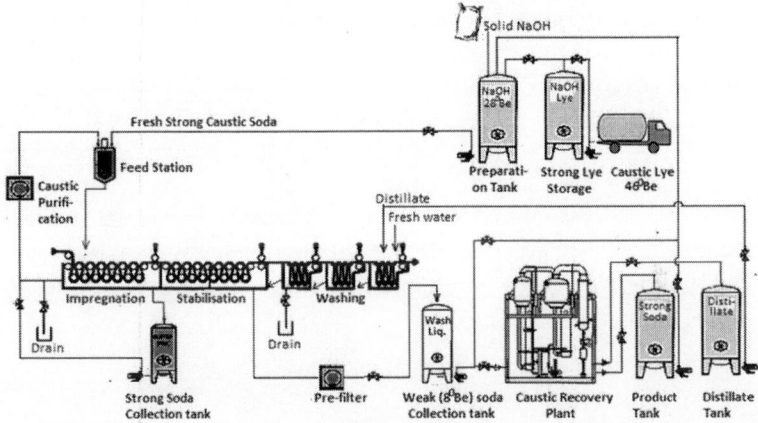

Figure 2.11. Flow Diagram of Mercerising Caustic soda Preparation, Circulation and Recovery

1. Caustic soda lye purchased from chemical suppliers/solid caustic soda.
2. Strong caustic soda for mercerisering operation (270–320 g/l or 28–32°Bé).
3. Weak caustic soda (50 g/l) in stabiliser and from washing units in counter current washing process.

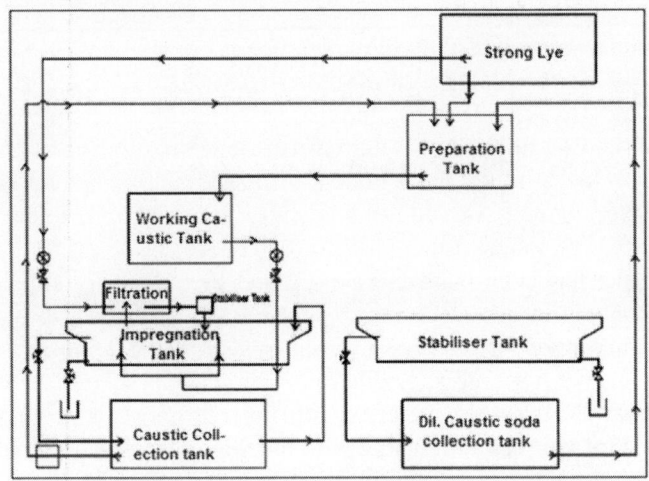

Figure 2.12. Caustic soda circulation in a merceriser

In the modern machines like Benninger Dimensa caustic soda strength is automatic controlled, but still you need a caustic soda circulation and recovery system for efficient usage of NaOH and to avoid any wastage.

Generally the following circulation system is used: caustic lye or solid caustic soda is dissolved to merceriser impregnation strength and fed to the impregnation tank through filtration. (If necessary dissolution is done in ice cold water as the caustic soda dissolution/dilution is an exothermic process. This caustic is circulated to keep the strength of caustic constant and cleaning (see Figure). The volume is replenished from the dissolution tank. The weak caustic soda solution from stabiliser, washing, etc., are controlled to get a caustic strength of 8°Bé, which is concentrated by caustic recovery system to mercerising strength which is recirculated. Alternately, weak caustic (8°Bé) is concentrated to a lower strength than merceriser strength and is used for dissolving solid caustic soda or diluting caustic lye instead of water, whereby steam consumption in caustic recovery plant can be reduced.

2.1.6   Hank mercerising

Yarn mercerising generally entails rolling a 54-inch long (the length of one loop) hank weighing about 500 g a number of times between two adjustable rollers. The yarn is moved by the turning of the rollers, with penetration of the alkali, application of tension and rinsing occurring automatically. In this, one cycle takes about 3–5 min, and 4–8 kg can be treated at one time. In the latest machines, all operations are automated, including control of the alkaline solution's concentration, temperature, the addition and recovery of the alkali, along with application of tension on the yarn and rinsing. An advantage of hank mercerisation is that during the yarn's absorption of the alkali, treatment can be carried out without tension, and so the alkali solution is able to penetrate the inside of the yarn sufficiently, and after the fibres have swelled sufficiently, any level of tension can be applied and the yarn rinsed for removal of the alkali. This allows the production of goods with satisfactory mercerisation effects. Uneven mercerising due to uneven tension is an unavoidable problem in current methods of mercerisation.

Yarn that has been hank mercerised and then dyed for finishing into solid-colour fabrics can result in a barre effect. Hank mercerised pre-dyed yarn is mainly used for products with narrow stripes or a checkered design when barre is not noticeable. Another problem in hank mercerisation is that conventional rinsing after mercerisation is insufficient, and without neutralisation through separate rinsing with hot water, the remaining alkali can cause problems.

2.1.7   Cheese mercerisation

Carrying out mercerisation, scouring, bleaching, and, in some cases, dyeing, along with oiling or sizing, with the yarn in cheese form results in a major rationalisation, which can raise productivity and reduce costs. However, mercerisation in cheese form can only be expected to achieve half-mercerisation, and not the same degree of evenness as hank mercerisation or other types of mercerisation. One problem is how to limit the difference in shrinkage between the inside and the outside of the cheese. Important factors in cheese mercerisation are – the adjustment of the twisting and the density of the winding of the yarn, the size of the take-up tube, the thickness of the layers during the winding, and finally the alkali concentration and temperature during the treatment.

2.1.8   Single end mercerisation

This method, also called, cone-to-cone, or cheese-to-cheese, involves taking up yarn into a cheese or cone shape, and, with one machine per cone, conducting alkali penetration, rinsing (with hot and cold water), neutralisation, rinsing again, and drying in consecutive order with the correct scheduling, and then taking up into a cheese or cone form. The yarn speed in this being approximately 450 m/min, the productivity per machine is low, and the equipment costs are high, because the sequence is automated, it only requires a very small number of staff. The mechanism for conducting mercerisation with these machines involves three revolving rollers– two squeezing rollers which are pressed tightly together and a third roller placed, at a certain distance, more or less parallel to these two. Yarn is lined-up in parallel from one end to the other of the third roller, which is removed from the nip space of the two squeezing rollers, and moved in a spiral perpendicular to the roller, during which time the alkali penetrates, tension is applied and rinsing (both with hot and cold water) and neutralisation occur. A problem in this form of mercerisation is the relative difficulty of controlling the tension on the yarn as it is introduced, and differences in the level of tension between machines and between cheeses or cones can occur easily and lead to patchy dyeing. Other problems relate to yarn breakage, yarn overlap, and yarn skewing. In general, this method of treatment requires two-fold yarn with a yarn count less than 60, and it is unsuitable for the treatment of yarn with fine yarn counts higher than this. In general, this method of treatment requires two-fold yarn with a yarn count less than 60, and it is unsuitable for the treatment of yarn with fine yarn counts higher than this.

Normally, 400 or so yarns are wrapped around a beam or a ball with a wrapper and 8–10 of these beams or balls are set in a stand. Yarn is unreeled from the balls or the beams at the same time and lined-up in ropes made with light twisting, which are mercerised continuously in a manner similar to that of roller mercerisation of fabric.

The equipment used in this method looks like a row of soapers, and each treatment bath is driven separately, tension on the yarn is controlled, and the shrinkage due to swelling during absorption of the alkali and the level of strain after this can be adjusted freely. This type of equipment can produce a large amount of yarn of consistent quality and so this method is suitable for the production of mercerised yarn for use in knits, and the treatment of fine yarn that is two-fold yarn with a yarn count of around 100–110 is also possible. One problem concerning the type of equipment used is the separation of the yarns in the rope after it is dried at the conclusion of the mercerisation process, and the way in which the yarns are unwound is very important for ensuring the smoothness of the operation.

2.1.9   Warp mercerisation

While tow mercerisation involves the treatment of yarn lined-up in rope-form, in warp mercerisation yarn is wound onto a beam and fed into a machine with the same system as in a slasher-sizing machine. Mercerisation takes place with sheets of separate threads, and the machinery used can be exactly the same as that in tow mercerisation. Thus, in the warp-beam method and the tow method, only the handling is different. Problem, however, is that during treatment the breakage of a single thread can lead to major difficulties, and so if the yarn is not of very good even quality, industrial implementation of this method is difficult.

2.1.10   Chainless/roller mercerising machine

In the classic roller mercerisation the fabric is fed between the rollers with permanent positive guidance. The shrinking process is prevented as far as possible using a special roller arrangement in the mercerising compartment, as the fabric adheres to the rollers. However, if the contraction forces are greater than the adhesive and friction forces between fabric web and roller, the selvedges start to shrink and thicken. Chainless mercerisation systems cause selvedge thickening of approx. 6–9%.

This method of mercerisation involves running fabric through a number of rollers without the use of a clip stenter is also called roller mercerisation.

The machine used has a number of stainless rollers, or stainless and rubber rollers, of a relatively-large diameter tiered zigzag in close contact to each other inside a long trough, with the lower tier designed to submerge in alkaline solution for mercerisation.

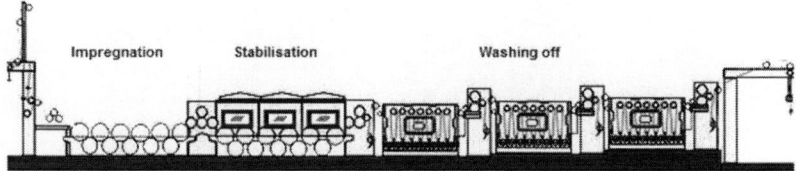

Figure 2.13. Schematic diagram of a chainless merceriser

The absorption of alkaline solution and fabric swelling take place as fabric sequentially glides through the surface of these rollers, and, although this movement from roller to roller in close contact with them reduces the width wise contraction to a minimum, the resulting fabric expansion remains within a limited range, thus displaying the mechanism of mercerisation at fixed length.

Figure 2.14. A chainless merceriser

To achieve maximum effects, the scouring of the liquor must firstly be carried out under tension. The installation of effective expander rollers in the first rinsing section (See Figure) (stabilisation section) of chainless fabric mercerising machines is therefore important. The fibre swelling of the cotton in caustic soda liquor is a fundamental pre-requisite for successful mercerisation. This arrangement is equivalent to the chain section of the chain mercerising unit (see Chain Mercerising section)

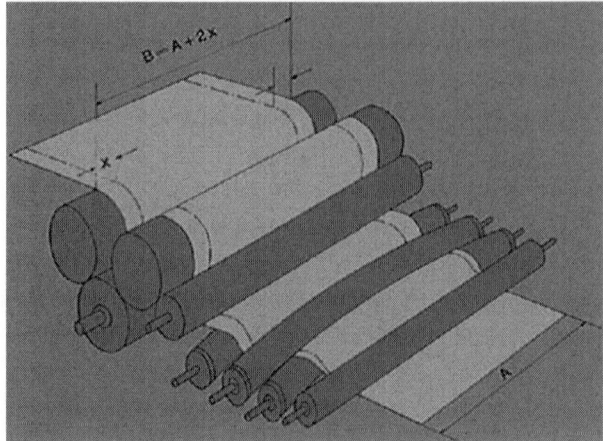

Figure 2.15. Expansion rolls arrangement and the increase
of width of the fabric in stabilisation unit

A similar device is used for the removal of most alkali following this initial stage of alkali penetration and fabric swelling, and an open-width soaping machine for further removal and neutralisation. Therefore, the machinery required is extremely concise and the cost is low, in comparison with the chain mercerisation. However, this method is subject to a considerable number of constraints due to inflexible width wise control over fabric depending on the kind and use. Mercerising of pet and cotton blends poses a little problem to the processor with respect to dimensional stability. This problem looks larger when the PET content is much low since the dimensional stability to the fabrics of higher PET content can be achieved by heat setting which is not the case with lower polyester. Roller mercerisation is not suitable for these sheer plain weaves. This type of machine is widely in use in India, Europe, in contrast to its scarce usage in Japan.

2.1.11   Chain mercerizing machine

In order to make up for the shortcomings of the roller mercerising machine, a clip stenter is used for post-mercerisation treatment, in which a width wise tension is applied. The critical alkali content is showered-off (removed) from the fabric during the passage of the fabric through the stenter, followed by thorough alkali removal and neutralisation using an open-width washing machine. In practice a heavy padding mangle is used for the application of alkaline solution using a 2 dip, 2 nip mangle, while allowing the fabric to have sufficient time for penetration and swelling of the cellulose in a timing cylinder, instead of undergoing an operation using so many rollers and so much solution as in roller/chainless mercerisation, to ensure reduced use of the alkali.

Figure 2.16. Chain merceriser

In chain mercerising systems the fabric is fed freely in the impregnation and reaction area. To compensate for the resulting loss in width, the fabric is tensioned width-wise using a stenter frame. The selvedge thickening is less due to the even pre-shrinkage. However, in many cases the loss of width cannot be entirely compensated. Unsatisfactory residual shrinkage figures in the finished product are the consequence.

Since the chain mercerising machine operates at an extremely high speed of 120–200 m/min, a clip stenter is commonly used after two consecutive treatments of alkali application/penetration. It is a device of considerable size, capable of holding, while maintaining a width wise tension, 70–90 m fabric at a speed of 120 m/min, or 117–150 m at 200/m, so that sufficient time is allowed, approximately 35 s for polyester/cotton blends, and 45 s for 100% cotton, between the initial application of alkali solution and the subsequent start of showering the alkali of the fabric. Furthermore, thorough removal of the alkali is ensured in this stenter stage, through repeated showering of low concentrated alkali and vacuum treatment.

The efficiency of the vacuum treatment will be most influential in the removal of alkali, especially in cases of using heavy cotton weaves, deficiencies in alkali removal makes the showering, even in an increased amount, an ineffective flow over the fabric surface and allows the fabric to be released from the stenter while still immature, resulting not only in incomplete setting of the width wise dimension but also in fluctuations in the dyeing stages that follow. Moreover, in the case of sheer cotton weaves, sufficient application of alkali solution will be important, since the relationship between controls over tensile strength for the obviation of crease production during the timing cylinder stage, controls over the fabric width on the stenter and the amount of alkaline solution required is extremely delicate.

Considering the points mentioned above, the performance of chain mercerising machines developed to date seems hardly satisfactory. The removal of the remaining alkali after the stenter stage barely comes into question in terms of the resulting mercerisation effect, however, a crucial watershed will be

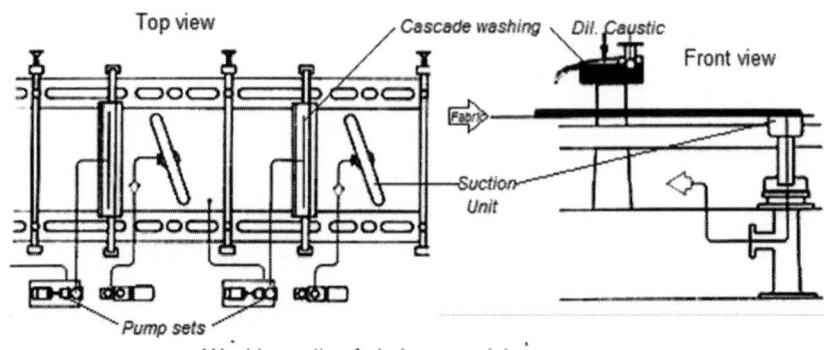

Figure 2.17. Washing arrangement on stenter chain of chain mercerising

whether the remaining alkali can be reduced to less than 3°Bé before the fabric leaves the stenter. On the chain, there are a series washing units in which the dilute alkali is pumped from the next unit and dropped on the fabric on the chain through cascades and as the fabric moves the caustic solution is sucked from below the fabric by suction unis (see Figure above). After the chain section the fabric enters normal washing units where they are hot, warm, cold washes and neutralisation takes place.

Disadvantages of chain mercerising

- The inherent disadvantage of chain mercerising is–as the force for keeping the material under tension acts mainly on the outer edges and the line of force diminishes towards the middle, a greater elongation takes place at the edges than at the middle of the fabric.

- The warp density, threads per cm are uniform before stentering and after stentering it is less at the edges than at the middle of the fabric.

- It is also noteworthy to remember that the control over shrinkage in warp way is better when compared with weft way.

- In other words weft tension control is standard and not satisfactory.

2.1.12   Merceriser with combined chainless–Chain combination

There are mercerisers with chainless impregnation combined with short chain portion for width stabilisation. The idea is to take the advantage of both type of mercerisers into one. Benninger Dimensa is the best example for this. It has the

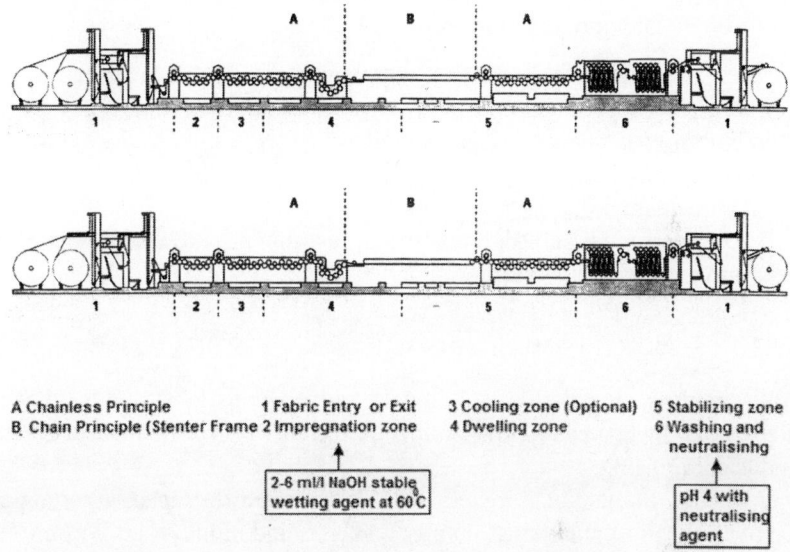

A Chainless Principle   1 Fabric Entry or Exit   3 Cooling zone (Optional)   5 Stabilizing zone
B Chain Principle (Stenter Frame 2 Impregnation zone   4 Dwelling zone   6 Washing and
                                                                   neutralisinhg

2-6 ml/l NaOH stable
wetting agent at 60°C

pH 4 with
neutralising
agent

Figure 2.18. Schematic drawing of Benninger chainless merceriser

impregnation section and dwelling section and final washing zones in chainless format and the stabilisation section in chain format.

The width loss in the impregnating zone is therefore less than on clip mercerisers. Due to the stenter frame the width can be corrected. The washing out of the caustic soda is completed in the subsequent stabilisation compartment using the principle of roller mercerisation. In this way fabric width shrinkage is prevented. The use of the controls selvedge stretching unit is new. Selvedge thickening produced in the roller mercerisation compartment is effectively corrected with this unit.

Selvedge thickening and control

The selvedge thickening also depends very heavily on the weaving construction and the yarn.

- The more complex the weaving construction, the greater the selvedge thickening (lowest thickening with L1/1, highest thickening with jacquard)

- Plied yarns cause greater selvedge thickening
- The coarser the yarn, the greater the selvedge thickening
- The greater the width difference between grey width and finished width, the greater the selvedge thickening (grey width losses of 12–17% produce high selvedge thickening)
- The selvedge thickening is predominantly to be seen at the edges of the fabric web and reduces to 0% towards the middle (after a distance of approx. 20 cm from the selvedge).

There are machines developed with mechanical selvedge stretching unit integrated into the machine. The required selvedge expansion may be adjusted both mechanically and also automatically. When the wales thickening is corrected on the edges of the fabric web in advance resulting in a varying weight per square meter in the treated area.

With mechanical selvedge control an instrument is available that effectively prevents selvedge thickening on woven fabric and knitwear.

2.1.13   Batch mercerisers

In this method, an alkaline solution is padded onto fabric which is then rolled-up, and when padding is completed the alkali is removed through continuous cold rinsing.

Although the use of the method is not common in Japan, a certain degree of application, including in knits, can be found in India and Western Europe. Despite costs for facilities being remarkably low, it is not an interesting method except for some special cases, as quality management and productivity remain problematic.

Figure 2.19. A batch merceriser

Still, for the growing cases of carrying out alkali reduction for the polyester side of cotton/polyester blends to achieve both the mercerisation of cotton and the alkali reduction of polyester in a single treatment, the application of this cold batch method is particularly interesting as a device that can combine the two separate stages which would otherwise raise facility problems.

2.1.14   Mercerising knitted fabrics

Knitted fabric is mercerised in tubular and open width form. Of late the open width knitted mercerising is becoming more popular.

The fabrics are impregnated in a short impregnating zone followed by dwelling in the chainless section of the machine. A minimum tension is exerted on the fabric during transportation in the chainless dwelling zone. The tension is controlled by the ratio metered AC drives located at the fabric transportation drums. The dwelling zone is followed by a vertical return pin chain/stenter section, where the fabric is uniformly stretched during the gradual removal of the caustic content. This means an initial stabilisation of the fabric already begins in the chain zone followed by further stabilisation in the chainless section of the machine. Fabrics stabilised in this manner are subsequently washed and neutralised in washing compartment. The combination of the chain and chainless principles ensure the dimensional stability of the knitted fabric. The dimensions set during the processing are fixed and memorised permanently.

The true mercerisation of knitted machine was a problem since after the padding, chiefly the knit goods shrink very considerably and if the stabilising does not take place under tension (inflation of the tube), it is effectively

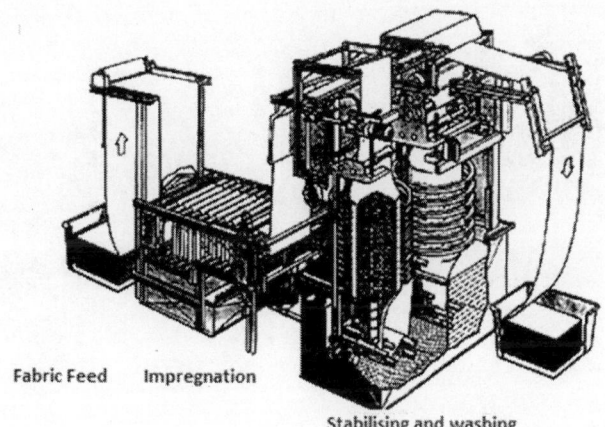

Fabric Feed   Impregnation

Stabilising and washing

Figure 2.20. Schematic diagram of a tubular merceriser (Dornier)

only causticised. Hence, width tension of 5–10% across the desired final width and longitudinal tension is 20–30% was always a problem in case of tubular fabric was always a problem. Dornier machines gives a most efficient mercerisation with stretched the tube using expandable "cigars" during stabilisation and also rinsing with same type cigars, which were expandable in the diameter. The longitudinal tension is given by a series of rollers.

There are many types of mercerising machines available in the market. One of them is mercerised in the tubular form in tandem with a continuous bleaching range in rope form. One of such machine is shown in the Figure below. Nowadays the demand for mercerising has been reduced in the market especially after the introduction of pad batch dyeing of cotton knotted materials in the open width form which gives a near to mercerised dyed effect.

Tubular mercerising Machine

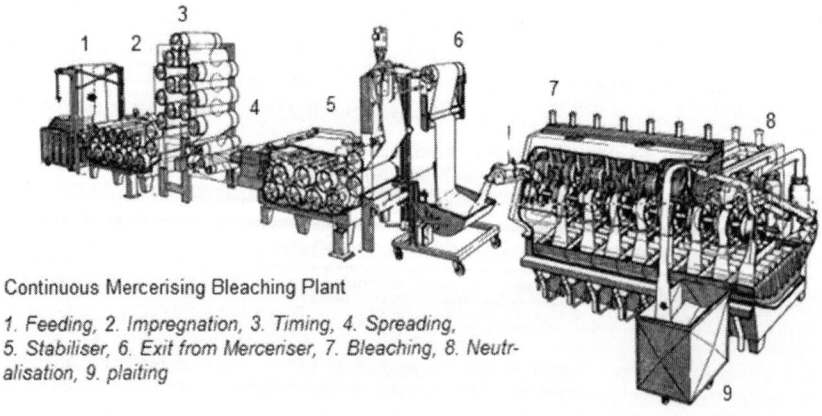

Continuous Mercerising Bleaching Plant

1. Feeding, 2. Impregnation, 3. Timing, 4. Spreading,
5. Stabiliser, 6. Exit from Merceriser, 7. Bleaching, 8. Neutralisation, 9. plaiting

Figure 2.21.

(a) Cotton yarn

Table 2.5. Parameters of mercerising of knitted material

Items	Quantity/time
Caustic soda	28–30°Bé (48.2–52.6°Tw)
Suitable wetting agent	4–8 ml/l
Temperature	15–25°C
Preliminary tensioning	25 s
Shrinking	40 s
Stretching	50 s
Squeezing	20 s
Hot water	30 s
Cold water	25 s
Squeezing	15 s

(b) Piece goods

Table 2.6. Normal mercerising/causticising process

Form	Woven and knit goods			
Machine	**All mercerising and or causticising machine**			
Type of treatment	**Mercerising**	**Causticising machines**		
Substrate	Co, PES/Co, FI, PES/FI	Co, PES/Co, FI, PES/FI	CV, PES/CV	
Guide recipes				
Caustic soda °Bé	26–30	18–22	6–7	6–7
Mercerising wetting agent ml/l	4–6	6–10		
Wetting/scouring agent			2–4	2–4
Liquor temperature °C	14–18	20–25	20–25	20–25
Time s	40–60			
Pick-up %		80–100	80–100	80–100

Notes:

The subsequent rinsing operations must adapted to the machine conditions.
Causticising: Sample working method:

- impregnate
- plait down or roll-up
- batch 10–60 min at room temperature
- wash in open width or rope form

at 40–50°C for CV and PES/CV
at 80–90°C for Co, PES/Co, FI and PES/FI.

Table 2.7. Cold causticising

Form	Woven
Substrate	Co, PES/Co, FI, PES/FI
Machine	Pad batch machine and OW washer
Type of treatment	Causticising and alkaline shock on OW washer
Guide recipe	
Caustic soda solid	22–23°Bé
Mercerising agent for hot reductive causticising	25–30 ml/l
Working method	
- Saturator	
Impregnate dry goods	40–60
Liquor temperature °C	20–30
Pick-up %	100
– Batch	
At room temperature	4–16 h
– *Wash-off*	110–140
With alkali shock	5–20
Compartment 1 and 2	Water at 90–100°C
Compartment 3 and 4	Water at 90–100°C + 10 g/l soda ash and suitable scouring agent
Compartment 5 and 6	Water at 90–70°C
Compartment 7	Cold water and neutralisation

Notes:
1) Impregnation liquor = 22–23°Bé caustic soda; all feed liquors = caustic soda 25°C.
   When setting the liquor with caustic soda compensate for any diluting heat by adding ice is advisable to use soft water (max. 6.25°Bé) for all treatment liquors.
2) If the goods have been submitted to previous wet treatment, e.g., desizing and scouring, essential to dry them before causticising.
3) Alkali shock–necessary to improve absorbency. Metering of strengthening bath required.

2.1.15   Stenter mercerising (SM) process

Stenter mercerisation (SM), is a new type of processing technique for knit and woven goods of cotton and its blends with modal or polyester fibres. A basic operation of wet processing (mercerising) can now be carried out completely different with the SM process. For knit goods in particular, this results in new processing possibilities. It leads to articles of a high quality standard. The mercerising process is done on a normal stenter with slight changes in the machine. As a strong caustic soda is used on the machine all parts which are coming into contact has to be made of ss. The brush trim in the punching may only contain polyamide brush hairs. The squeezing rubber rollers should be resistant to 23–25°Bé NaOH. The process includes padding at room temperature, stretching the width on chain, and drying at 130°C, followed by scouring. This method of mercerising can achieve very good dimensional stability with a high uniformity of fabric structure and high shine. The strength of caustic soda, temperature and level of caustic soda has to be maintained and controlled as in the case of other type of mercerising. After the padding, chiefly the knit goods shrink very considerably. The feed phase (before the drying system) is carried out under stronger longitudinal tension. The width tension should be applied in such a way that the fabric is also pinned down in a heavily tensioned state, but without producing clip defects. When drying, the following rules of thumb should be observed for the dimensional stability and correct setting of the knitted geometry: width tension of 5–10% across the desired final width plus 3 cm for selvedge cutting (single jersey). The longitudinal tension is 20–30%. The adjustment of the width and longitudinal tension is a matter of experience and depends on the final dimensional stability, knitted geometry and final weight per square metre. The fabric appears

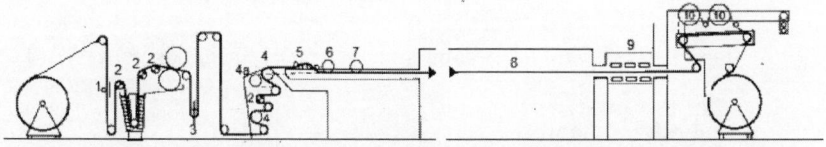

Mercerization on the stenter frame by Sandoz. 1 = stretcher; 2 = expander roller; 3 = floating roller; 4 = drawing roller; 4a = drawing roller lowered for direct fabric feed; 5 = finger spacers; 6 = shrink roller; 7 = rotary brush; 8 = drier; 9 = cooling zone; 10 = cooling drum.

Figure 2.22. Stenter Mercerising

yellow as it gets completely dried. The fabric can be plaited (knitted fabric) or batched and further scoured using also the caustic soda left on the fabric whereby the original colour and shine and other properties of the mercerised fabric is obtained. The method is not very well practiced now.

Advantages of the SM process

SM process (stenter mercerising process) makes it possible to set selected dimensions for problematic fabrics (knit goods–flat knit goods and cut circular knitted fabric) during mercerisation which are readily reproducible. Compared to conventional mercerising of knit goods in cut open form, this offers the following advantages:

- Regular surface area

  No thickening at the edges, i.e., a uniform, geometric setting of the stitch structure over the entire width and length of the goods.

- Dimensional stability
  - Improved stretch and deformation behaviour in the dry state
  - Improved creasing behaviour in the wet state enabling reproducible setting of the desired low weight or achievement of the prescribed final width

  Optimum setting of dimensional stability is obtained with the appropriate finish.

- Dye yield

  Like the conventional mercerising processes, the SM process also effects a considerable increase in the dye uptake capacity.

  Machine conditions for the use of the SM process:
  - Wherever the purchase of a mercerising range is not worthwhile for production reasons
  - Wherever it is preferred to purchase a stenter (instead of a mercerising range) due to its versatile application possibilities, e.g., for applying the SM process, finishes, curing/fixing operations.

  The SM process comprises two stages of treatment:
  Stage 1 = stenter mercerisation
  Stage 2 = finishing for optimum properties of the goods with regard to

- Handle
- Dimensional stability
- Sewability.

Table 2.8. Parameters of Stenter Mercerising

Form	Woven and knit goods	
Machine	Stenter	
Substrate	Cotton, cotton/modal, PES/Co, grey or bleached goods	
Type of treatment	Silicate alternative	Lubricating agent alternative

Guide recipes		
Caustic soda °Bé	23	22
Mercerising wetting agent ml/l	50	50
Sodium silicate 38 °Bé ml/l	30–50	
Suitable lubricating agent ml/l		5–10
Sequestering agent ml/l	2–4	
Working method–Padding		
Impregnation temperature °C	20–30	
Impregnation time s	5–10	
Drying °C	120–130	

Notes:

1) Washing-off: as hot as possible and neutralise, open-width washer, winch, overflow, etc.

2) The width and lengthwise tension depends on the article:

   Remarks on setting the pad liquor:

   − If possible use soft water (up to maximum 6.25° Bé)

   − Maintain the above order of addition

   − Dilute caustic soda to the largest possible volume with water; compensate for any diluting heat by adding ice

   − Dilute mercerising wetting agent and sodium silicate with cold water 1: 1 or

   − Dilute lubricating agent with cold water 1: 10, then add mercerising wetting agent liquid and mix.

3) Pad liquor = 23°Bé caustic soda, feed liquor = caustic soda 25°Bé

4) The lubricating agent alternative is particularly suitable for a soft handle.

5) Lubricating stability in caustic soda has to be checked.

2.1.16   Caustic recovery

The explanation of mercerisation is not complete without the explanation of caustic recovery. In a mercerising plant, as explained earlier the strong caustic soda is applied on the fabric and after the mercerisation process is over the caustic soda is washed-off from the fabric. Wash waters from the mercerising process contain the bulk of the caustic soda used during fabric or yarn mercerising. The economy in the mercerisation cost is more depends on the reuse of the washed liquor (caustic soda) effectively. There may be or may not be the same processes where the weak caustic soda can be used, but these processes may not consume the whole weak liquor. The caustic soda from such wash waters is considered quite suitable for use either in other wet-processing treatments requiring dilute NaOH or for reuse in mercerisation after concentrating dilute wash waters by the process of evaporation. Some of the methods of reuse are given below:

1. Use the liquor for preparing merceriser strength of caustic soda (26°Bé) from caustic lye (46°Bé)

2. Use the liquor for dissolving solid caustic soda to merceriser strength

3. Evaporate to merceriser strength and reuse as such.

The evaporators used are generally designed to concentrate the wash liquors by effecting the evaporation of the water in two or three stages. They are technically known as double-effect or triple-effect evaporators. Figure below shows a triple-effect evaporator schematically.

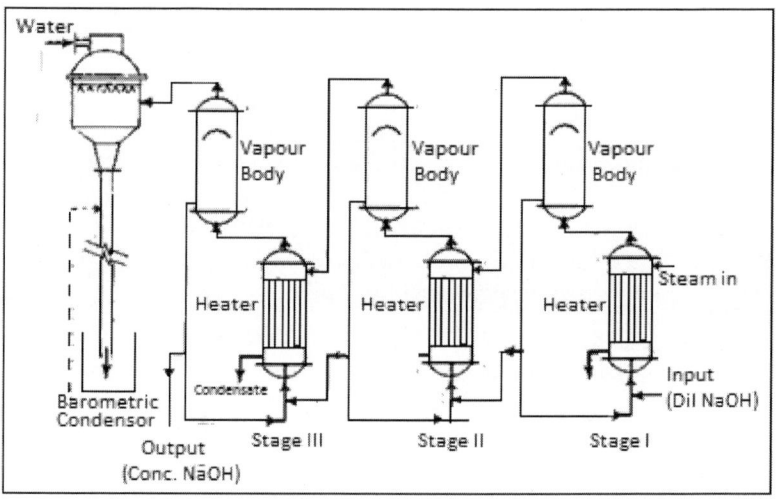

Figure 2.23. Typical 3 Stage Caustic Recovery Plant

Usually the evaporation is done at a lower pressure than atmospheric so that the boiling point of the liquor is reduced. This is done using a barometric condenser (See figure 2.33). The evaporation is done in stages (usually 3, some evaporators are designed for 2 stages also) where by the pressure in each stage is adjusted so that each stage the evaporation takes place at a lower temperature than the previous one. Thus the vapour produced in one stage (vapour body) can be used in the next evaporator. The condensate from the may be used in any process where hot water is used (Boiler, washing, etc.).

Checking of effectivity of mercerising

- Microscopic analysis

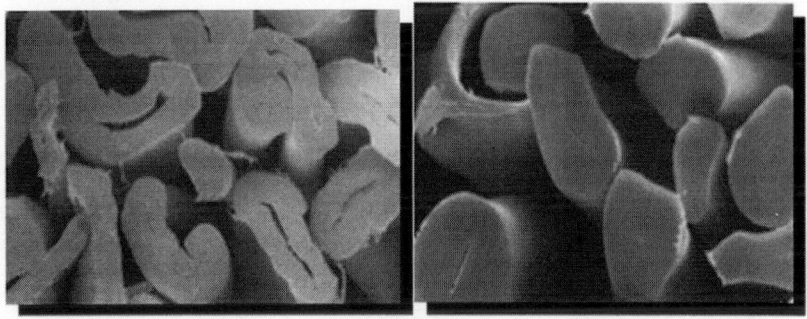

Unmercerised	Mercerised

Figure 2.24.

- Barium activity number

treatment with Ba(OH)₂-solution, determination of the use of Baryt-solution by titration with 0.1 N HCl.

$$\text{Baryt number} = \frac{(a-b)\times 100}{(a-c)}$$

a = consumption of HCl for Ba(OH)$_2$–solution (blank sample)
b = consumption of HCl for Ba(OH)$_2$–solution of the merc. sample
c = consumption of HCl for Ba(OH)$_2$–solution of the not merc. sample.

2.1.16.1   Other analysis method for mercerising process

Determination of NaOH concentration

Procedure
Take approximately1 l of the bath solution from the caustic measurement device.

Titration for caustic solution:
Pipette 1 ml or 10 ml of this solution into a conical flask. Mix it with approx. 50 ml of distilled water. Titrate with 1 N/1 M hydrochloric acid or 1 N/0.5 M sulphuric acid against phenolphthalein until the colour becomes colourless.

Calculation:
1 ml 1 N HCL = 40 mg NaOH 100%
Conversion factor

Factor for determination of NaOH concentration
10 ml solution titrated with 1 N HCL solution

Consumption in ml HCl 1 N × factor 4.0 = g/l NaOH 100%

Table 2.9.

Degrees Baume	Degrees Tw added	Weight %	Y - spec. weight	1 kg contains NaOH [g]	1 litre contains NaOH [g]	Degrees Baume	Degrees Tw added	Weight %	Y - spec. weight	1 kg contains NaOH [g]	1 litre contains NaOH [g]
1	1.4	0.59	1.007	5.9	6.0	26	44.0	19.7	1.220	196.5	239.7
2	2.8	1.18	1.014	11.8	12.0	27	46.2	20.6	1.231	206.0	253.6
3	4.4	1.85	1.022	18.5	18.9	28	48.2	21.6	1.241	215.5	267.4
4	5.8	2.50	1.029	25.0	25.7	29	50.4	22.5	1.252	225.0	281.7
5	7.2	3.15	1.036	31.5	32.6	30	52.6	23.5	1.263	235.0	296.8
6	9.0	3.79	1.045	37.9	39.6	31	54.8	24.5	1.274	244.8	311.9
7	10.4	4.50	1.052	45.0	47.3	32	57.0	25.5	1.285	255.0	327.7
8	12.0	5.20	1.060	52.0	55.0	33	59.4	26.6	1.297	265.8	344.7
9	13.4	5.86	1.067	58.6	62.5	34	61.6	27.7	1.308	276.5	361.7
10	15.0	6.58	1.075	65.8	70.7	35	64.0	28.8	1.320	288.3	380.6
11	16.6	7.30	1.083	73.0	79.1	36	66.4	30.0	1.332	300.0	399.6
12	18.2	8.07	1.091	80.7	88.0	37	69.0	31.2	1.345	312.0	419.6
13	20.0	8.78	1.100	87.8	96.6	38	71.4	32.5	1.357	325.0	441.0
14	21.6	9.50	1.108	95.0	105.3	39	74.0	33.7	1.370	337.3	462.1
15	23.2	10.30	1.116	103.0	114.9	40	76.6	35.0	1.383	350.0	484.1
16	25.0	11.06	1.125	110.6	124.4	41	79.4	36.4	1.397	363.6	507.9
17	26.8	11.84	1.134	118.4	134.0	42	82.0	37.7	1.410	376.5	530.9
18	28.4	12.60	1.142	126.0	145.0	43	84.8	39.1	1.424	390.6	556.2
19	30.4	13.50	1.152	135.0	155.5	44	87.6	40.5	1.438	404.7	582.0
20	32.4	14.35	1.161	143.5	166.7	45	90.6	42.0	1.453	420.2	610.6
21	34.2	15.15	1.171	151.5	177.4	46	93.6	43.6	1.468	435.8	639.8
22	36.0	16.00	1.180	160.0	188.8	47	96.6	45.2	1.483	451.6	669.7
23	39.0	16.91	1.190	169.1	201.2	48	99.9	46.7	1.498	467.3	700.0
24	40.0	17.81	1.200	178.1	213.7	49	102.8	48.4	1.514	484.1	732.9
25	42.0	18.71	1.210	187.1	226.4	50	106.0	50.1	1.530	501.0	766.5

Determination of caustic on the fabric

Procedure
1. Sample after padding pick-up
2. Sample after padding pick-up and NaOH on the fabric (ML)
3. Sample after padding and NaOH on the fabric (MS)
4. Sample after padding and NaOH on the fabric.

The above-mentioned samples were cut from the fabric when the machine was running in production. To take or cut the sample, the machine has to be stopped (this must be an emergency stop and not a normal stop, as otherwise an accurate result will not obtained).

The samples were cut with a sharp cutter of approx. size 10 × 10 cm
- The sample is immediately transferred into a closed conical flask.
- Determine the WET weight of the sample.
- Determination of the amount of NaOH on the fabric.

Take the sample, put it into the conical flask and dip it in distilled water (the sample must be completely covered with water). Keep the sample for approx. 30 min prior to titration. The flask should be stirred every 10 min so that the alkali in the fabric will be mixed properly with the water.

Titration

Add 3–4 drops of phenolphthalein (0.1% solution) to the flask solution (with the sample)

Titrate with 1 N or 0.1 N hydrochloric acid (HCl) until the pink colour disappears, add a few drops of phenolphthalein while shaking the flask to check whether the pink colour appears. After titration, remove the sample from of the flask (caution: including any loose yarn) and dry the fabric until it is completely dry.

Determine the dry weight of the fabric

Grams of 100% NaOH per kg of fabric

$$= \frac{\text{Consumption HCl} \times 40 \text{ 1 N HCl} \, (\times 40.1 \text{ N HCl})}{\text{Dry weight of the fabric}}$$

Determination of NaOH on the fabric (reverse titration)

Procedure

Put a cutting of fabric into an Erlenmeyer flask filled with distilled water. Add a few drops of methyl orange and 20 ml of 0.1 N HCl. The colour of solution must be pink, otherwise add another 10 ml of 0.1 N HCl. Boil this solution for five minutes, allow it to cool and titrate it with 0.1 N of NaOH until the colour changes towards orange. Afterwards rinse the fabric sample, dry it and weigh it.

Calculation

Grams of 100% NaOH per kg of fabric $= \dfrac{b \times 4}{a}$

Where

a = dry weight of fabric.

b = quantity in ml of 0.1 N HCl minus the quantity in ml of 0.1 N NaOH.

Determination of selvedge thickening by fabric weight (g/m²)

Method: Comparison of weight between selvedge/centre/selvedge

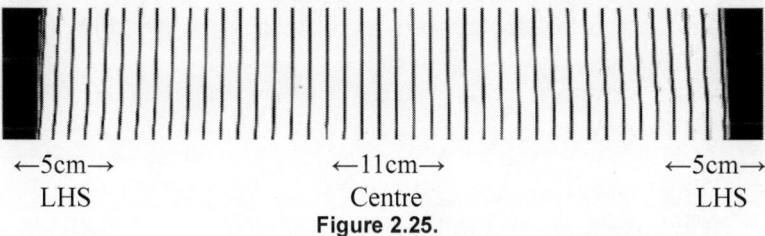

←5cm→   ←11cm→   ←5cm→
LHS   Centre   LHS
Figure 2.25.

Cut samples at the place shown in the diagram and weigh.

Calculation

$$\frac{\text{Weight in the centre} - \text{weight at the side} \times 100}{\text{Weight in the centre}}$$

Example

Side: 150 g/m², Centre: 140 g/m²

$$\% = \frac{150\text{g/m}^2 - 140\text{g/m}^2 \times 100}{140\text{g/m}^2}$$

$= 7.1\%$ thickening.

2.1.17   Ammonia mercerisation

Treatment of cotton fibre with liquid ammonia produces similar effect that of mercerisation developed by Coats in 1960. Like mercerisation with caustic soda, ammonia produces changes in the structure of cellulose – reorganisation of the crystalline network with cellulosic chains rotating and translating around their axle – giving a better accessible network to reagents. The process was then introduced in the 1970s, originally in the USA, as a finishing treatment for denim fabric and other heavy cotton fabrics, as the treatment causes interesting effects on these articles, in particular an excellent dimensional stability to washing and tumble-drying, a pleasant soft handle and improved smooth-drying properties. The liquid ammonia process is increasingly being used as a pre-treatment for shirt, blouse and dress materials made from 100% cotton which are subsequently given an easy-care finish.

The differences between the two types of mercerisation are as follows:

1. Penetration of ammonia into the fibre and its elimination are nearly instantaneous (between 15 and 25 s). Therefore, the treatment is very fast.
2. Ammonia is recoverable and reusable after a purifying distillation. Caustic solution after mercerisation, especially in grey form, is soiled and it creates pollution problems.
3. Water consumption is reduced to less than half in ammonia mercerisation.
4. Ammonia is a natural substance and the process may be considered as ideally eco-friendly.
5. It gives reagent-free textile goods.

The liquid ammonia treatment is widely accepted for yarns used in sewing threads, and special fabrics like Denims, Corduroys, chambrays, Pillow material, linen, jute and blends of cellulose with PET and/or nylon. Among various amines, the liquid ammonia appears to be unique in its swelling action on cellulose and its effect on crystal structure. Anhydrous liquid ammonia being smaller molecule, penetrates cellulose rapidly and complexes with hydroxyl groups of cellulose after breaking hydrogen bonds in crystalline regions and increases the distance between cellulose chain in crystallites.

2.1.17.1 Process and theory

The process consists of treatment of the fabric by impregnation with liquid ammonia at its boiling temperature (–33°C), thus causing the cellulose fibres to swell rapidly. The ammonia is then volatilised by guiding the fabric over a heated drying cylinder and any residual traces are removed from the material in a subsequent steaming zone. The expelled ammonia is liquefied once again and reused. The fabric is treated in various finishing stages, e.g., boiling-off (bleaching is carried out after the ammonia treatment), mercerising, solid-shade dyeing, or colour-woven.

In cotton yarn, the cellulose polymer chains are supposed to be in stretched semi-rigid chains stabilised by intramolecular hydrogen bridges and intramolecular forces (Van der Waals forces). As a result the chains arrange themselves into supramolecular ordered sectors, whereby the crystalline areas alternate with somewhat less distinct ordered areas in the fibre axis. The ordered states are arranged in fibrils which are separated from one another by long chain interference zones (micro voids) along the fibre axis. The liquid ammonia is able to swell the fibrils in such a way that the ordered states in the fibrils are also affected, including the crystallites (intrafibrillar and intracrystalline swelling). This is principally due to the higher basicity of the ammonia molecules in comparison to water which makes it possible to split the dense hydrogen bridge network into crystalline zones. If inter- and intra-molecular interactions are split in oriented cellulose fibres due to the effect of a swelling agent, the fibre can be shrunk to a greater or lesser degree. The main cause of this is the disorientation of fibrils and molecular chains during the swelling process. If the fibre is kept at a constant length during this process, a shrinking force develops. The extent of shrinkage and the shrinking force is dependent on both the orientation of the fibre structure and the fibrillar and crystal structure, as well as on the swelling conditions (swelling agent, temperature, tension).

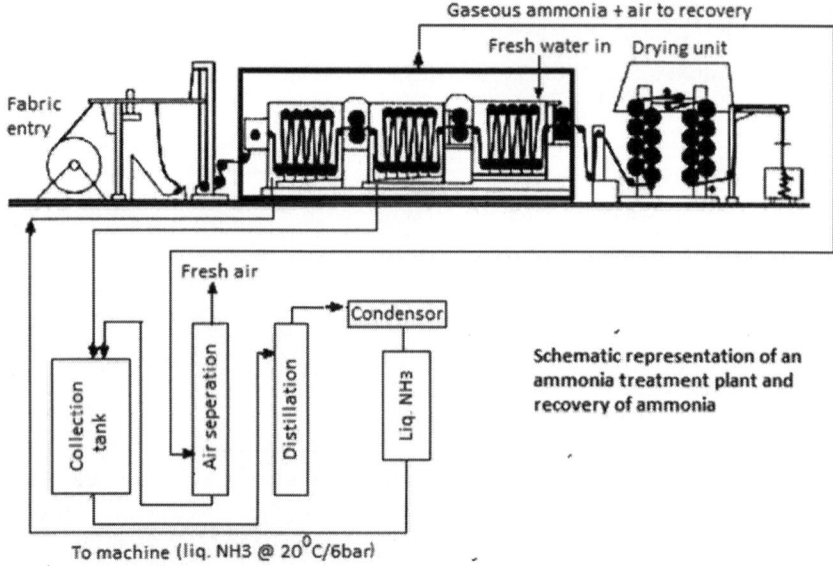

Figure 2.26.

The ammonia mercerising changes the fibre properties compared to NaOH mercerising as follows:

The lustre produced by ammonia mercerising is less than caustic soda mercerising.

The increase in dye affinity and the change in appearance by conventional mercerising are almost retained by ammonia mercerising also.

Ammonia, a weak alkali, does not damage the fibre in contrast to the very aggressive caustic soda. Mechanical properties like abrasion resistance, tensile and tearing strength are improved. Ammonia treatment increases tensile strength by 40% and the elongation at break by 2–3% of that of untreated cotton.

Less swelling by ammonia causes the fibre to become more plastic and malleable. The article life is increased and the new appearance is retained for a longer time. The difference in swelling also explains the more brilliant aspect of the fibre after caustic soda treatment and the satin gloss after the ammonia treatment.

Two processes known as "PROGRADE" for yarn mercerising and "SANFORSET" for woven and knitted fabrics are widely used.

2.1.18　　Prograde process–J. & P. Coats Ltd. (Scotland)

In this process, the yarn is treated with liquid ammonia at its boiling point of –33°C for less than a second (0.7 s), followed by immersion of yarn under tension in hot water (93°C) for about 0.1 s where it is stretched 10–30% as measured against ammonia-swollen length. Ammonia mercerising achieves a 40% increase in tensile strength, improved lustre, etc. The yarn after ammonia treatment is stretched and ammonia is removed by hot water washing. The yarn is wound on a spool and dried continuously with hot air. The method, developed primarily for sewing thread, is claimed to be one thousand times faster than traditional yarn mercerising processes. Prior moisture treatment (up to 30% of dry weight) improves the effects, especially uniform dye pick-up.

But in case of fabric mercerising is done with continuous lengthways tension. The process consists of impregnation of goods with liquid ammonia at atmospheric pressure (i.e., at the boiling temperature of –33°C) followed by the removal of ammonia by evaporation or washing.

2.1.19   Sanforset process

This process represents a combination of exclusive liquid ammonia processing and controlled compressive shrinkage to provide wash and wear characteristics without strength loss on cotton denims along with an unusual soft and supple handle. Monforts, Germany has successfully developed Sanforset treatment range.

After entering process, the material passes over five pre-drying cylinders then through a cooling station. The drying cylinders reduce the moisture content of the material below the humidity, while the cooling station takes care of the cooling of the material. Then the material passes into actual treatment chamber where it is impregnated with liquid ammonia in a trough. The treatment chamber is kept under slight vacuum to prevent ammonia from escaping. Finally it is squeezed-off in a padder and the ammonia is driven-off in two felt calendars. In this heat treatment 90–95% of ammonia is recovered. The rest which is chemically bound to the cellulose is removed in a steaming compartment consisting of a pre-dwelling zone and a steaming zone. The entering and take-off compartments are sealed by locks. An after dwelling zone and take-off zone complete the process. The evaporated ammonia is led to a recovery unit where it is compressed, cooled and liquefied to be led to a storage tank. The chemically bound ammonia is removed from the material separately, by dissolving in water and reusing it as chemical or manure. It can also be heated until it escapes and is oxidized. However, care must be taken to see that no pollution occurs.

Advantageous of ammonia mercerisation

- Less pollution
- Higher productivity

- Blends can be mercerised
- Jute and linen can be mercerised
- Evenness and uniformity of swelling is comparatively better
- Comparatively softer feel
- Very suitable for heavy weight fabrics like denims
- Moisture regain is comparatively better
- Absorbency is comparatively better
- Fabrics will not shrink during domestic laundering
- Wash and wear properties improved
- Better CRA's if finished with resins
- Good shrinkage stability
- High smoothing capacity
- Almost same mercerising effect as compared to caustic soda mercerisation.

Disadvantages of ammonia mercerisation

- Degree of swelling is less
- Higher shrinkage
- Lower lustre
- Lower colour yield
- Very costly
- Ammonia recovery even though not a problem with the machinery developed, leakages will do great harm to the personnel working.

2.1.20   Weight reduction (deweighting) of polyester

When polyester fibres are given an alkaline treatment with caustic soda in a hot medium (hydrolytic action) the surface of the fibre is peeled-off. The desired handle modification which results is silk-like. A pleasant effect of deweighted goods is their improved absorbency. Although the deweighting of PES fibres produces a loss in weight (finer titre) which can be controlled by the amount of alkali, temperature and treatment time, it also results in loss of strength which must be taken into consideration. It is therefore necessary to make a compromise and produce a fibre which is as silky soft as possible without losing too much of its strength. The attainable deweighting effect depends to a large extent on the type of PES fibre. It is therefore advisable to carry out tests in the lab in each case to determine the acceptable degree of deweighting, weight and strength losses.

2.1.20.1   Discontinuous deweighting

In most cases deweighting is carried out on the relevant dyeing machine (winch, overflow, HT beam, etc.). To accelerate alkalisation, treatment can be carried out under HT conditions or a cationic accelerator can be applied, which permits a drastic reduction in the caustic concentration. At the end of treatment an anionic surfactant is added to the liquor to stop the peeling process and to disperse the peeled fibre residues.

Table 2.10. Guide recipe and process for weight reduction of polyester

Form	Woven and knit goods
Machine	Winch, overflow, HT beam, etc.
Liquor ratio	10:1–20:1
Substrate	PES Monofil, texturised staple fibres–grey, scoured/ relaxed/and or pre-set goods
Type of treatment	Exhaust
Guide recipes	
Caustic soda solid	5–10 g/l
Cationic accelarator[1]	2–4 g/l
Suitable surfactant[2]	2–3 g/l
Working method	
	Heat within 30 min to boil
	Treat to boil for 30 min
	Cool to 80°C
	Add surfactant[2]
	Rinse warm and cold
	Neutralise with acetic or formic acid

[1]e.g., Lyogen PAL liquid.
[2]e.g., Sandopan TFL liquid.

2.1.20.2   Semi continuous and Continuous deweighting

Interesting alternatives are the continuous methods. Apart from the treatment on pad-steam ranges, cold causticising by the pad batch method is particularly suitable for deweighting 100% polyester articles.

Table 2.11. Pad batch method (cold)

Form	Woven and knit goods
Machine	Pad batch range
Substrate	PES Monofil, texturised staple fibres–grey, scoured/ relaxed/and or pre-set goods
Guide recipes	
Caustic soda solid	22–23°Bé
Sequestering agent	25–30 g/l
Saturator	Impregnate dry goods
Liquor temperature °C	20–30
	80–100%
Batch	24–48 h at room temperature
Wash-off	With soft water 60–80°C
	Neutralise with 0.5–1 g/l formic acid concentrated
	Rinse warm and cold
	on tension free washing machines or dyeing machines

Table 2.12. Continuous Steaming method

Form	Woven and knit goods
Machine	Pad team, Combi steamer, conveyor, U box, etc.
Substrate	PES Monofil, texturised staple fibres–grey, scoured/ relaxed/and or pre-set goods
Type of treatment	Steaming method, medium dwell times
Guide recipes	
Caustic soda solid	50–200 g/l
Mild reducing agent	5–10 g/l
Suitable wetting agent	20–30 ml/l

Working method

Saturator	Impregnate dry goods
Liquor temperature °C	20–30
Pick-up	80–120%
Steam	10–20 min at 120°C
Wash-off	PES–60 –80°C, PES/Co 90–100°C
	Neutralise with 1–2 g/l formic acid concentrated

2.1.21   Continuous reductive causticising process (CRC process)

This process is a new, simple, economical, rapid process for the pre-treatment of Co and PES/Co woven fabrics. In the CRC process the goods are desized, scour boiled, reduction bleached, given high dye up take capacity in a single treatment operation.

The CRC process is particularly suitable for pre-treating Co woven fabric prior to printing, especially African style prints, but also for pre-treatment of dyed goods in general. For white articles, pale brilliant and pastel shades or for complete removal of cotton husks it is advisable to bleach the goods as well, either with sodium hypochlorite or with a cold and/or hot peroxide bleach. On PES/Co fabrics special effects with a soft, supple handle can be obtained by deweighting the PES component. No long hot treatments should be given to PES/Co fabrics, e.g., on pad roll ranges, because the deweighting effect would be too severe.

Advantages of the CRC process

- Time saving
- Saving of water, chemicals and energy
- Increase in productivity
- Avoidance of possible bottlenecks in production (e.g., for washing and mercerising)
- Good degree of desizing (in difficult cases preliminary desizing is advisable)
- Good absorbency of the goods
- Adequate brightening/bleaching effect (comparable with that of a reductive bleach)
- Good dye yield of the goods (comparable with that of cold causticising or mercerising).

Table 2.13. Continuous one bath one stage CRC process

Form	Woven goods		
Substrate	Co, PES/Co–grey or desized	Co grey or desized	Co, PES/Co grey or desized
Machine	Pad-steam NT/HT all OW steamers, short to Med. dwelling	Pad roll, (hot batching) long batching times	Pad batch, (cold batching) long batching times
Type of treatment	Hot reductive causticising–one-stage causticising/ scour boiling		
Guide recipes			
Caustic soda °Bé	22–23	22–23	22–23
CRC aid chemical[1]	25–30	25–30	22–30
Working method			
Saturator			
Impregnate dry goods			
Liquor tmperature °C	20–30	20–30	20–30
Pick-up	100%	100%	100%
Steam			
Time min	5–15 (1–2 HT)	60–120	6–8 h
Steam	100–102 (NT), 120–130 (HT)	90–100	Room temperature
Wash-off	At 90–95°C, Neutralise	At 90°95°C, Neutralise	At 90–95°C, Neutralise

[1]e.g., Sirrix CRC liquid.

Notes:

1) PES/Co: Only on steaming ranges with short to medium dwell times, observe HT steam conditions.

2) Impregnation liquor: caustic soda 22–23°Bé, all feed liquors = caustic soda 25°Bé. When setting liquors with caustic soda compensate for any diluting heat by adding ice. Soft water (maximum 6.25°Bé) should be used for all feed liquors. .

3) If the goods have been submitted to previous wet treatment, e.g., desizing and scouring, it is essential to dry them before causticising.

Table 2.14. Continuous one bath two stage CRC process

Form	Woven goods
Substrate	Co, PES/Co–grey or desized
Machine	Pad batch/pad-steam NT combined long/ short dwelling
Type of treatment	Hot reductive causticising–one-bath causticising/scour boiling
Guide recipes	
Caustic soda	22–23°Bé
CRC agent[1]	25–30 ml/l
Working method	
Saturator	
Impregnate dry goods	
Liquor temperature °C	20–30
Pick-up	100%
Batch	
At room temperature	1–3 h
Steam	
Temperature	100–102°C
Time min	1–2
Wash-off	At least 90°C, Neutralise

Notes

1) Impregnation liquor: caustic soda 22–23°Bé, all feed liquors = caustic soda 25°Bé.

   When setting liquors with caustic soda compensate for any diluting heat by adding ice.

   Soft water (maximum 6.25 Bhé) should be used for all liquors.

2) If the goods have been submitted to previous wet treatment, e.g., desizing and scouring, it is essential to dry them before causticising.

   Pad batch: see also "Cold causticising with the CRC formula".

2.1.22   One bath causticisation and bleaching (CBF process)

The CBF process (caustification blanchiment a froid) is an economical, combined causticising and bleaching process with low time and energy consumption.

This process is generally used for pre-treatment of PES/Co blend fabrics to be dyed or printed. The CBF process is particularly suitable for finishers who have no mercerising and/or steaming units. The CBF process is an economical, combined causticising and bleaching process with low-time and energy consumption.

Advantages of the CBF process

- Time saving
- Saving of water, chemicals and energy
- Increased productivity
- Good degree of desizing
- Perfect removal of cotton seed husks
- Good degree of whiteness for dyeing and printing
- Increased dye yield (comparable with that of cold causticising or mercerisation)
- Coverage of dead and immature cotton.

It can be done on pad batch range (impregnation trough and batching device, e.g., combined with singeing machine), open width washer (if necessary rinsing can also be carried out on the jig).

Table 2.15. One-bath causticising and bleaching or continuous one-bath one-stage CBF process

Form	Woven goods
Substrate	PES/Co–Grey goods
Machine	Pad batch–Impregnation trough and batching device
Type of treatment	One bath causticising and bleaching
Guide recipes	
Caustic soda	22–23 °Bé
Mercerising wetting agent	6–8 ml/l
Wetting/scouring agent (caustic stable)	12–15 ml/l

Sodium silicate 38 °Bé	15–30 ml/l
Hydrogen peroxide 35%	20–30 ml/l
Working method	
Saturator	
Impregnate dry goods	
Liquor temperature °C	20–30
Pick-up	90–100%
Batch	
At room temperature	9–12 h
Wash-off	As hot as possible with alkali shock, rinse

Notes:

1) The bath must be cooled to 20–25°C before adding the peroxide.
2) Impregnation liquor caustic soda 22–23°Bé, all feed liquors of caustic soda at 25°Bé.

2.2   Scouring

Alkaline scour boiling is a cleaning process for cellulosic fibres using large quantities of caustic soda at high temperature as well as special chemical products to remove the natural impurities in cotton and other cellulosic fibres, free them from troubling substances and make them absorbent. At the same time vegetable contamination swells perfectly and is softened for rapid decolouration during bleaching. Alkaline scour boiling is an intermediate process after enzymatic desizing and before peroxide bleaching.

Even if the quantity of impurities is not so high it is essential to eliminate them because of their negative effect in subsequent finishing processes. Fats and waxes prevent penetration of the water-soluble dye and mineral substances can cause precipitation of the dye or during bleaching in the presence of iron annoying catalytic damage. It is also essential to eliminate vegetable residues as well as the proteins and other substances present in the fibre.

There are various scouring agents used in processing of fabrics, out of which the alkaline scouring agents are the most widely used worldwide. Scouring agents can be generally classified as follows:

Alkali: NaOH, KOH, Na_2CO_3, liquid ammonia, sodium metasilicate, sodium silicate, sodium phosphate, trisodium phosphate, tetrasodium phosphate, sodium tripolyphosphate, borax, etc.

Surfactants: Anionic activator, non-ionic activator.

Emulsifiers

Organic solvents

(a) **Chlorine system:** Carbon tetrachloride, Trichloroethylene, Perchloroethylene, Methylchloroform, Trichloromethane, Fluorine.

(b) **Hydrocarbon system:** Benzene, Industrial gasoline, White spirit, Solvent naphtha.

The appropriate type of scouring agent generally depends on the kind of fibre; fabric type i.e., woven or knitted, thick or thin, texturised or non-texturised and the extent and type of impurities present in the fibre. The quality of scour boiling is measured above all by the absorbency; without absorbency there is no dyeing and no print.

Impurities to be removed from the fibres by scouring include fats and waxes, pectin and related substances, minerals and heavy metals, amino acids or proteins, lubricating or knitting oils, etc. A scouring recipe should contain chemicals which help in removing each impurities present in the fibre/fabric to be treated. The impurities and the methods of removal with the chemicals involved for the same are given in the table below:

Table 2.16.

Impurities	Methods of removal from fibres
Fats and waxes	Generally, alkali ($NaOH/Na_2CO_3$) and surface active agents but removal may not be 100%. In difficult circumstances a solvent and mixture of surfactant may help
Minerals and heavy metals	Either by the action of acids to make into soluble salts (demineralisation), but restrictions in using mineral acids which is best suited, due to action on the machineries or by using sequestering agents which will trap the metal ions in water
Pectins and related substances	Solubilised by the action of alkali, by saponification. Alkali doubly help by swelling the fibres which further facilitates the removal of the same
Amino acids and proteins	Producing by converting into soluble sodium salts by the action of an alkali
Lubricating and knitting oils	Knitting oils usually is either water soluble or self-emusifiable. Wetting and detergents can further help in emulsification

The table implicates the importance of alkali and surface active agents in scouring recipes. Manufacturers offer a mixture of organic solvents and surface active agents and other ingredients as branded scouring agents.

2.2.1   Chemicals for alkaline scour boiling

Apart from the caustic soda and water the chemicals used for alkaline scour boiling are *surfactants* and *organic sequestrents*. Certain sulphur or sugar-based reducing agents can also be used for reductive decolouration.

2.2.1.1   Surfactants

All the liquids show surface tension at the liquid–air, liquid–liquid and liquid–solid interfaces. The energy accumulated in the surface molecules of a liquid (water in most textile processing) such as van der Waals forces, dipole bonds, etc., are felt as surface tension. This interfacial surface tension resists water to enter the textile material. It is necessary to reduce this surface tension, which is about 0.073 N/m (Newtons per metre) has to be reduced as much as possible to enable water (and thereby the chemicals added) to act effectively in a scouring process. Surface active agents, wetting agents, detergents and washing-off agents reduce the surface tension of water. When added to water, these agents more or less cover the liquid surface. They are not as strongly attracted to the inner water molecules as the water molecules were previously, hence the surface tension of water is reduced to as much as 0.030 N/m in the presence of a surface active agent. Thus they are very important in the scouring process and find use as wetting agents, softeners, detergents, emulsifiers and defoaming agents, to name a few applications. It is available in the market in many forms like membrane, micro-emulsion, liquid crystal, liposome, gel, etc.

Surfactant molecules contains two groups with contrasting solubility, which is termed as amphiphilic. When molecules of this structure are introduced to an oil–water interface, they align themselves at that interface, with the hydrophilic group being solubilised into the aqueous phase and the hydrophobic group being solubilised into the organic phase. Not all amphiphiles display such activity: only amphiphiles with hydrophilic and lipophilic properties are likely to migrate to the surface or interface. If the amphiphilic molecule is too hydrophilic or too hydrophobic, it stays in one of the phases.

Surfactants reduce the tensions on the surface of the fibre allowing the water, caustic soda and other substances to penetrate to the heart of the fibre in order to degrade and solubilise the impurities in the textile material.

Surfactants are classified by their ionic character:

- The *anionic* group
- The *non-ionic* group

- The *amphoteric* group
- The *cationic* group.

Diagrammatically, surfactants have a linear end (hydrophobic, lipophilic, attraction for oils) and a round end (hydrophilic, attraction for water).

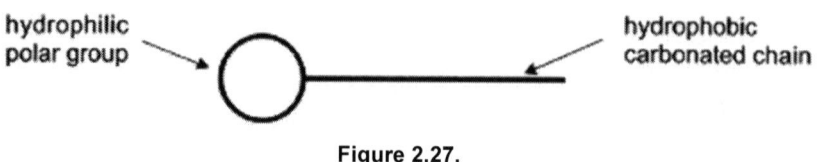

hydrophilic polar group   hydrophobic carbonated chain

Figure 2.27.

Other properties of surfactants

- Stability to caustic soda and scope of application
- 0–2°Bé NaOH–surfactants for discontinuous processes or desizing; anionic or non-ionic.
- 0–8°Bé NaOH–surfactants for continuous and discontinuous processes, desizing, scour boiling and bleaching; anionic or non-ionic.
- 0–18°Bé NaOH–surfactants for concentrated scour boiling with all-in reinforcement bath; anionic.
- 18–32°Bé NaOH–surfactants for mercerising and causticising; anionic.

Wetting-out

The wetting out properties are the ability to eliminate surface tensions very rapidly enabling the water or aqueous solution to penetrate rapidly to the heart of the fibre. The wetting effect depends on the pH as well as the temperature of the bath. It is measured in seconds with round swatches of grey cotton (hydrophobic) and a progressive concentration of wetting agent.

Behaviour of surfactant in a liquid

As the concentration of surfactant in a solution is increased it becomes energetically favourable for the individual molecules, or monomers, to combine together to form large aggregates, or micelles, which shield the hydrophobic components from the solution. It's a group of surfactant molecules associated in a cluster.

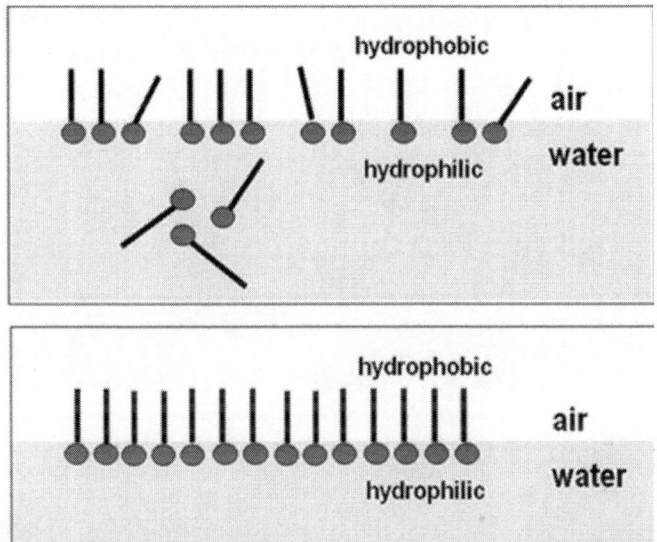

Figure 2.28. Schematic representation of action of surfactant

- Surfactant molecules migrate to the surface with the hydrophobic "tails" pointing outwards and the hydrophilic groups remaining in the water.
- When the interface is completely covered with surfactant molecules, minimum interfacial tension is attained.
- The concentration at which no further reduction of surface tension occurs is known as: "Critical Micelle Concentration" (CMC).

In case of a surfactant molecule with linear structure, for e.g., sodium dodecyl sulphate (SDS), a planar monolayer of surfactant molecules is formed at the interface. With more active shearing and mixing, these types of surfactants are able to form oil-in-water (O/W) and water-in-oil (W/O) dispersions by forming droplet structures called micelles of one phase dispersed within the other.

- Surfactant molecules added at the CMC do not migrate to the surface, but remain in the water, forming clusters known as "Micelles".
- The micelles form the substance reserves in the solution.

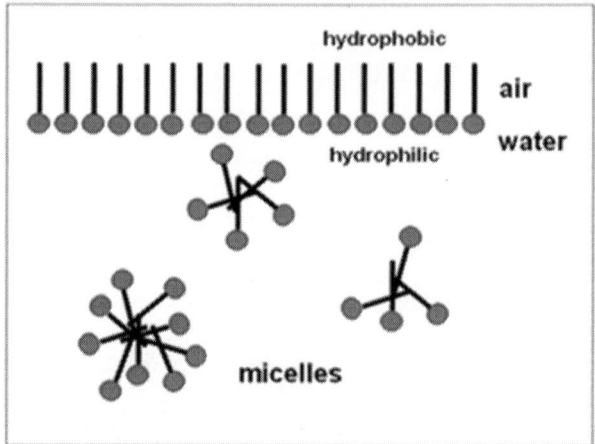

Figure 2.29.

The concentration above which micelle formation becomes appreciable is termed as the critical micelle concentration (CMC).

At low surfactant concentration the surfactant molecules arrange on the surface. When more surfactant is added the surface tension of the solution starts to rapidly decrease since more and more surfactant molecules will be on the surface. When the surface becomes saturated, the addition of the surfactant molecules will lead to formation of micelles. This concentration point is called critical micelle concentration.

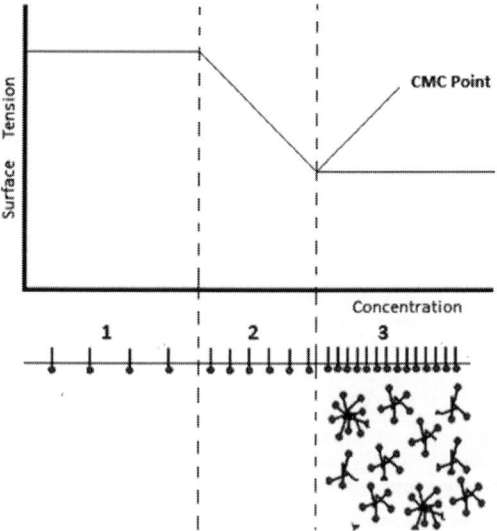

Figure 2.30. Critical micelle concentration

At very low surfactant concentration only slight change in surface tension is detected. Addition of surfactant decreases the surface tension drastically At CMC point, surface becomes saturated and the addition of surfactant molecules do not effect on the surface tension.

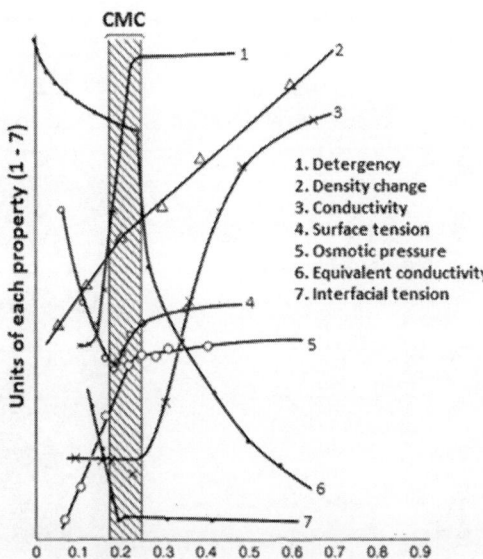

Figure 2.31. Sodium Dedecyl Sulphate (%)

Factors affecting CMC

(1) Number of carbon atom

Increasing hydrophobic part (number of carbon atom) will result in a decrease of CMC. In aqueous medium, the CMC of ionic surfactants is approximately halved by the addition of each CH_2 group. For non-ionic surfactants this effect is usually even more pronounced. This trend usually continues up to about the C16 member. Above the C18 member the CMC tends to be approximately constant. This is probably the result of coiling of the long hydrocarbon chains in the water phase.

(2) Thermal agitation

Micelle formation is opposed by thermal agitation and CMC's would thus be expected to increase with increasing temperature but not always.

(3) Addition of electrolytes

With ionic micelles, the addition of simple electrolyte reduces the repulsion between the charged groups at the surface of the micelle. The CMC is, therefore, lowered.

(4) Addition of organic molecules

Organic molecules may influence CMC's at higher additive concentrations by virtue of their influence on water structuring. Sugars are structure-makers and as such cause a lowering of CMC, whereas urea and formamide are structure-breakers and their addition causes an increase in CMC.

Micelles containing more than one surfactant often form readily with a CMC lower than any of the CMC's of the pure constituents.

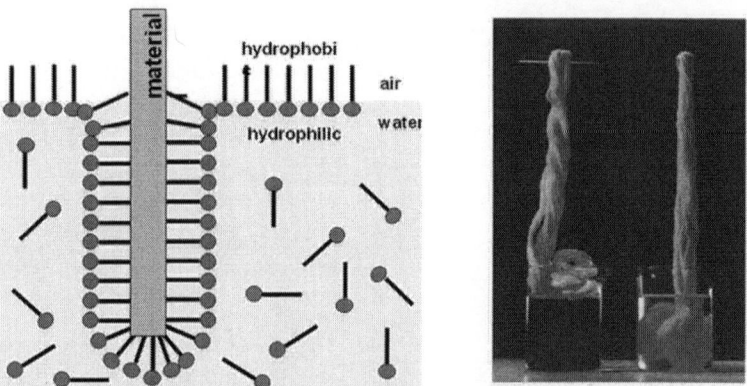

Figure 2.32. Wetting out of a material in presence of a surfactant

- If we disregard the swelling characteristics of the fabric, the difference between wetted and non-wetted woven goods is that liquid rather than air fills the hollow areas between the fibres.
- For the liquid to fill the fine hollow areas between the fibres, the liquid has to penetrate the fabric; it has to form an interface between the liquid and the fabric surface.
- The forming of an interface and the suppression of the air are the two most important elements of the wetting process.
- The wetting of non-ionic surfactants slows down at temperatures above the cloud point due to surfactant insolubility.
- The aim of each washing is to remove foreign substances on the fabric by using an aqueous detergent solution.
- A part of the solid soil will be removed from the fibre by mechanical action.
- It is more difficult to remove impurities which are water insoluble. With surfactants the adhesion between fat and/or oil and the fibre surface will be eliminated.

• If the soil from the fibre is present in the washing solution, the oil and fat have to be emulsified and the solid soil dispersed.

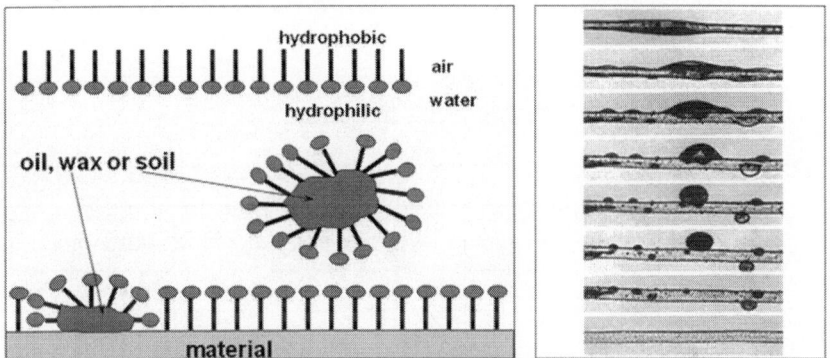

Figure 2.33. Removal of fat, oil, wax from fabric by surfactant

Foam formation

One of the intrinsic properties of surfactants is the formation of foam in aqueous solution. Depending on the chemical composition of the surfactant, its concentration, the pH and temperature of the bath, this foam can be more or less abundant. As a rule this surface foam causes considerable problems in textile applications, e.g., in jets or in impregnation baths. The addition of an antifoam or deaerating agent avoids these problems. It goes without saying that the antifoam used must be free from silicone (risk of spots) or a rubber solvent (attack on the rollers).

Absorbency

Absorbency is the most important parameter of the pre-treatment of cellulosic fibres, without which it would be impossible to dye or print. The surfactant in combination with caustic soda makes it possible to obtain a certain absorbency which depends on several factors:

– The concentration of caustic soda
– Scour boiling temperature
– Scour boiling time
– Type of textile material.

Surface active agents can be anionic, cationic, non-ionic and amphoteric. But basically anionic and non-ionic agents or their blends are mainly used in textiles scouring. A high quality synthetic detergent provides a good balance with wetting, cleaning, emulsifying, dispersing and foaming properties, thus providing it a good cleaning ability. Anionic have a negative molecular charge, cationic positive and non-ionic have no charge. In the case of amphoteric, there are both positive and negative charges. Anionic and non-ionic surfactants provide most of the industrial surfactant requirements and are the most common.

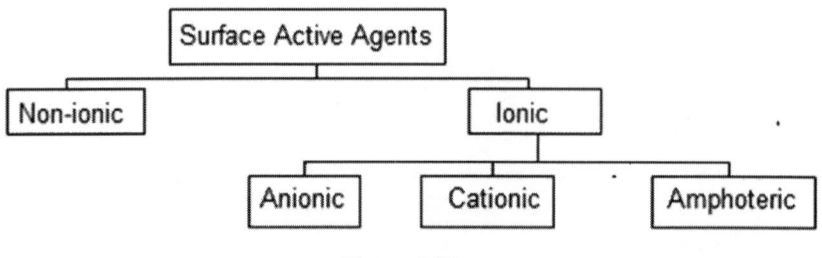

Figure 2.34.

Anionic surface active agents.

Anionic surfactants are dissociated in water in an amphiphilic anion, and a cation, which is in general an alkaline metal (Na^+, K^+) or a quaternary ammonium. This class includes alkyl benzene sulphonates (detergents), (fatty acid) soaps, sodium lauryl sulphate (foaming agent), dialkyl sulphosuccinate (wetting agent), lignosulphonates (dispersants), etc., as indicated below.

1. *Non-nitrogenous*
   • Phosphonium compounds
   • Sulphonium compounds, etc.
2. *Ampholytic*
   • Amino and carboxy
     – Non-quaternary
     – Quaternary
   • Amino and sulphuric ester
     – Non-quaternary
     – Quaternary
   • Amino and alkane sulphonic sulphuricacids
   • Amino and aromatic sulphonic acids.
3. *Anionic surfactants*
   • Carboxyl
     – Carboxyl joined directly to hydrophobic group
     – Carboxyl joined through an intermediate linkage
   • Sulphates
     – Sulphate joined directly to hydrophobic group
     – Sulphate joined through an intermediate linkage

- Alkane sulphonic acids
  - Sulphonic group directly linked to hydrophobic group
  - Sulphonic group joined through an intermediate linkage
- Alkyl aromatic sulphonic acids
  - Hydrophobic group joined directly to sulphonated aromatic nucleus
  - Hydrophobic group joined indirectly to sulphonated aromatic nucleus
- Miscellaneous anionic hydrophilic groups
  - Phosphates and phosphoric acids
  - Persulphates, thiosulphates
  - Sulphonamides
  - Sulphamic acids, etc.

This is one of the major groups used in textiles. This includes soaps, synthetic scouring agents. The soaps are more effective when the numbers of carbon atoms are more in their molecule or simply higher molecular weight. Soap with high molecular weight (C_{12}-C_{18}) is only suitable for use when scouring can be carried out at or near the boiling point of the solution. The instability of soap in acid solutions is the main drawback. Synthetic detergents are surfactants developed to overcome the drawbacks of soaps. They are salts of strong acids and when pure they are practically neutral. In contrast to soap they are less liable to hard water precipitation and washing-off is much easier.

Detergents	Chemical Structure
Sodium Stearate	C17H35COONa
Dipropyle naphthelene sulphonates	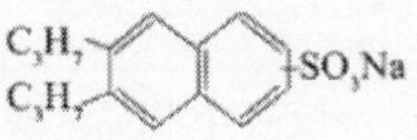 C_3H_7 — C_3H_7 — SO$_3$Na
Alkyl benzene sulphonates	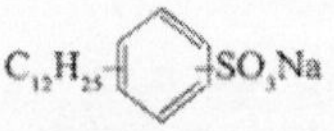 $C_{12}H_{25}$ — SO$_3$Na
Alkyl sulphates	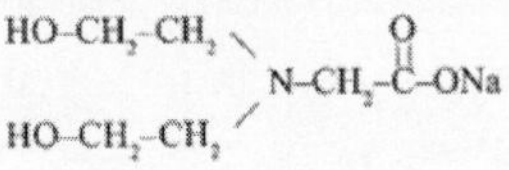 HO–CH$_2$–CH$_2$ \ N–CH$_2$–C–ONa HO–CH$_2$–CH$_2$ /
Alkane sulphonates	C 7H33CO-N-C2H4SO3Na
Phosphate esters	R-O-(C2H4 O)x -PO 3Na2 R = Octyl or nonyl phenol or fatty alcohol

Figure 2.35. Anionic surfactants

Cationic surfactants

As explained earlier the cationic detergents are rarely used in scouring formulations, but it finds useful in other areas like levelling, softening, retarding of dyeing, water repellency, emulsifying etc. in textile processing. Cationic surfactants are dissociated in water into an amphiphilic cation and an anion, most often of the halogen type. Many are nitrogen compounds, such as fatty amine salts and quaternary ammoniums. The standard classification of these surfactants is based on their dissociation in water. The hydroxide used to neutralise the acid is of great importance, because of the hydrolysis reaction which takes place in water. With very alkaline hydroxides, e.g., NaOH or KOH, the pH of the soap aqueous solution is very high. This increases cleaning power but can damage skin. Selecting the soap cation controls the balance of cleansing action and solubility. They generally have the hydrophobic part of the molecule as organic ammonium or pyridinium compound containing one or more hydrophobic residues and an example is

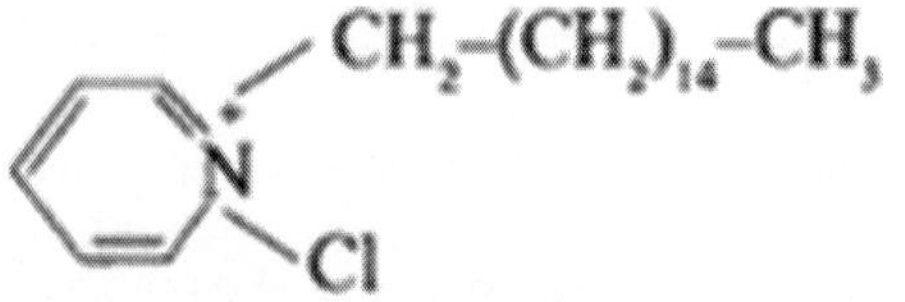

Cetyl pyridinium chloride

Cationic surfactants may be classified as follows:

1. Amine salts (primary, secondary and tertiary)
    Amino group joined directly to hydrophobic group
    Amino group joined through an intermediate linkage
2. Quaternary ammonium compounds
    Nitrogen joined directly to hydrophobic group
    Nitrogen joined through an intermediate linkage
3. Other nitrogenous bases
    Non-quaternary bases (e.g., guanidine, thioronium salts, etc.)
    Quaternary bases.

Examples of cationic surfactants

$$R - CO - NH - CH - CH - CH_2 - N\,(CH_3)_2 - CH_2 - COO^-Na^+$$
Alkyl amido betaine

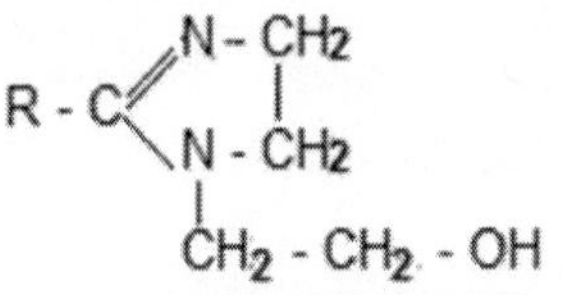

Hydroxyethyl Imodazoline

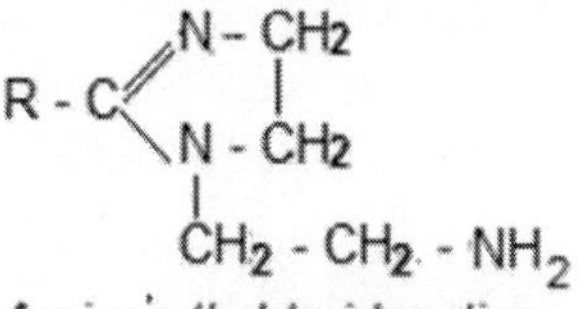

Amino ethyl Imidazoline

Non-ionic surfactants

After anionic surfactants, the major surfactants used in textile industry is non-ionic surfactants. Main non-ionic detergents are ethylene or propylene oxide condensates and these are made by the reaction of fatty acids, alcohols and alkyl phenols, fatty amide, etc., with them. Non-ionic surfactants do not ionize in aqueous solution because their hydrophilic group is of a non-dissociable type, such as alcohol, phenol, ether, ester or amide. A large proportion of these non-ionic surfactants are made hydrophilic by the presence of a polyethylene glycol chain, obtained by polycondensation of ethylene oxide.

Examples of non-ionic detergents are follows:

Ethoxylated primary alcohols	$(R\text{-}O\text{-}(C_2H_4O)_n H)$,
Ethoxylated thioethers	$(CH_3\text{-}S\text{-}C_2H_4\text{-}(C_2H_4O))_n$,
Ethoxylated fatty acids	$(R\text{-}CO\text{-} (C_2H_4O)_n\text{-}OH)$,
Ethoxylated fatty amides	$(R\text{-}CO\text{-}NH\text{-}(C_2H_4O)_n H)$.

As explained earlier many of the non-ionic surfactants are ethylene oxide condensates. Ethylene oxide itself is not very stable due to its peculiar triangular structure which can be seen in the figure below, where one can see that the angle of the bonds and bond distance are different than normal.

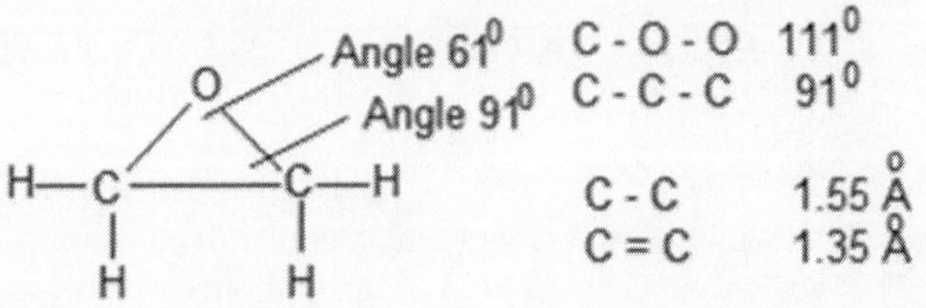

Angle 61^0	C - O - O	111^0
Angle 91^0	C - C - C	91^0
	C - C	$1.55\ \overset{o}{A}$
	C = C	$1.35\ \overset{o}{A}$

Some of the non-ionic surfactants can be produced by the following reactions from ethylene oxide.

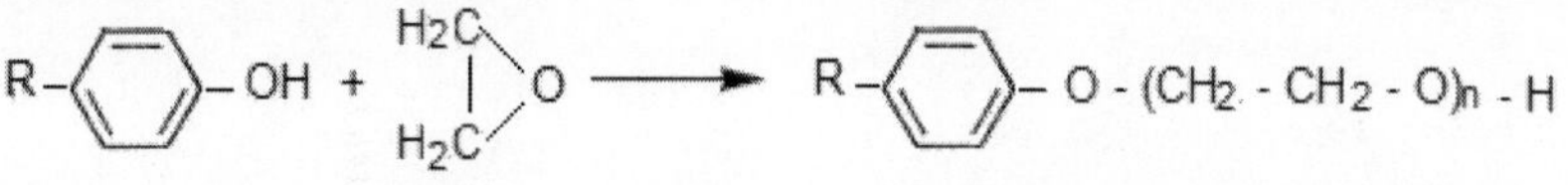

Alkyl phenol   Ethylene oxide   Alkyl phenol ethoxylate

(R = H, octyl, nonyl, dinonyl, dodecyl, bisphenol A)

$$HO - CH_2 - CH_2 - O - (CH_2 - CH_2 - O)n \; H$$

Ethylene oxide   Ethylene glycol   Polyethylene Glycol (PEG)

$$HO-(CH_2-CH_2-O)_n H + R-CO-H \longrightarrow R-CO-O-(CH_2-CH_2-O)_n H \quad OR$$

$$R-CO-O-(CH_2-CH_2-O)_n-CO-R$$

Polyethylene Glycol (PEG)   PEG Monoester or PEG Diester

$$R - O - (CH_2 CH_2 - O)_y - (CH_2 - CH - O)_x - H$$
$$\qquad\qquad\qquad\qquad\qquad CH_3$$

Propylene oxide   Ethylene oxide   Alcohol   Alcohol Alkoxylate

Ethylene oxide condensates are the major group in non-ionic surfactants and their application more or less controlled by the hydrophilic–lipophilic balance (HLB). When the HLB is low it serves only as a wetting agent, as the HLB increases it can be used as a detergent and at further higher HLB it can act as a solubiliser.

Non-ionic surfactants do not contain an ionisable group and have no electrical charge contrary to anionic or cationic surfactants. A special character of these surfactants are the phenomenon of cloud point which is the temperature at which a 1% solution become cloudy or insoluble. Normally, larger the number of ethylene oxide molecules in the product, higher is the cloud point. This phenomenon is a problem in the textile processing. There are methods to prevent this to blend the non-ionic with an anionic such as a soap, a sulphonate or a phosphate. The cloud point of non-ionic surfactant solution can be depressed by the addition of an electrolyte like common salt, Glauber's salt, etc. They are miscible with all surfactants in any proportions, and are good wetting agents and rewetting agents, emulsifiers, etc.

Amphoteric surfactants

Amphoteric surfactants may be cationic, anionic or non-ionic depending upon the pH of the aqueous solution. A typical amphoteric surfactant can be represented as

For example, an amphoteric surfactant of the structure $Cl^-R-N^+H_2-CH_2-CH_2-COOH$ in acidic medium it will act cationic where as another one with structure $R-NH_2-CH_2-CH_2COO^-Na^+$ in alkaline medium will act as anionic. Some of them are not stable at higher temperature and hence not suitable for high temperature scouring. There major use is in the scouring of silk wool, etc., and as a dyeing assistant to prevent chafing, crack marks and crowfeet effects. They are also used as lubricating agent, corrosion inhibitor, wetting agent and protective colloid in the dyeing of protein fibres.

The main component in a scouring recipes are scouring agents which constitute the most important group of detergent components. Surfactants used in scouring agents include:

- Carboxylic acids and salt
- Sulphuric acid derivatives
- Sulphonic acids and salts
- Alkoxylated alcohols
- Alkanolamides
- Ethoxylated fatty acids.

2.2.1.2   Sequestering agents

Sequestering agents are added in the alkaline scouring baths to avoid the precipitated mineral hydroxides settling back on the fabric by chelating or fibres which will affect the subsequent processing and especially dyeing. Sequestering agents or chelating agents are negatively charged and are capable of forming strong ring structures with the metal ions present in hard water and in pectins of cotton. The positively charged metal ions, particularly Fe^{3+} and Ca^{2+} are readily available for reaction with any negatively charged anion such as OH^- or CO_3^{-2} and insolubilise soap in the fibre which may disturb subsequent operation. This problem is much more acute when scouring is carried out in continuous process involving padding bath where liquor ratio is much lower than the batch process. Thus, the functions of the chelating agents in the soap and detergent

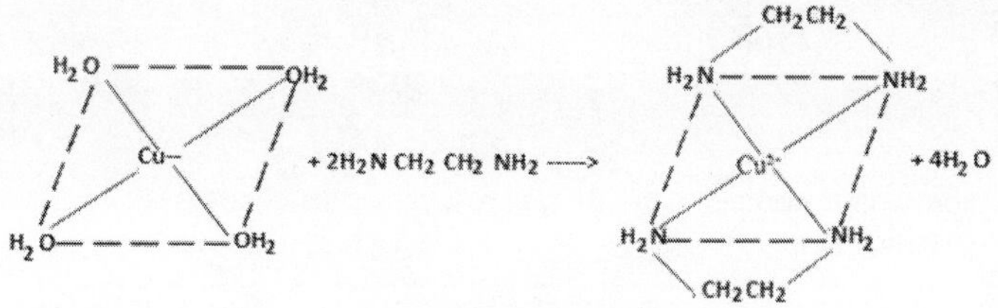

Figure 2.36. A typical chelate structure

formulations are for the prevention of film and scum formation, precipitation of hard water, calcium and magnesium inhibition of foaming properties, clogging of liquid dispersions, haze turbidity in liquid solutions, and rancidity and oxidation that cause discolouration of formulation. Sequestering power is influenced by pH of the scouring bath. At a given pH, the quantity of sequestering agent to chelate a given metal varies. Organic chelating agents are more frequently used in textile formulations.

Table 2.17. Some of the sequestering agents

Name of sequestering agents	Chemical structure
EDTA–Ethylene diamine tetraacetic acid	
HEDTA–N-(2-Hydroxyethyl) ethylenediamine tetraacetic acid	
DTPA–Diethyltriamine pentaacetic acid	
DHEG–N, N-Di (2-hydroxyethyl) glycine	
DPTA–Diaminopropanol triacetic acid	
NTA–Nitrilotriacetic acid and salts	
ATMP–Aminotri-methylene phosphonic acid	

Name of sequestering agents	Chemical structure
HEDP–1-Hydroxyethyledene-1,1-diphosphonic acid	$\underset{\underset{PO_3H_2}{|}}{\overset{\overset{OH}{|}}{CH_3-C-PO_3H_2}}$
EDTMP–Ethylenediamine tetramethylene phosphonic acid	$\underset{H_2O_3P-H_2C}{\overset{H_2O_3P-H_2C}{}}N-CH_2-CH_2-N\underset{CH_2-PO_3H_2}{\overset{CH_2-PO_3H_2}{}}$
DTMP–Diethylene triaminopentamethylene phosphonic acid	$\underset{H_2O_3PH_2C}{\overset{H_2O_3PH_2C}{}}N-CH_2-CH_2-N-CH_2-CH_2-N\underset{CH_2PO_3H_2}{\overset{CH_2PO_3H_2}{}}$ $\underset{PO_3H_2}{\overset{CH_2}{|}}$

Some of the sequestering agents works both as sequestering agent and peroxide stabiliser.

Testing the capacity of sequestering agent for chelating

Anti-precipitant effect

The organic sequestrant can avoid any insoluble precipitation by chelating (masking) metallic cations. For example, the addition of a chelating agent can avoid the precipitation of copper hydroxide $Cu(OH)_2$ when adding caustic soda to the cupric solution (see below).

Without   With Seque-
Sequestrant   strant

Figure 2.37.

Inhibiting effect on calcium

Test: Add a complexing agent to a bath containing 25°e calcium, 25°e magnesium, 20 g/l sodium carbonate, 20 g/l potassium sulphate, 20 g/l potassium chloride. Adjust the bath to pH 11 with NaOH and hold at the boil for 1 h, and then filter-off hot. The tendency to form calcium and magnesium precipitates is very high in an alkaline medium at 100°C.

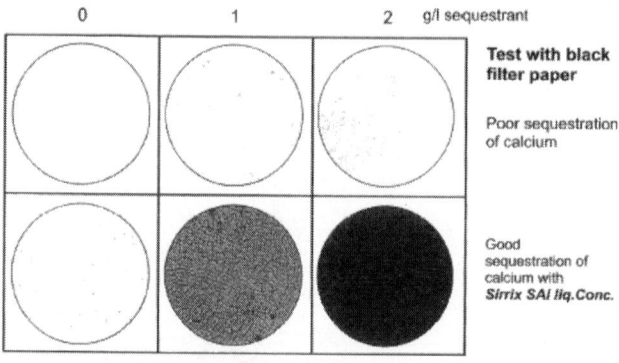

Figure 2.38.

2.2.1.3   Builders

Scouring in kiers is a process of old age. Builders are generally added to the kier boiling bath to increase the activity of soap or detergents. Builders are generally salts such as borates, silicates, phosphates, sodium chloride, sodium sulphate, etc. Sodium metasilicate ($Na_2SiO_3 \cdot 5H_2O$) also acts as a detergent and buffer and assists other chemicals to penetrate into the lignin mass of cellulose materials. In general, the function of the builders is to drive the soap from water phase to fabric/water interface and consequently increase the concentration of soap on the fabric. It is inexpensive, but its addition in the scouring bath increases the ash content of cotton.

2.2.1.4   Fibre protecting reducing agents

Air or oxygen inside the kier can help in the formation of oxycellulose when boiled under high pressure (Kier). In kier operation, the pressurisation is done only after removing the air inside the kier by mere expelling by steam during the heating up operation. But is not necessary that all the oxygen is expelled, especially some air pockets may be present inside the rolls which can cause oxycellulose formation and tendering or cause problems during further processes (e.g., dyeing). To overcome these problems, a small quantity of mild reducing agent such as sodium bisulphite or even hydrosulphite is added to the kier liquor.

2.2.1.5 Mild oxidising agents

During scouring process in the absence of any oxidizing agents, higher temperature can create reducing atmosphere, especially in close units (e.g., Jet, Steamer), due to chemicals or the cellulose (fibre). This may not be a problem in a grey fabric, but when one is scouring striped fabric this can cause problems like bleeding, staining of adjacent and layers (in batched form) etc. Azoic dyes may be mechanically transferred while the vat dyes may be reduced to its soluble leuco state in presence of alkali and impurities removed from the cloth and in case of reactive dyes it can dullen the shade or even reduce the depth or unevenness in case of any colour. For such types of fabrics mild oxidising agents like Ludigol, Resist Salt L, etc., may be added to the scouring bath to resist the reducing properties of dyes. Chemically, these products are sodium salt of nitrobenzene sulphonic acid. Alternatively, such fabrics containing coloured threads can be scoured in open kier, on winch machine, jigger or continuous machine with boiling soda ash liquor (2.5% o.w.f.) instead of caustic soda to reduce the mark-off to a minimum.

2.2.1.6 Solvents

Solvents are necessary to remove residual fats, waxes and oils which are not fully removed by alkali. The modern scouring agents usually contain some of the solvents to tackle, especially waxes which is difficult to remove. High boiling solvents like cyclohexanol, methylcyclohexanol, tetrahydronaphthalene, decahydronaphthalene, etc., are usually preferred to withstand the scouring temperature.

2.2.1.7 Emulsion scouring

Emulsion scouring is mainly applied for synthetic and woollen materials which contain lot of oils like lubricants, coning oils, knitting oils, antistatic agents, etc., which support in the manufacturing operations. Most of these oils are water soluble or self-emusifiable, but may not be completely removed. Increasingly sophisticated spin finishes are used and long group of compounds are known to serve as fibre lubricants, mineral oils being used since long. Mineral oils, tar stains, grease marks are not scourable, especially after heat setting (in case heat setting is done before scouring operation. Local stains are removed by spot cleaning but usually it should not be allowed to dry, otherwise halo mark can result. This is often not practical and hence in such cases emulsion scouring is a preferable choice. The spot cleaning chemical is usually an emulsified solvent. They are usually organic solvents containing emulsifier dissolved in them and hence become self-emulsifiable. The solvent must be capable of dissolving the hydrocarbon impurity and the detergent must be capable of emulsifying both the hydrocarbon and the solvent. Other alternative methods are that the powerful stain removers may be applied from the bath along with wetting agents cum detergents at high temperature and solvent scour. One such formulation which has been widely used for polyester/cotton blended fabrics for removing unsaponifiable waxes from cotton portion apart from lubricants and conning oils from the yarn is emulsion scouring. The mechanism by which emulsion scouring occurs is attributed to the hydrophobic portion of the micelle being saturated with the solvent. The micelle, when loaded with solvent, is more effective in solubilising oily impurities from the fabric than is the detergent alone, provided both solvent and emulsifier are properly selected and used. The critical feature of such a solvent is its KB value, which is a measure of its solvency for a particular oily soil.

Table 2.18. Example of an emulsion scouring recipe

Quantity	Unit	Additives
20–50	g	Essential oil
20–10	g	Perchloroethylene
40	g	Emulsifier (e.g., nonyl phenol HLB 13–15)
100	g	

2.2.1.8   Solvent scouring

Solvent scouring is much better environmental friendly than chemical scouring, mainly due to reduced water pollution, reduced energy cost and consumption apart from effective removal of impurities. Solvent preparation gives excellent results in terms of uniformity, reproducibility and high absorbency. But it may not be useful for all fibres especially cotton. This method is suitable for synthetic fibres which are more clean fibres and probably wool. Solvent mainly can remove the lubricants, knitting oil from synthetic hosiery, certain sizes, oligomer from polyester etc. In case of wool, it naturally felts less in solvent media because of the absence of moisture and is particularly useful in wool milling and application of shrink-resist resins to wool.

The solvent can be absorbed into the fibre web and can give problems in further steaming operations. But the main problems in this process are the recovery of the solvent and related issues. The solvent can affect the rubber sleeves of the bowls and cause peeling-off due to the dissolution of binders and as such the rubber can be attacked by the solvent thus making serious problems for the mangles in subsequent processing. However, solvent scouring is gaining more importance due to the general non-availability of huge amount of water for wet processing anywhere and presently this method is an alternative to reduce the water consumption and more eco-friendly.

The solvent selected should be recoverable, which makes the process environmental friendly. The usual solvent employed are the same as the dry cleaning solvents like chlorinated hydrocarbons, e.g., tetrachloroethylene (perchloroethylene), trichloroethylene and 1,1,1-trichloroethane. Trichloroethylene is not generally used because of its stripping action of dyed polyester materials if the temperature should rise above 30°C. As mentioned earlier the stability of solvents to recovery by distillation or by adsorption process. The recovery of stabiliser in the solvents is also equally important. The stabilisers recommended for different solvents are shown in table below:

Table 2.19. Solvent stabilisers

Solvent	Stabiliser
Trichloroethylene	Substituted diamines
Perchloroethylene	Crotonaldehyde dialkylhydrazones
1,1,1-Trichloroethane	Mixture of acetonitrile and nitromethane or
	Dimethoxyethane and 1.4 Dioxan
Halogenated hydrocarbons	Triethyleorthoformate, or
	Phosphoric acid or
	Benzotriazoic

In certain processes "booster solvents" are added to perchloroethylene. Soil removal in chlorinated hydrocarbons can be improved by the addition of solvent detergents, e.g., monoethanolamine, alkylbenzene sulphonate, alkyl poly (glycol ether) and alkyl pyridine chloride. Such products are usually mixtures, e.g., dodecylbenzene sulphonate (13%), nonyl phenol along with 5 ethylene oxide units (5%), isopropylalcohol (5%), water (7%) and chlorinated hydrocarbon (70%). Chlorinated hydrocarbons containing such products are capable of dissolving water and hence enhances detergency.

2.3   Batch wise process

2.3.1   Alkaline scouring

Boiling out is the treatment of cellulose under strong alkaline conditions. Raw cotton contains a large number of foreign substances such as hemicelluloses, proteins, lignins, pectins, fats, waxes, natural dyes and seed husks. These are partly water-soluble, and are only removable by an alkaline process. In some cases an acid treatment is necessary. Seed husks and cotton waxes can only be eliminated by longer alkaline boiling or kier boiling. This process is important to improve the hydrophilicity (a must for continuous dyeing and printing). A boiling process is also useful to reduce the danger of a catalytic damage in a subsequent peroxide bleach.

2.3.1.1   Kier boiling

This method is very seldom practiced, other than special cases. A good scouring process, this method is not much followed due to its high utility costs. Fabric in rope form or in the form of batched rollers after soaking in the scouring chemicals are loaded in a vertical reaction vessel, called Kier. (The essential feature of the kiering process is the treatment of a large quantity of fabric, typically several tonnes, over a period of some hours in a steam-heated pressure vessel. A typical kier has the form of an enclosed vertical cylinder with a relatively small opening at the top through which the fabric rope is loaded and unloaded. At one time fabric was loaded manually, with operatives actually standing in the machine; nowadays mechanical loading devices are employed.) The vessel is filled with the liquor and boiled under pressure in a closed kier or under atmospheric pressure in an open kier. Liquor is circulated through a heat exchanger where it is heated to the required temperature. The process must be a carefully controlled operation in order to avoid channelling, i.e., the formation of channels through which the liquor may flow in preference to the load.

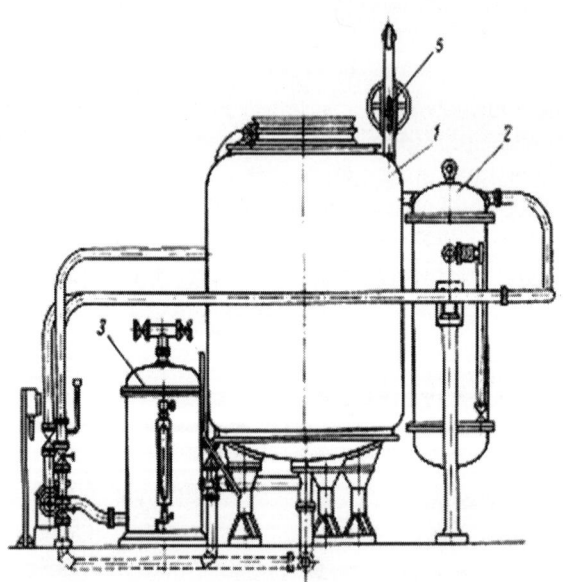

1. Kier, 2. Heat exchanger, 3. Buffer, 4. Circulation pump, 5. Air vent

Figure 2.39.

In a typical open kier boiling process–10–20 g/l sodium hydroxide for 4–6 h at 95–98°C. After completion of the boil it is still important to avoid contact with the air and it is therefore advisable to run in hot rinsing water as the alkaline liquor is drained-off.

Whereas in pressure boiling alkali requirements are lower than open boiling–5–10 g/l sodium hydroxide at a pressure of 200 kPa (30 lbf/in²) and a temperature of 130°C for 2–4 h. Little or no degradation of the cotton fibre. After a pressure boil the liquor continues to circulate whilst cooling to 90°C, before dropping the liquor gradually and giving several rinses, hot and then cold.

Notes

1. Pressure boiling helps in better removal of fats and waxes than in atmospheric scouring and achieve excellent absorbency, but it can give a rough and harsh handle.

2. Pressure boiling produces a degree of whiteness that is often adequate for dyeing medium to dark shades, even without further bleaching.

3. Undue movement of the fabric during processing must be avoided, because contact with the kier walls is liable to cause abrasion marks (chafing) on the goods.

4. Normally, the fabric is loaded in to the kier in dry state or wetted out (with liquor) and the hot liquor is introduced from the bottom so that the hot liquor wets out or otherwise, the heat drives the air trapped to the top and allows it to escape through the open valve at the top of the dome cap.

5. Even after the air escaped the liquor is heated further and the valve is closed only after confirming the area above the fabric is filled with steam. Any air trapped inside or even above the fabric can tender the fabric under pressure. Atmospheric boiling is less risky.

2.3.2   Acid scouring

Acid cracking is the treatment of cellulose under strongly acid conditions. Raw cotton contains many times high concentration of alkaline earth ions (Ca, Mg) and heavy metal ions (Fe, Cu, Mn) which have a negative influence on the process security and the final results (e.g., catalytic damage, precipitation on machines and fabrics). These impurities can be better removed in the acid conditions than in the alkaline conditions.

Problems occur when too many impurities are brought into the peroxide bleaching process. Main of these are as follows:

(a) Precipitations and scales in the bleaching machinery and on the fabric (mainly insoluble calcium salts)

(b) Low-bleaching effect

(c) Danger of catalytic damage

Inorganic acids due to its corrosive property, formation of hydrocellulose, difficult to wash-off, damage of fibre due to the residual acid (H_2SO_4) are normally not recommended. Organic acids have advantages but cause problems like smell, volatility (acetic acid), formation of alkali salts (formic acid), costly, lcium calcium salt precipitation (oxalic acid). Hence normally formulated acid cracking agents are used. The formulated acid cracking agents do not show such disadvantages:

- Non-corrosive for steel
- No formation of insoluble salts
- Practically neutral salts when used for neutralisation
- Additional dispersing and cleaning power (threshold-effect).

Additional complexing power and inhibition of precipitations in changing pH-ranges.

Requirements for dis-continuous (batch wise) applications

- Low-foaming (for jet machines)
- Good dispersing properties
- Good Emulsifying properties
- Pumpable formulation (for automatic dosing type machines).

Desizing and alkaline boiling-off (CO, CO/PES)

If the goods are difficult to desize, different desizing methods can be combined, for example, enzymatic desizing and a subsequent alkaline treatment with ammonium persulphate.

Table 2.20. Guide recipe

Details	Discontinuous	Continuous (pad roll, pad-steam, J-box
Caustic soda solid	10–12 g/l	50–60 g/l
Suitable detergent/SA agent	1–2 h/l	5–10 g/l
Ammonium persulphate	3–7 g/l	3–5 ml/l
	1–3 h at 95°C	Impregnate at 50°C and pick-up 100% store for 30–90 min at 90–95°C

Table 2.21. Alkaline treatment (boiling-off)

Discontinuous (CO and linen)	Yarn and piece goods. Winch 20:1 Jig 3:1–5:1	Piece goods pressure Kiers 4:1
Caustic soda solid	6–12 g/l	5%
Suitable scouring/wetting agent	0.3–2 ml/l	0.2–0.6%
Time	1–3 h	4-8 h
Temperature	95–100°C	Pressure 1.5–2 atm

Jigger process

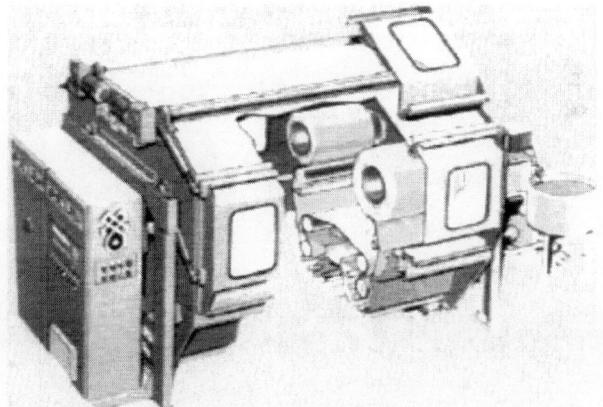

Figure 2.40. Jigger

Fill machine and load fabric add in two ends

Table 2.22. Guide recipe

Quantity	Unit	Chemicals
3	%	Caustic soda solid
0.2	%	Wetting agent
3	%	Hydrogen peroxide 50%
0.3	%	Scouring agent
0.15	%	Sequestering agent (optional)

Run 2 ends in cold, and then raise temperature to 85–90°C
Run 4 ends (90–120 mins)
Drain
Hot rinse

Neutralise with 0.5 gpl acetic acid at 60°C.
Peroxide removal with 0.15% peroxide killer at 55°C cold wash.

Notes:

1. Greater amount of chemicals apply for shorter liquor ratio.
2. After the alkaline treatment, rinse as hot as possible to remove emulsified cotton waxes from the goods.
3. After rinsing it is advisable to acidify with 1–2 ml/l concentrated hydrochloric acid at 30–40°C to clear insoluble calcium salts from the goods.
4. Finally rinse again to neutralise the goods.

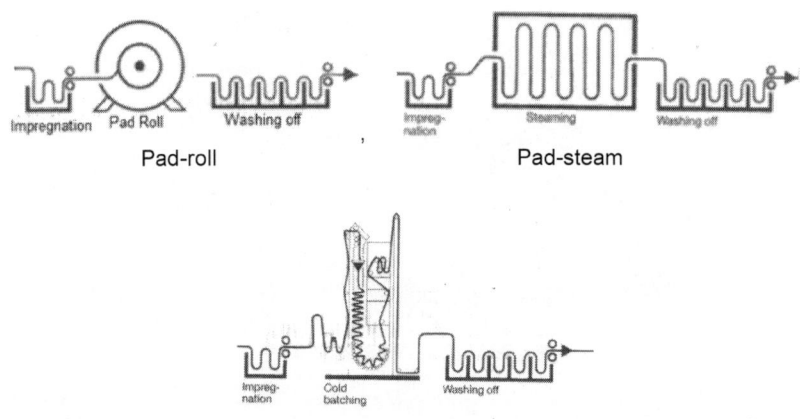

Pad-roll   Pad-steam

J - box

Figure 2.41.

Table 2.23. Guide recipes

Continuous for cotton only	Piece goods			
	Pad-roll O/W	Pad-steam O/W	J-box O/W and rope	HT steamer O/W
Caustic soda solid	40–50 g/l	50–90 g/l%	50–70 g/l	50–70 g/l
Suitable detergent	1.5–5 g/l	5–8 g/l	1.5–2 g/l	3–5 g/l
Impregnation	60–80°C	50–60°C	50–60°C	60–80°C
Liquor pick-up	80–100%	80–100%	80–100%	80–90%
Steaming temperature	90°C	103–105°C		142°C
Reaction time	60–90 min	60–90 s	1–2 h	1 min

Notes:

1. If the goods are padded wet on wet, the padding liquor should be 5–8 times stronger.
2. For cotton/PES, viscose/PES, the amount of alkali is adjusted to the cellulose fibre content of the blend.
3. After the alkaline treatment rinse as hot as possible to remove maximum emulsified.

2.3.3   Continuous scouring

Pad batch enzymatic cracking–pad-steam bleach

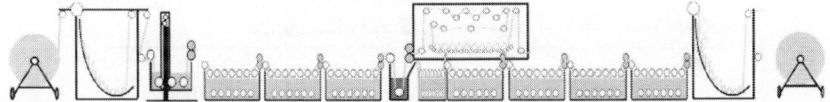

Figure 2.42. Pad batch enzymatic cracking (scouring)

Table 2.24. Guide recipe

Quantity	Unit	Chemical
2–5	ml/kg	Wetting agent
1–3	ml/kg	Sequestering agent
2–5	ml/kg	Enzyme

Process: Impregnate at 60°C, pick-up 100% &and batch for 6–24 h.

Table 2.25. Pad-steam peroxide bleaching

Quantity	Unit	Chemical
2–5	ml/kg	Wetting agent
6–10	ml/kg	Peroxide stabiliser
1–3	ml/kg	Scouring agent
12–25	g/l	Caustic soda solid
20–40	g/l	Hydrogen peroxide 50%

Process: Impregnate cold, pick-up 100%, 15–20 min steaming at 98°C.

Pad-steam enzymatic cracking and pad-steam bleach

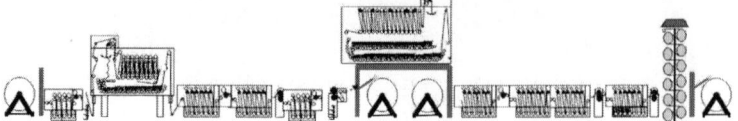

Figure 2.43. Pad-steam enzymatic cracking (scouring) and bleaching

Table 2.26. Guide recipe

Quantity	Unit	Chemical
2.0–4.0	ml/kg	Suitable enzyme
1.0–2.0	ml/kg	Wetting agent
2.0–4.0	ml/kg	Scouring agent

Process: Impregnate at RT, pick-up 100% dwell for 8–10 min.

Combined alkaline treatment and bleaching

    1.   *Rapid peroxide bleach for cotton, polyester/cotton, viscose/polyester*

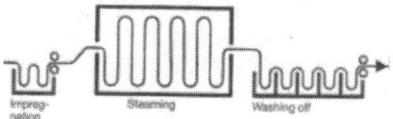

Figure 2.44. Pad steam

Table 2.27. Guide Recipe for Pad-steam peroxide bleach

Continuous	Pad-steam
Detergent liquid or scouring agent	5–8 ml/l
Dispersing agent + White spirit	6–10 ml/l
Stabiliser	1–2 g/l
Sodium silicate 38°Bé	10–30 ml/l
Caustic soda	10–15 g/l
Hydrogen peroxide 35% by weight	20–40 g/l

    2.   *Method for cotton containing husk (kitties)*

Pad the above recipe at 20–30°C at 70–80% pick-up–Batch–Store for 3 h–Pad again the same liquor–Steam continuously for 60–90 s at

102°C or 140°C–After treat in an open width soaper–Rinse as hot as possible–Acidify–Rinse–Dry.

3. *Method for cotton free from husks and polyester/cotton blends*
    Pad the above liquor at 20–30°C and at 70–80% pick-up–Steam continuously for 60–90 s at 100 or 140°C–Rinse as hot as possible–Acidify–Rinse–Dry.

Neutral scouring (enzyme scouring)

Neutral scouring is more important in the present scenario because of following reasons:

- Shorter process time, time saving, higher production in given time
- Saving in cost–Less water (30–60% water saving in pre-treatment), steam usage
- Cost saving in effluent treatment due to low BOD/COD/TDS (less than 50% compared to conventional process, more life to reverse osmosis (RO) membrane and shorter effluent treatment
- 1–2% weight gain compared to conventional methods
- Less damage to the substrate–better quality
- Better handle due to less usage of chemicals like caustic soda. Less finishing required
- Less fibrillation due to clean surface.

Pectin connects the waxy material to cellulose of the primary wall. In the conventional manner, high amount of caustic and high temperature is used to break this pectin linkage to free the waxes which subsequently are emulsified with the help of emulsifiers present in the bleach bath.

However enzymes destroy the morphology of cuticle structure by digesting the pectin's and releasing the waxes which does not have any affinity to the cellulose. This free wax is then emulsified easily with the help of emulsifier and thus improved wettability is obtained.

Table 2.28. Guide Recipe

Quantity	Unit	Bath additions
0.5–0.8	% OWM	Wetting agent
0.35–0.5	% OWM	Enzyme
0.1–0.3	% OWM	Emulsifying agent

OWM–on weight of material

Run for 20 mins at 60° C–then raise temperature to 90–95°C and run for 15 min and then drain.

Note: The quantity of Lenetol GC 333 required depends on the process and substrate.

The whiteness of the fabric may not be as good as the conventionally scoured fabrics but can be used for medium and dark shades. But for light bright shades it is better to follow up with a light peroxide bleach to improve the whiteness.

Peroxide   2.0–4.0% owm
Caustic flakes   0.5–1.0% owm
(to maintain pH 10.5–11 during bleaching)
Run for 20–30 min at 98°C.

Table 2.29. Comparison of conventional and enzyme scouring

Property	Neutral scouring	Alkaline scouring
BOD, COD, TDS	Low	High
Weight loss	Low	Medium
Handle	Soft	Coarse
Wettability	Good	Good
Mote/Kitty removal	Low	Medium
Brightness	Low	Medium
Productivity	High	Lowe
Energy, water consumption	Low	High
Waste water	Low	High
Total cost	Medium	High

Example of treatment in jigger for woven and subsequent dyeing (CRODA process).

Desize by pad batch

(GSM 320, 16's × 12's count).

Table 2.30. Guide Recipe

Quantity	Unit	Chemicals
4	g/l	Suitable wetting cum scouring agent
1.5	g/l	Dispersing agent
3	g/l	Enzyme
1.5	g/l	Sequestering agent

Pad at 70°C and batched overnight, then loaded on jigger.

Strong hot wash (85–90°C) × 2 ends (drain after each end) and then started dyeing.

Dyed Navy, Black, Brown shades for confirmation of the results.

Continuous pre-treatment

Process: Pad batch enzymatic cracking –pad-steam bleach

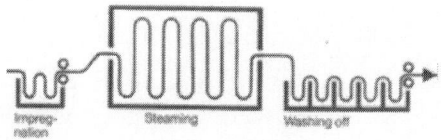

Recipe

Table 2.31. Pad-steam peroxide bleach

Quantity	Unit	Bath additions
2.0–5.0	ml/kg	Suitable wetting cum scouring agent
6.0–10.0	ml/kg	Stabiliser
1.0–3.0	ml/kg	Sequestering agent
15.0–25.0	g/kg	Caustic lye 50%
20.0–40.0	ml/kg	Hydrogen peroxide 50%

Process: Impregnate cold, pick up 100%, 15–20 min steaming at 98°C

Recipes for alkaline boiling off/scouring in various machines

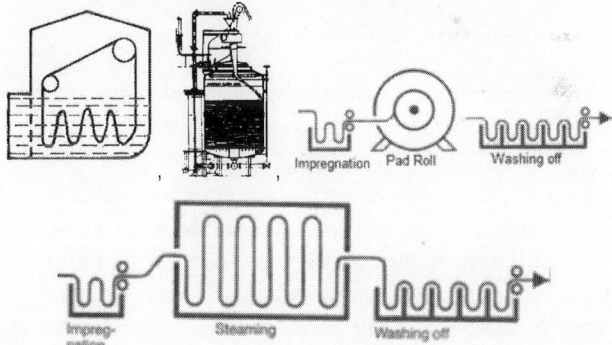

Chemicals	Winch 20:1, Jigs 5:1, Yarn, Piece goods	Piece goods, pres-sure Kiers	Pad-roll open width	Pad-steam open width	J-box open width and rope	HT steamer open width
Caustic soda solid g/l or %	6–12	5%	40–50	50–90	50–90	50–70

Suitable combined* detergent/ emulsifier g/l or %	0.3–1.0	0.2–0.6%	3–5	5–8	1.5–2	3–5
Conditions of treatment	1–3 h at 95–100°C	4–8 h at 1.5–2 atm pressure	1. Impregnate at 60–80°C, 80–100% pick-up. 2. Steam at 90°C for 60–90 min	1. Impregnate at 50–60°C, 80–100% pick-up. 2. Steam at 103–105°C for 60–90 s	1. Impregnate at 50–60°C, 80–100% pick-up. 2. Steam in J-box 1–2 hrs (time of stay)	1. Impregnate at 60–80°C, 80–90% pick-up. 2. Steam at 142°C at 3 atm for 1 min

Notes:

1. Caustic soda: If goods are padded as wet-on-wet. Then the padding liquor has to be 5–8 times stronger. The calculations are given elsewhere in the book.
2. The detergent scouring agent should be powerful with good emulsifying, dispersing, detergent, solvent actions combined.
3. After boiling-off goods can be washed neutralised rinsed and dried.

2.3.4   Preparation of cotton knitted goods

Scouring of knitted goods are little different than scouring of woven goods. The scouring of hosiery goods are easier because they do not contain size, the removal of which is one of the most difficult part in preparation process, hence easier to prepare. Hosiery yarns whether carded or combed yarns are used, knit goods contain fewer seeds and less twist which makes the scouring can be done more effectively. Moreover, these goods better penetration of the liquor because of the low-twist of the yarn and porous structure.

The only disadvantage of the knitted goods scouring processing is that the material has to be handled carefully without tension to avoid distortion of the loop structure. There will be some tension whatever the case may be but it should be within the range of recovery of the structure by further processing steps like steering, compacting, etc.

The scouring can be done in both tubular and open-width form. Even though tubular form of scouring is more popular (mostly because the scouring and dyeing is continually done in soft flow machines), the open width form of scouring is gaining over tubular scouring of late, especially after the introduction of pad batch and control dyeing of reactive dyes in open-width form.

Scouring of knitted good on Jet/soft flow

Grey cotton goods are normally scoured by boiling-off in strongly alkaline medium, since cotton husks swell under these conditions to a degree that permits them to be removed by subsequent hydrogen peroxide bleach. Both open width and tubular fabric can be treated in jet/over flow machine.

Treatment procedure

Ready for dyeing (RFD)→ Demineralisation–Scouring/partial bleaching.

Full bleach → Demineralisation–Scouring/full bleaching–Optical whitening.

Demineralisation is a process by which the heavy metal contamination in the fibre is removed. The removal of these metal ions help in achieving a better whiteness. Demineralisation can be done as a separate process or a combined process along with scouring process. In case of full bleach it is better to do the demineralisation as separate bath taking into consideration of the higher whiteness that has to be achieved.

A demineralisation prior or after an alkaline process (scouring or bleaching) leads to an excellent removal of heavy metals and hardness from the fabric. Demineralisation prior to an alkaline process helps in the removal of heavy metals results in a more fibre protecting bleach. Neutralisation/Demineralisation in the rinsing bath gives additional extraction, low residual hardness and an even neutral pH-value.

Even though, strictly speaking there is no bleaching in a scouring process, it is customary to do a partial bleaching during the scouring process to see that all the naps, natural colours, etc., are removed to get a clean base for dyeing in case of RFD. In case of dark shades a white base may not be necessary, one can decide to do away with the addition of bleaching agent during scouring process, but it is always better to do a partial bleaching which will add to better absorbency and brightness of the shades.

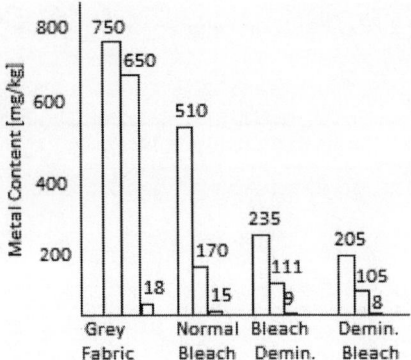

Figure 2.45. Removal of heavy metals in different process

Basic whiteness requirements:

Base whiteness degree for RFD: 70–75 Berger.

Base whiteness degree for full white: 82–88 Berger.

A full pre-treatment procedure for knits on circulating machines like soft flow can be summarised as follows:

Scouring in tubular or open-width form

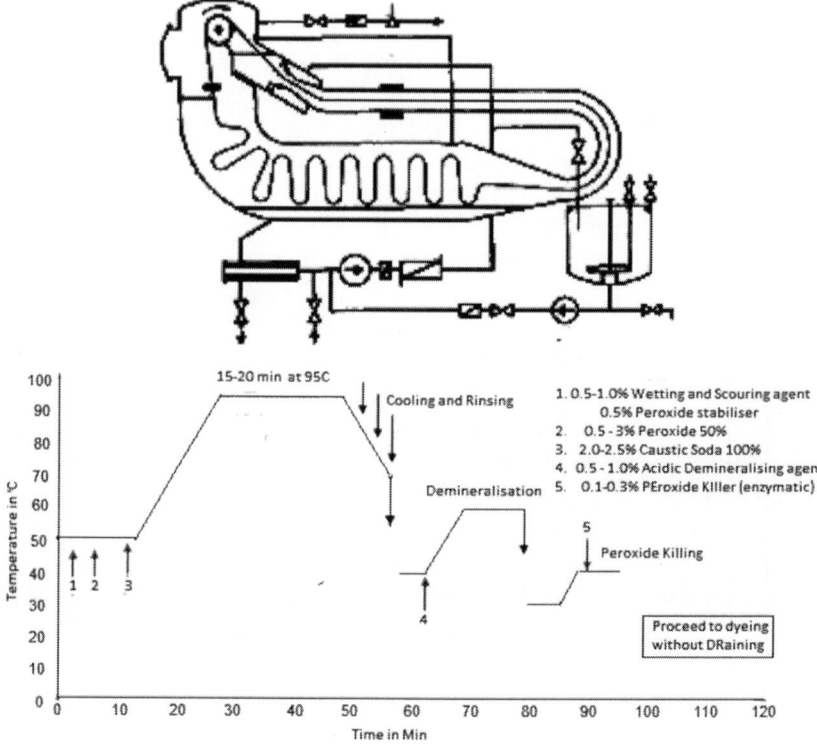

Figure 2.46. Scouring at atmospheric pressure (95°C)

Table 2.32. Guide recipe

Quantity	Unit	Bath additions
0.5–1	%	Wetting and scouring agent
0.5–1.5	%	Lubricant
0.4	%	Peroxide stabiliser
0.8–1.5	%	Caustic soda
1.5	%	Hydrogen peroxide 50%

Demineralisation, which may not be a must in the RFD, but it is suggested to do to avoid the interference of heavy metals in dyeing, is combined with neutralisation to save time. It should be noted that the demineralising agent is normally a strong acid

Notes:
Quantity of peroxide has to be decided as per the whiteness requirement. For example, it is better to keep a higher whiteness for a pale or light shade.

It is suggested to use lubricating agents to avoid rope marks and crease marks during the scouring process.

HT scouring
HT scouring has many advantages. In a processors point of view saving time is money and reducing cost. Every minute saved is higher production and saving. HT scouring was introduced to get a better scouring in shorter time in the steps of HT continuous bleaching. The process is based on the principle– scouring/bleaching speed can be increased by increasing the temperature, thus the process time can be clearly shortened.

The main advantages of HT scouring/bleaching are:

• Economical scouring process

• Saving of time

• Saving of water

• Saving energy

• Higher degree of absorbency and whiteness.

The process of bleaching at atmospheric pressure can be followed except the temperature has to be raised to 105–110°C, in a closed system. For the same whiteness requirement the amount of pperoxide and caustic soda can be reduced, after doing trials. It should be noted that the dangers of pin holes, weight loss, etc., are higher when the temperature is raised higher. But careful controls of process parameters can always give good results without any undue problems.

Notes
1. The lubricating agents reduce the risk of weight loss and rope marks and crease marks.

2. The temperature can be kept at 105°C also, the holding time may be fixed as per trials, but as a thumb rule the holding can be reduced 1 min per degree reduction from 110°C

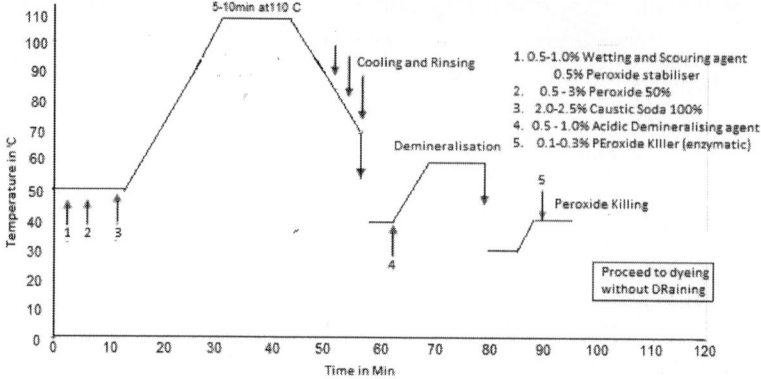

Figure 2.47. Scouring at 110°C for RFD

As there are risks of pin holes due to the presence of heavy metals it is always better to do the demineralisation process separately before scouring process if the presence of heavy metals is higher. In such cases neutralisation after scouring has to be done as an additional process using a core neutraliser. Alternately an alkaline demineralising agent can be employed along with the scouring bath.

PART III

Pretreatment of cellulosics - bleaching

3.1 Bleaching of cellulosics

Bleaching of textiles has been practiced for a very long time. Even today fabrics produced from the natural cellulosic fabrics are bleached to improve whiteness. The only difference is that the bleaching processes have changed enormously with the development of various types of bleaching agents. In spite of these processes whiteness would have been impossible without optical brighteners. However, the degree of whiteness is not only the main criterion in evaluation of bleaching, but also absorbency. Besides, cellulosic fibres consist of motes, naps, pectin, wax, fragments of seed coat.

The bleaching process aims at achieving several general goals:

- A high and uniform absorptivity of fabric for water, dyestuffs and finishing agent, achieved by the removal of hydrophobic impurities from the natural fibre.

- A sufficiently high and uniform degree of whiteness, which is stable to storage. The fabric should not be damaged with the degree of polymerisation remaining high.

- Of the two methods of bleaching, oxidative and reduction bleaching, the latter processes predominate in the treatment of wool whilst the former process is most frequently used for cotton.

There are two chemically different bleaching processes:

Oxidative bleaching: There are two systems in this processes which are as follows:

(a) Peroxide systems–With hydrogen peroxide, sodium peroxide, sodium perborate, potassium permanganate, peracetic acids and other acids.

(b) Chlorine systems–Sodium hypochlorite, bleaching powder, lithium hypochlorite, sodium chlorite, chloramines, isocyanural trichloride.

Reductive bleaching: Bleaching with sodium hydrosulphite, stabilised hydrosulphite preparations and other sulphoxylates, sulphur dioxide, acidic sodium sulphite, sodium bisulphites.

The choice of chemicals depends on the required degree of whiteness, on technological and ecological aspects, on the machinery and on economic aspects. Overdosing of the bleaching chemicals, insufficient temperature regulation, too long bleaching times, existence of catalysts, insufficient stabilising, etc. may lead to damaging of fibres.

3.1.1   Hypochlorites

(Redox potential: 1400–1550 mV)
1 l chlorine bleaching lye contains about 150 g active chlorine.

Table 3.1. Concentration active chlorine required for different processes

Process	Concentration active chlorine	Temperature	Time
Discontinuous	1–2 g/l	20–25°C	30–120 min
pad batch dwelling	2–6 g/l	20–25°C	10–120 min

pH = Approx. 9.0–11.5
Additions = Buffering substances (0.5–2 g/l) and protection against corrosion ($NaNO_3$) is necessary.
Antichlor = with H_2O_2.
Chlorination of soda ash solution to produce hypochlorous acid involves a somewhat complicated sequence of reactions, which begins with the formation of hypochlorite and bicarbonate in two steps, as follows:

$$Na_2CO_3 + Cl_2 \rightarrow NaOCl + NaCl + CO_2 \qquad \text{--------(1)}$$

$$Na_2CO_3 + CO_2 + H_2O \rightarrow 2NaHCO_2 \qquad \text{--------(2)}$$

On further chlorination, hypochlorite is first decomposed without liberation of carbon dioxide:

$$NaOCl + Cl_2 + H_2O = NaCl + 2HOCl \qquad \text{---------(3)}$$

On further chlorination, the bicarbonate is decomposed and hypochlorous acid is formed with liberation of carbon dioxide:

$$NaHCO_3 + Cl_2 = NaCl + CO_2 + HOCl \qquad \text{----------(4)}$$

This reaction, however, is very inefficient because the formation of chlorate ion and chloric acid preponderates. In the production of hypochlorous acid, the chlorination should be stopped at the end of reaction (3) when the solution begins to effervesce as a result of the liberation of carbon dioxide.

The available chlorine is then present largely in the form of free hypochlorous acid. Thus reactions 1, 2 and 3 may be combined as follows:

$Na_2CO_3 + Cl_2 + H_2O = NaCl + NaHCO_3 + HOCl$

Such solutions decompose very quickly and must, therefore, be utilised immediately.

Hypochlorous acid (HOCl), the active ingredient for bleaching, is an unstable compound known only in water solutions.

Heat of formation

$Cl + O + H + aq = HOCl$ aq.   29.93 + 31.6 + 29.78 kg-cal/mol

Heat of solution

$Cl_2O + aq. = HOCl$ aq.   9.44 kg-cal/mol

Heat of neutralisation

NaOH aq.+ HOCl aq.   9.88 kg-cal/mol

Dissociation constant

$K = 20°C - 1.04 \times 10^{-6}$

$K = 30°C - 1.18 \times 10^{-6}$

Hypochlorite provides a convenient form of handling chlorine, particularly for use in the concentrations required for textile preparation processes. Commercially, they are available as a solution of sodium Hhypochlorite (NaOCl) or as bleaching powder. The latter is normally a mixed salt with a formula approximating to $3Ca(OCl)ClCa(OH)_2.5H_2O$.

The product strength of hypochlorite is generally expressed as available chlorine content

$NaOCl + 2HCl \rightarrow NaCl + H_2O + Cl_2$

It undergoes self-decomposition as follows:

$2NaOCl \rightarrow NaCl + NaClO_2$

$NaOCl + NaClO_2 \rightarrow NaCl + NaClO_3$

$3\ NaOCl \rightarrow 2\ NaCl + NaClO_3$

Another reaction, which is sluggish, is

$2\ NaOCl \rightarrow NaCl + O_2$

The relative proportion of above reaction depends on the pH. Maximum decomposition was observed at pH 7. A major portion of hypochlorite decomposed results in chlorate formation. This chlorate formation reaction predominates almost all the hypochlorite decomposition on the alkaline side, where the oxygen evolution is practically nil, while on the acidic side, oxygen evolution predominates over chlorate formation. The best conditions for the storage of sodium hypochlorite are about 2N NaOH at 0–5°C and away from light. As pH of sodium hypochlorite decreases from 12 to 4.6 all of the sodium hypochlorite is converted to hypochlorous acid and while below a pH of 4.6 chlorine gas is liberated. Hence bleaching in the acidic side is not carried out 4. Following disadvantages of chlorine bleaching as:

- This method is low priced bleaching, but whiteness of the cloth obtained is not satisfactory.
- Cellulosic fabrics may be damaged to a large extent as compared to hydrogen peroxide bleaching.
- Corrosion resistant equipment is necessary.
- There is always unpleasant smell in the factory.

The commercial products used in textiles generally having the following available chlorine content.

Sodium hypochlorite liquid % w/w 12.5 (g/l ca. 150)

Bleaching powder, normal % w/w ca. 35.0 (g/l ca. 65.0)

The efficiency of hypochlorite bleaching depends on many factors.

Effect of pH

Hypochlorite entirely different in acidic, neutral and acidic pH. The most suitable pH I between 9 and 11, where proper formation of active ions OCl⁻ or a complex of OCl⁻ with HOCl is formed which are responsible for the bleaching action of hypochlorites as shown below:

$$HOCl \rightleftharpoons H^+ + OCl^-$$

$$HOCl + H^+ + Cl^- \rightleftharpoons Cl_2 + H_2O$$

To maintain the pH at 9–11, 5 g/l Na_2CO_3, which acts as a buffer.

In the acidic pH range say below 5 the bleaching actin is reduced due to the liberation of chlorine which can happen in case of sodium hypochlorite and in case of bleaching powder it is much faster. At the pH range of 1–2 the whole hypochlorous acid is converted into chlorine. In the neutral pH range, the presence of hypochlorite ion and hypochlorous acid is increased and bleaching action is increases, but more effective at higher pH as mentioned earlier, where the hypochlorite is more stable.

A large excess alkalinity does not, stabilise 15% NaOCl solutions any more than a slight excess (0.5% NaOH), but if the alkalinity drops below 0.04% NaOH, or a pH of about 11, decomposition becomes more rapid.

Effect of temperature

Even though hypochlorite is stable at even boil at pH values 11, the bleaching of cellulose is carried out at temperatures below 40°C (1–1.5 g/l active chlorine for 1 h). This is mainly due to degradation of cellulose along with bleaching. When the pH is buffered at 11 the speed of bleaching is increased at the rate of around 2.3 times for every 10° raise in temperature but it will be very difficult to control the degradation of cotton, but below 40°C the bleaching process is very safe.

Concentration of certain catalysts

Copper, nickel, and cobalt are powerful catalysts of decomposition. Of these, copper is the one generally present to some extent and should be kept as low as possible, not over 0.0001% (one part per million) in the finished (15%) bleach. In this connection, it is important to observe every precaution to avoid copper contamination from pipe lines, vessels and valves which may be used in handling the chlorine. Such lines and valves must be kept internally free from moisture to prevent active corrosion by the chlorine.

Effect of concentration of available chlorine

It has been found that half-life of the 25 gpl solution is decreased over 98% when the storage temperature is 60°C as compared with 25°C also the half-life at 25°C is decreased about 56% when the concentration is changed from 25 to 50 gpl. The commercial solution of sodium hypochlorite contains 14–15% available chlorine, compared to 35–36% in bleaching powder. The concentration of hypochlorite in the bleaching bath generally varies from 1 to 3 g/l available chlorine. The optimum bleaching conditions, however, depend on the degree of discolouration of the cloth and thus the temperature and time of reaction should be adjusted according to the requirement. The concentration of hypochlorite solution is normally estimated by means of standard thiosulphate or arsenite titration.

Effect of light

The influence of light on the stability can easily be demonstrated by dividing a sample of bleach, putting one portion in an amber glass bottle and the other in a colourless glass bottle and exposing both to sunlight, other conditions being the same. In a test of this kind, the half-life of a 200 gpl bleach was reduced to about 7 days in the colourless bottle as compared with a normal 44 days in darkness.

Effect of electrolyte

On adding sodium chloride in hypochlorite bleaching bath, it increases the bleaching activity in the beginning due to the production of nascent chlorine.

The addition of chloride moves the equilibrium to the left-hand side, but after, say 15 min, when the new equilibrium is reached, the sudden stimulated effect is ceased and bleaching resumes its normal course and hence action of salt is only temporary and confined to the first stage of bleaching.

$$2Ca(OH)_2 + 2Cl_2 \rightleftharpoons Ca(OCl)_2 + CaCl_2 + 2H_2O$$

$$HOCl + NaCl = 2Cl + NaOH$$

The most stable solutions are those of low hypochlorite concentration, with a pH of 11 or higher, low copper and nickel content, and stored in the dark at low temperatures. Iron and alumina seem to have slight effect on stability.

3.1.2   Bleaching powder

Bleaching powder is calcium hypochlorite. Bleaching powder is important as a bleaching agent because of the stable powder, hassle free transportation compared to sodium hypochlorite which is less stable and transportation is hazardous. Bleaching powder acts as a bleach when it is dissolved in water. But a solution of bleaching powder contains many other chemicals other than calcium hypochlorite $[CaO(ClO)_2.4H_2O]$ like calcium chloride $(CaCl_2)$, Lime $[Ca(OH)_2]$, and possibly free chlorine, calcium chlorate, calcium carbonate, etc. For bleach solutions, bleaching powder is mixed with water (bleaching powder do not fully dissolve in water) and allowed to dissolve the maximum and the supernatant liquor is separated and taken as bleach liquor. It may be 12°Tw and shall contain 3.9 g/l available chlorine and a pH of about 11.05 and can be diluted as required and used. The bleaching action is due to the hypochlorous acid formed by the decomposition of calcium hypochlorite. This bleaching agent is not much used in textile due to its pollution problem, hazardous effect on human being, contamination with calcium, etc.

3.1.3   Sodium chlorite

Formula $NaClO_2$;

Molecular weight 90.5

(Redox potential: 1040–1200 mV)

Physical forms: powder, 50–80% or liquid, 25–30%

The sodium chlorite bleach is widely used. Thanks to its good bleaching effect on cellulosic fibres, including bast fibres, and on fully synthetic fibres.

- As a bleach for cotton and regenerated cellulose and their blends with fully synthetic fibres.

- As an intermediate bleach for bast fibres which are difficult to bleach (e.g., linen, jute, etc.).
- As a bleach for full white on synthetic fibres and their blends with cellulosic fibres.

In contrast to sodium hypochlorite, sodium chlorite bleaches virtually only the fibre impurities and the natural dye and is therefore gentler to the fibres.

The sodium chlorite bleach is carried out in an acid medium (pH 3–4.5) at 80–90°C. The lower the pH and the higher temperature, the greater is the bleaching effect. However under these conditions unpleasant chlorine dioxide is formed and corrosion of the bleaching equipment occurs. As the pH of the bleaching liquor plays an important role in this connection an organic acid, preferably formic acid, is added to the long liquor and buffer salts are added to control the pH during bleaching. In the pad roll process activators are used instead of acid. This causes a pH shift to the weakly acid region at temperatures above 70°C which counteracts to rapid release of chlorine dioxide.

Bleaching with sodium chlorite requires a treatment time of an hour or more and depends on the type of process. For a good cleaning effect and to produce better absorbency a brief boil with soda ash during washing-off is recommended. If there are still chlorine residues in the goods after bleaching and washing it is advisable to carry out an antichlor treatment.

3.1.4   Chlorite bleaching–Theory

Chlorine dioxide gas is poisonous and explosive and is also very corrosive to metals in the aqueous medium. Chlorite bleaching is therefore often carried out in an exclusive room in the dye house that is very well-ventilated. The machines are fabricated from special stainless steels that have a high proportion of molybdenum or titanium.

Alternatively bleaching is done in equipment made of stone, PTFE (Polytetrafluoroethylene, Teflon) coated steel or wood.

For protection of the stainless steel metal of the machines, it is not uncommon to add sodium nitrate in a quantity equal to the chlorite that moderates decomposition of the chlorite and inhibits corrosion of metals.

The pH of around 4 ± 0.2 required for bleaching is maintained with buffers or as termed in industry activators, like sodium acetate or sodium dihydrogen phosphate (NaH_2PO_4) latter is usually preferred because it improves whiteness of goods. Neutral or slightly acid chemicals that liberate acid on heating are also used occasionally. Organic esters like ethyl lactate or titrate and their ammonium salts are also suitable for this purpose.

Sodium chlorite is an oxidizing agent containing chlorine as available chlorine and it bleaches cellulose on the acidic side at the boil without degrading it. Sodium chlorite decomposes predominantly to sodium chlorate than into gaseous oxygen.

$$3NaClO_2 \rightarrow 2\ NaClO_3 + NaCl$$

$$NaClO_2 \rightarrow NaCl + O_2$$

The decomposition products of sodium chlorite over a pH range 1.6–5.95 at 40–80°C include sodium chloride, sodium chlorate, chlorine dioxide and oxygen. Chlorine dioxide formation is main reaction and is maximum at pH 2.5–3.5.

$$5ClO_2 + 2H^+ \rightarrow 4CO_2 + Cl^- + 2OH^-$$

Oxygen evolution is maximum at pH 2.4

$$ClO_2^- \rightarrow Cl^- + {}^+O_2$$

Chlorate formation commences below pH 4.8

$$3ClO_2^- \rightarrow 2ClO_3^- + Cl^-$$

It is found that the amount of sodium chlorite decomposed decreases with increasing pH. At 100°C in fairly acidic solutions (at pH 4) most i.e., about 95% of the chlorite is decomposed within 2 h of boiling. The optimum condition for using acidified chlorite solution for bleaching should be those that promote maximum formation of sodium chloride, minimum formation of chlorine dioxide and minimum formation of chlorate. At pH less than 4, chlorine dioxide evolution is excessive which leads to loss of the oxidizing power of the solutions. This loss can be prevented by controlling pH in the range 4–7. Disadvantage of sodium chlorite is evolution of chlorine dioxide, which is toxic in nature and has corrosive properties. Organic acid like acetic acid or formic acid used for activating chlorite solutions result in the evolution of chlorine dioxide gas. A reduction in corrosion in bleaching unit is possible by the addition of sodium nitrate to bleach liquor. Various acid liberating agent like organic ester e.g., diacetin, ethyl lactate or triethanol amine hydrochloride have been recommended. Magnesium chloride, zinc nitrate and aluminium sulphate are found to be effective activator.

The advantages of sodium chlorite have attracted wide attention despite its high cost.

1. Sodium chlorite decomposition is not catalysed by metal ions, nor is degradation of cellulose brought about.

2. Due to lower weight loss and incomplete removal of waxes, the hand of chlorite bleached fabric is softer compared to hypochlorite or peroxide bleached fabric.

3. The bleached fabrics have low residual alkaline, which facilitates removal of chemical residue.

Disadvantages of sodium chlorite

1. The degree of whiteness of the cloth is good permanent whiteness and it is very effective in removing neps/motes from cotton fabrics.
2. There is a slight danger of damage for cellulosic fabrics, negligible danger for synthetic fabrics.
3. There is danger to the corrosion of equipment. Processing methods are limited to discontinuous and semi-discontinuous procedures.
4. Chlorite bleaching is applicable to all materials including synthetic fibres and mixtures of all these fibres.
5. There is no danger of damage of cellulose with sodium chlorite bleaching.
6. The vigorous alkaline treatments employed in kier boiling are not necessary in using sodium conditions of chlorite bleaching, hardness of water has little harmful effect and therefore low ash content is obtained on bleached cotton.
7. Sodium chlorite is rather insensitive to the presence of metal ions as iron with peroxide.
8. Since the bleaching is done at acidic pH there are possibility of the release of chlorine dioxide-with consequent toxicity and corrosion problem often lacking hydrophilicity.
9. Mildest bleaching process, causing minimum fibre damage and good degree of whiteness.
10. Chlorite bleaching is more expensive.
11. Peroxide bleaching also more or less as versatile as sodium chlorite except the case of say, acrylics. The advantage of chlorite bleaching is best suited for blend is applicable to peroxide also in most of the cases.
12. Even at acidic pH of 4–5 there are possibilities of release of chlorine dioxide which is corrosive and toxic.

Table 3.2. Conentration of sodium chlorite for different processes

Process	Concentration (NaClO$_2$ 80%)	Temperature	Time
Discontinuous	1.5–5 g/l	70–85°C	1–3 h
Pad roll/pad-steam	10–25 g/l	85–100°C	20 min–4 h

pH = 3.5–4 (with formic, acetic or oxalic acid).
Additions = Buffering substances (0.5–4 g/l) and protection against corrosion (NaNO$_3$) are necessary.

Cold pad chlorite systems rely on generation of chlorite dioxide in the roll and, in order to cope with occasional irregularities in formation, it is usual to provide a fan at or near floor level at the pad mangle and the still age. If fumes develop in the pad box, this should be emptied at once and rinsed out well before proceeding. Chlorite formaldehyde pad liquor has limited life, and a cheap way of ensuring continuity is to use two storage tanks alternately. More economical use of chlorite is possible if it is applied by proportionate feed, in a wet-on-wet pad box, to cloth that has been rinsed well, if the components are metered separately from two tanks, more concentrated mixtures can be used safely. It is important to prepare the pad liquor carefully, by mixing each component roughly with plenty of water in a bucket and add them in the order shown into the bulk of the water, with efficient stirring. It is imperative that the temperature be kept below 30°C. Most of the brass or copper parts be allowed to come into contact with the permanganate–persulphate–chlorite solution, since they catalyse decomposition of persulphate, causing generation of gas. Stainless steel padding boxes are not corroded by alkaline chlorite solutions.

After padding with chlorite, the batches should be covered with polythene, several layers of wet cloth, the outer layer being saturated with weak bisulphate. Peroxide bleached batches require hot soda boil for maximum whiteness. Cloth bleached with chlorite formaldehyde may retain fumes and must be unrolled from under the surface of bisulphate solution. Cloth treated with permanganate persulphate chlorite needs washing-off in hot bisulphate, oxalic acid or neutral hydrosulphite, together with sequestering agent. This is particularly suitable where an alkali scald conferring wettability follows, or where a size is present which coagulates in acid solutions.

Activation of $NaClO_2$ by formaldehyde was carried out at alkaline medium (pH 10) to avoid the troubles associated with evolution of chlorite dioxide. Based on the reheats obtained, combined desizing scouring and bleaching of loom state cotton fabric could be achieved by treating the latter at 70°C for 60 min with aqueous solution containing 3 gpl $NaClO_2$, 0.5 gpl HCHO, 2 gpl wetting agent at pH 10 using a material to liquor ratio 1:20.

Loom state (starch sized) cotton fabric was treated with sodium chlorite/potassium permanganate co-oxidant under a variety of conditions. It was found that the combined bleaching effect relies on the permanganate concentration, pH of the bleaching medium, and duration of the bleaching treatment. The bleaching effect of chlorite is assessed by the whiteness index. Loss in fabric weight wettability, copper number, carboxyl content and tensile strength of the treated fabric is favoured at acidic pH (4–6) using 0.01% and at alkaline pH (8–10).

Increasing the duration of the treatment from 15 to 120 min is also accompanied by an enhancement in the bleaching effect of both chlorite and chlorite permanganate co-oxidant enhances the bleaching effect significantly.

Based on the results obtained, new bleaching formulation could be developed for combined desizing, scouring and bleaching of loom state cotton fabric. The treatment method based on these formulations involves treatment of loom state cotton fabric with a solution containing 0.3% chlorite, 0.2% wetting agent at pH 4–6 using 0.01% permanganate for 60 min or at pH 8–10 using 0.04% permanganate.

Table 3.3. Physical properties of sodium chlorite

	NaClO$_2$ 26% w/w	NaClO$_2$ 80% w/w
Appearance	Yellowish clear liquid	White free flowing powder
Sp. gravity	1.23	
Solubility in water @200C g/l	Soluble in all proportions	ca. 400
NaClO$_2$ content g/kg	260	800
NaClO$_2$ content g/l	317–320	

3.1.5   Open-width bleaching using sodium chlorite

In this process, the goods are padded in open-width with a solution of sodium chlorite and activator (Special auxiliary for adjusting the pH for chlorite reaction), heated to the required reaction temperature and then suitably stored until the bleaching agent is spent. Whilst the impregnating liquors have a weakly alkaline reaction and are completely stable and inactive, activator provides the required acidity for bleaching during heating-up of the liquor.

A mixture of wetting agent and activator is dissolved in hot water and thoroughly mixed. The sodium chlorite is then either sprinkled in or added in dissolved form. The fabric is preferably padded at 40–50°C but may also be padded at room temperature. In this case, it is advisable to employ a somewhat higher concentration of activator. Impregnation is best carried out in a roller vat. Squeezing rollers are fitted at the end of the impregnating trough to remove the residual liquor. It is of advantage to run small idling rollers of approximately the same diameter on the upper guides of the impregnating trough. These rollers press against the goods and their repeated slight squeezing ensures effective and uniform wetting. The impregnating trough may be charged with wet or dry goods.

When bleaching dry goods, the initial bath concentration and that of the feeding addition are the same, i.e., the liquor carried over by the goods is replaced by a solution of all the chemicals in the same concentration. In the majority of cases, however, the goods are submitted to wet treatment before bleaching (desizing, mercerising) so that the bleaching plant is usually charged with wet goods.

In the cases of wet goods, the padding liquor gradually becomes diluted as the water contained in the goods is displaced by it. The displacement is, however, incomplete; it depends on the type and size of the trough and to a certain extent also on the type of goods. It lies between 60 and 90%.

Example

In open-width bleaching, a desized cotton nettle fabric requires 2.3% sodium chlorite.

At a squeezing effect of 90% after impregnation, and an assumed displacement of 85%:

$$\frac{2.3}{0.9 \times 0.85} = 30$$

i.e., the impregnating liquor should contain 3% sodium chlorite or 30 g/l sodium chlorite.

Dilution with the water introduced by the goods must be balanced out by a concentrated feeding addition, the exact concentration depending on the following factors:

1. The amount, in grams per litre, of sodium chlorite in the impregnating liquor = a

2. The amount of liquor in the goods after squeezing at the end of the impregnating trough = b

3. The water content of the goods to be bleached in % = c.

The concentration is calculated according to the following formula:

$$\frac{a - b}{b - c} \text{ g/l sodium chlorite}$$

The concentration of the other chemicals is calculated in the same way.

It follows from the formula that the sodium chlorite concentration of the feeding addition is inversely related to the factor b - c, i.e., the difference between the water content of the goods before and after padding with sodium chlorite. The concentration of the feeding addition should not be greater than 250 g/l sodium chlorite as otherwise there is a danger that the auxiliaries will no longer dissolve because of salting out. The goods should therefore be squeezed thoroughly before being entered into the impregnating trough.

The concentrated feeding addition should be entered at the front end of the impregnating trough, over its entire width, through a suitable distributor. In this way, the movement of the goods ensures uniform mixing of the solution added.

If the flow of concentrated feeding addition is controlled to maintain the liquor in the trough at the same level, constant concentration of the chemicals is obtained provided that goods of the same quality are involved.

After impregnation, the goods are heated to the required reaction temperature. This is best carried out with saturated steam. The speed of the passage should be adapted to the steam conditions and the thickness of the goods.

Infra-red methods of heating are not recommended.

When bleaching mercerised goods, the residual alkali must be completely removed by souring and washing before the goods are padded with sodium chlorite. If this is not done, bleaching may be uneven or may be completely ineffective. The bleaching time is dependent on the temperature of the goods:

80°C	3–3 1/2 h
90°C	1 1/2–2 h
100°C	3/4–1 1/4 h

The goods are best washed in an open-width washing machine with 4–6 boxes. To obtain good absorbency, it is important that the first two boxes should be at 85–90°C. The third box and, if necessary, the fourth should be at 40–50°C and the last box cold. The goods must on no account be cooled with cold water before being entered into the first box.

Chlorite bleaching advantages

a. The cotton suffers only a very slight loss in weight (usually 2–3.5%) and its excellent absorption is not affected.

b. Seed husks are completely removed with minimum damage to the fibres.

c. Bleaching results are entirely uniform and a soft, full handle is obtained.

d. This process is also suitable for bleaching fabrics from bast, regenerated and synthetic fibres and their blends. In this case, different concentrations and possibly combination with other bleaching agents may be necessary.

Chlorite disadvantages

a. It is carried out under acidic conditions which gives rise to a major disadvantage–the release of chlorine dioxide with consequent toxicity and corrosion problem. Hence its use is limited.

b. It should never be used on protein or polyurethane fibres.

Table 3.4. A typical sodium chlorite recipe

Quantity	Unit	Chemicals
1–2	% owf	Sodium chlorite (80%)
0.25–0.5	% owf	Sodium dihydrogen phosphate
0.1–0.25	% owf	Wetting agent
3.8–4.2	gpl	Formic acid to adjust pH to
1–2	gpl	Sodium nitrite
80–85°C		Temperature
2–3 h		Time

Discontinuous and continuous recipes with sodium chlorite

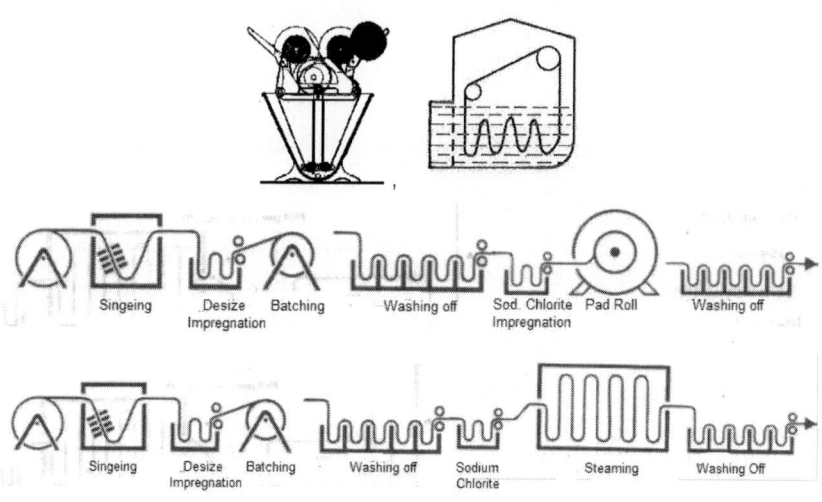

Can be followed for cotton, cotton/polyester yarn and piece goods.

Table 3.5. Guide recipes

Yarn and piece goods	Discontinuous	Continuous	Continuous
	Machines 10:1–20:1, Jig 5:1, Winch 20:1	Pad-roll	Pad-steam
Sodium chlorite 80% g/l	1–2	15–20	15–20
Formic acid concentrated g/l	1–2		

Buffer salt	1–2		
Wetting agent ml/l	0.5–1	0.5–1	4–8
Monosodium phosphate g/l		7	7
Disodium phosphate		4	4
pH adjusted to	3.5–4.0	6.5	6.5
Temperature of the bath °C	90–95		
Impregnate at °C		30–40	30–40
Steaming °C		90	102
Time min	60	60–90	60–90

Pad roll recipe

Washing off   Sod. Chlorite   Pad Roll   Washing off
Impregnation

Figure 3.1. Pad -roll

Table 3.6. Guide Recipe

Quantity	Unit	Chemicals
5–7	g/l	Sodium chlorite 80%
2–3	g/l	Sodium nitrite
2–4	g/l	Sodium dihydrogen phosphate
1–2	g/l	Wetting agent
x	g/l	Formic acid to adjust to pH 3.8–4.2

Temperature 85–90°C
Reaction time 1–3 h
After the treatment, washing with boiling water and dechlorination with 0.5–1% cold sodium bisulphate solution is carried out.

Bleaching of regenerated cellulose (rayon staple and rayon filament)

The bleaching process for these fabrics differs from that for cotton only with regard to concentration, which should be approximately 10 g/l sodium chlorite and 2 g/l catalyst. The storage time may also be reduced for bleaching of blended fabrics from cotton and regenerated cellulose fibres. The sodium chlorite open-width bleaching process is also suitable for this type of fabric. Even poor-quality cotton can be bleached until all husks are removed without in any way damaging the regenerated cellulose component. With blending ratios of 50:50, the addition of 20–25 g/l sodium chlorite and 2–3 g/l activator.

Bleaching of bast fibre (linen)

Because of the high content of foreign matter in bast fibres, satisfactory results cannot be achieved with one-stage bleaching processes; an alkaline treatment will at least be necessary. Where higher degree of whiteness is required, a combined bleaching process is used, e.g., a chlorite peroxide process. Linen union fabrics from pre-bleached linen yarn, on the other hand, can be bleached by a one-stage process with sodium chlorite. In such cases, the concentration of sodium Cclorite and activator (catalyst) should be raised to 40–50 g/l and 4–5 g/l, respectively.

Bleaching of synthetic fibres

Fully synthetic fabrics are not generally suitable for open-width bleaching because of their low-water absorption. Blends of these fibres with cotton or other cellulose fibres may, however, be bleached in this way. In this case, the method used for cotton may be used.

3.2   Hydrogen peroxide (H-O-O-H)

Molecular weight–34
   (Redox potential: 810–840 mV)
   Strength–35 or 50%
   It is generally agreed that the first stage is an ionisation is to form perhydroxyl ions (HOO⁻)

$$H_2O_2 \rightarrow HOO^- + H^+$$

The formation of the active perhydroxyl ions is favoured by alkaline conditions and so most hydrogen peroxide bleaching is carried out in these conditions. A number of competing reactions also occur, particularly in the presence of metallic catalysts, which may be summarised as a direct breakdown of hydrogen peroxide to water and molecular oxygen.

$$2H_2O_2 \rightarrow 2H_2O + O_2$$

This breakdown is most rapid in highly alkaline solutions. The molecular oxygen escapes from the bleach solution reducing the bleaching effect and the intermediates formed are very active and can cause fibre damage. To control bleaching a balance must be achieved and maintained, between activation and stabilisation.

Whilst the maximum bleaching activity from H_2O_2 is obtained generally at about pH 11.5, in practice this alkalinity level is altered to suit fibre sensitivity and bleaching process requirements. Examples: wool/cotton, amount of seed, etc. Sodium hydroxide, sodium carbonates are the alkalis generally used in case of cellulosic fibres, whilst ammonia and tetra sodium pyrophosphate are use while bleaching protein fibres.

3.2.1   Mechanisms

1. The symmetrical molecule of hydrogen peroxide H–O–O–H is activated by the addition of caustic soda which causes heterolytic splitting with formation of the peroxo–anion ^-O_2H responsible for bleaching.

$$\text{H-O-O-H} \rightarrow \,^-O_2H + H_2O$$

2. Hydrogen peroxide can be defined as a weak acid characterised by the dissociation constant K_a

$$\text{H-O-O-H} \leftrightarrow H^+ + \text{H-O-O}^-$$

Acid constant $K_a = 2.5 \times 10^{-12}$

The peroxo-anion formed can produce an oxidising or reducing reaction depending on the medium.

1. Oxidising reaction: $H_2O_2 + \,^-OH \leftrightarrow O_2 + H_2O + e^-$ (e = –0.08 V).
2. Reducing reaction: $H_2O_2 \leftrightarrow O_2 + 2H^+ + 2\,e^-$ (e = +0.7 V).

The peroxo-anion formed can react in three possible ways during bleaching:

$$\underset{\text{Coloured}}{-\overset{1}{C}=\overset{1}{C}-} \xrightarrow{\,^-O_2H} \underset{\text{Colourless}}{-\overset{1}{\underset{\overset{|}{\underset{H}{O}}}{C}}-\overset{1}{\underset{\overset{|}{\underset{H}{O}}}{C}}-}$$

— Oxidation of the natural dyes in the cotton; the chromophoric group is oxidised and becomes colourless.

— Formation of molecular oxygen; this is a loss of oxidising agent resulting in a reduction in the degree of whiteness obtained by the bleaching process.

$$^-O_2H \rightarrow HO^- + \tfrac{1}{2}O_2$$

The bleaching of textile fabric with hydrogen peroxide is dependent on several factors, such as pH, temperature, time, stabiliser type and presence of metallic impurities. The deposition of peroxide is related to the concentration of alkali and stabiliser and the temperature of the reaction. For practical reasons, peroxide bleaching is carried out at higher temperature 80–90°C for 2–4 h.

Hydrogen peroxide has the following advantages:

1. It produces a stable white colour.
2. It's reaction products are non-toxic and innocuous thus reducing pollution problems.
3. Cellulose is not degraded by it under optimum conditions.
4. Higher absorbency of finished goods is achieved.
5. Lesser tendency of after yellowing of goods, bleached with H_2O_2.
6. It is used for both open-width and rope for bleaching.
7. There is no danger of corrosion to the equipment as compared to the other processes.
8. Most important aspect is that it has no air pollution problem.
9. Universally applicable; long and short time, hot and cold processes.
10. Used for all states of treatment.
11. Stabilising (inorganic and/or organic) is necessary to control the degradation of peroxide.
12. Bleaching in an alkaline medium.
13. Activation by temperature, alkali and impurities.

The disadvantage is that the bleaching is slow unless high temperatures are applied. Hence energy costs are high. The presence of iron, nickel, copper, cobalt and lead hydroxides in the bleach may lead to catalytic decomposition of H_2O_2.

3.2.2   Physical and chemical properties of hydrogen peroxide in various forms

Table 3.7. Properties of H_2O_2 at different concentrations

	H_2O_2 % by weight			
	27.5	30	35	50
Appearance	Clear colourless liquid	Clear colourless liquid	Clear colourless liquid	Clear colourless liquid

Specific gravity at 20°C	1.101	1.114	1.131	1.195
Freezing point °C	ca. -22	ca. - 27	ca. - 34	ca - 52
H_2O_2 content g/kg	275	300	350	500
H_2O_2 content g/l	302	334	396	598
Active oxygen g/kg	129	141	165	235
Active oxygen g/l	143	157	186	281

Conversion factor for commercial hydrogen peroxide on weight basis

Table 3.8. Conversion factor for commercial hydrogen peroxide on weight basis

H_2O_2	Required amount			
Percent weight/weight	27.5%w/w	30%w/w	35%w/w	50%w/w
27.5% w/w	1	0.92	0.79	0.55
30.0% w/w	1.09	1	0.86	0.6
35.0% w/w	1.27	1.17	1	0.7
50.0% w/w	1.82	1.67	1.43	1

Conversion factor for commercial hydrogen peroxide on volume basis

Table 3.9. Conversion factor for commercial hydrogen peroxide on volume basis

H_2O_2	Required amount			
Percent weight/weight	27.5%w/w	30%w/w	35%w/w	50%w/w
27.5% w/w	1	0.91	0.76	0.51
30.0% w/w	1.1	1	0.84	0.56
35.0% w/w	1.31	1.19	1	0.66
50.0% w/w	1.97	1.79	1.51	1

Examples of calculation

1. In place of 12 kg of H_2O_2 35% w/w, what will be the equivalent amount of 50% w/w.

   Calculation = 12.0 × 0.7 = 8.4 kg of H_2O_2 50% w/w.

2. In place of 12 l of H_2O_2 35% w/w, what would be the equivalent amount of H_2O_2 50% w/w.

   Calculation = 12.0 × 0.66 = 7.92 l of H_2O_2 50% w/w.

3.2.3   Decomposition of hydrogen peroxide

Peroxide solution is very unstable in even weakly alkaline solutions and the decomposition is catalysed even by traces of certain metallic salts. Copper ions are more effective catalysts than iron and generally the catalytic activity increases with increase in alkalinity. On the other hand, the stabilising effects are evident for phosphates, pyrophosphates, silicates, tartarates and borates. The catalytic effects of the metal ions are extended even to the fibres and this effect is enhanced if the catalytic impurity is already present in fibres. To counteract this catalytic activity, various stabilisers buffer the pH. It has been found that sodium silicates are the best stabiliser for hydrogen peroxide. They are cheap, effective and have a detergent action and inhibit corrosion of metal material. In addition they provide a buffering action which suppresses the concentration of free sodium hydroxide leading to further reduction in fibre damage.

The major disadvantage of silicate stabiliser is that their use tends to cause formation of hard scales on the processing equipment. These interfere with the free running of the cloth through the equipment, abrade the cloth and reduce the efficiency of heat exchangers. There is a tendency for the deposition of silicate onto the fibre. This may result in the cloth having a poor handle problem with dyeing and printings that have been attributed to silicate decomposition.

Many sequestering agents have also been introduced in stabiliser formulations for bleaching.

Certain amides can accelerate the decomposition of hydrogen peroxide, when incorporated in the bleaching bath, thereby reducing the bleaching time. Urea, benzamide, formamide and N, N dimethyl formamide can be used, while among them, benzamide gives the maximum whiteness.

Direct breakdown of hydrogen peroxide to water and molecular oxygen. This decomposition is not the reaction which is required for bleaching because molecular oxygen (O_2) escapes from the bleach solution reducing the bleaching effect and cause fibre damage.

$$2H_2O_2 \rightarrow 2H_2O + O_2$$

The actual mechanism by which bleaching (=oxidation) occurs is essentially done by the following two stages. Activation by NaOH + Oxidation

Decomposition of peroxide in presence of caustic soda

- $NaOH \rightarrow Na^+ + OH^-$
- $H_2O_2 + OH^- \rightarrow H_2O + HO_2^-$
- $HO_2^- \rightarrow HO^- + (O)$
- Bleach inactive:

- $2H_2O_2 \rightarrow 2H_2O + O_2$
- Homolytic fission: catalytic damage
- $H_2O_2 \rightarrow 2HO^{.}$

3.2.4   Stabilisation of bleaches and bleach liquors

When using the term "stabilisation" a difference should be made between the stabilisation of bleaching agent and stabilisation or control of the bleach liquor.

Bleaching agents, e.g., hydrogen peroxide and sodium chlorite are "stabilised" during manufacture by addition of certain materials so that their concentration stays almost constant over long period of storage. This is stabilisation in true sense of the word.

But a bleacher uses "Stabilisation" in a different sense. In bleach liquors which contain hydrogen peroxide or sodium chlorite, bleaching occurs after activation i.e., by addition of alkali or acid as appropriate or by increased temperature. The bleaching activity should be regulated to prevent rapid, spontaneous decomposition of bleach and minimum damage to the fibre, wastage of bleaching chemical and undesirable side reaction.

This process of regulation or control is the "stabilisation" in the textile bleaching world. Actually stabilisation is not stabilisation but is a controlled release of H_2O_2 as per the requirement of the process.

3.2.4.1   Selection of bleaching agent

Table 3.10. Suitability of Bleaching agents for different fibres

Fibre	H_2O_2	NaClO2	$CH_3CO.$ OOH	Persulphates
Cotton	s,d	s	s	d
Flax	s,d	s	s	d
Jute	s	s	bs	bs
Wool	s	n	s	bs
Silk	s	n	s	n
Viscose rayon	s,d	s	s	d
Cup. rayon	bs	s	n	n
Polynosic rayon	s,d	s	s	d
Acetate	o,bs	s	s	n
Triacetate	o,bs	s	s	n

Polyamide	o,bs	s	s	n
Polyester	d,bs	s	s	d
Acrylic	n	s	s	n
Polyurethene	s	n	bs	n

s–suitable for bleaching.
d–suitable for desizing.
bs–bleaching under special conditions.
o–compatible for fibres shown and so may be used on blend.
n–not suitable.

3.2.5   Stabilisation of hydrogen peroxide

Bleaching agents, e.g., hydrogen peroxide, are "stabilised" during manufacture. In bleach liquors which contain hydrogen peroxide, bleaching only occurs after activation, e.g., by the addition of alkali or/and by increased temperature. This bleaching activity must be "regulated" to prevent rapid, spontaneous decomposition of the bleach and to minimise damage to the fibre, to avoid waste of bleaching chemicals as well as undesirable side reactions. This process of regulation or control is also called as stabilisation.

A stabiliser is selected after considering following criteria:
1. Stabilising effect under various pH, temperature, liquor ratio, water hardness.
2. Effect on absorbency.
3. Degree of bleaching.
4. Influence on fibre.
5. Economics.
6. Final ash content of the goods.
7. Effect on handle of the bleached goods.
8. Physical handing i.e., flow, solubility and pumpability.

3.2.6   Inorganic stabilisers

3.2.6.1   Sodium silicate

Sodium silicate may be used in a colloidal silicate (water glass) or as orthosilicate or metasilicate. The stability action is improved considerably by the presence of magnesium salts. Magnesium silicates and hydrated silicas (SiO_2) should be kept in colloidal form during their formation and bleaching, which is seldom fulfilled completely in plant conditions. Hence some depositions of silicate on to the fibre which affects the handle of the fabric and dyeing and printing problems are inherent, when silicate is used as stabiliser in processing.

Mean specification of standard qualities of sodium silicate

Table 3.11. Mean specification of standard qualities of sodium silicate

Baume'	Density g/ml	Weight ratio $Na_2O:SiO_2$		Dry substance %	100 g of sodium silicate contains %	
					Na_2O	SiO_2
37–40	1.36	1:3.3	ca. 35	8.5	26.5	
40–42	1.4	1:3.3	ca. 40	9.5	30.5	
48–50	1.5	1:2.6	ca. 48	13.5	34.5	
58–60	1.7	1:2.1	ca. 56	18.5	37.5	

Calculation of $Na_2O:SiO_2$ ratio

For the best bleached results, the $Na_2O:SiO_2$ ratio should be between 1.3: 1 and 1:1.6. To calculate $Na_2O:SiO_2$ ratio the recipe in g/l is multiplied by the Na_2O and SiO_2 contents (as fractions) of the sodium silicate and caustic soda used. For solid caustic soda the Na_2O content is about 77% or 0.77 as fraction, its SiO_2 content is of course, nil.

Example for calculation

What is the $Na_2O:SiO_2$ ratio of a pad-steam bleach containing 25 g/l sodium silicate 79° Tw and 5 g/l caustic soda?

Calculation

	g/l	Na_2O	SiO_2
Sodium silicate 75°Tw	25	25 × 0.88 = 2.2	25 × 0.29 = 7.25
Caustic soda	5	5 × 0.77 = 3.9	5 × 0 = 0
Total		6.1	7.25

Hence the ratio is 1:1.2.

The general relationship is as follows

$Na_2O:SiO_2$ ratio = R = {[Silicate] × Y + [Caustic] × Z} / [Silicate] × X

Where

X = SiO_2 content of silicate

Y = Na_2O content of silicate

Z = Na_2O content of caustic soda

And [Silicate] and [Caustic] are concentration of silicate and caustic soda.

Example

In the above example, what changes are required if sodium silicate 42°Bé is used? ($Na_2O = 0.106$, $SiO_2 = 0.26$).

Calculation

The require amount of 42°Bé silicate = 7.25/0.26 = 28 g/l.

Now, 28 g/l 42°Bé silicate provides 28 × 0.106 = 3.0 g/l Na_2O.

But the required Na_2O is 6.1. So caustic is added to give 6.1–3.0 = 3.1 g/l Na_2O.

This is given by 3.1/0.77 = 4.0 g/l NaOH.

So the recipe has to adjusted to 28 g/l sodium silicate 42°Bé and caustic reduced to 4 g/l.

Specific gravity and $Na_2O:SiO_2$ ratios of various silicates (20°C)

Table 3.12. Physical and Chemical properties   different silicates and silicates solutions

Sodium silicate	S.G.	°Tw	°Be	Mean weight ratio $SiO_2:Na_2O$	Mean % Na_2O	Mean % SiO_2
Metasilicate.$5H_2O$				1	24	29
Metasilicate anhydrous				1	50.5	49.4
Ortho silicate				0.5	60	30
Solid. polysilicate				2	28	56
Powders				3.3	19.5	64.5
Polysilicate solutions	1.7	140	59.7	2	18	36
	1.625	125	55.8	2	16.5	33.2
	1.6	120	54.4	2	16.1	32.2
	1.56	112	52.1	2	15.3	30.6
	1.5	100	48.3	2	14	28.1
	1.5	100	48.3	2.5	12.5	31.1
	1.475	95	46.7	2.65	11.5	30.7
	1.53	106	50.2	2.7	12.5	33.6
	1.48	96	47	2.85	11.2	31.9
	1.41	82	42	2.5	10.6	26
	1.36	72	38.4	3.15	8.5	26.6

Sodium silicate	S.G.	°Tw	°Be	Mean weight ratio SiO_2:Na_2O	Mean % Na_2O	Mean % SiO_2
	1.423	84.5	43.1	3.2	9.5	30.3
	1.375	75	39.6	3.2	8.6	27.6
	1.41	82	42.2	3.3	9.1	29.9
	1.398	79.5	41	3.3	8.8	29.3
	1.388	77.5	40.5	3.3	8.7	28.7
	1.35	70	37.6	3.3	8	26.4
	1.37	74	39.2	3.37	8.3	27.9
	1.33	66	36	3.65	7.2	26.3
	1.263	52.5	30.2	3.85	5.8	22.2

3.2.6.2   Polyphosphates

Polyphosphates like tetrasodium pyrophosphate ($Na_4P_2O_7$) (TSPP) and hexaphosphates are of interest as stabilisers in alkaline bleach baths. These are produced as a result of the condensation of orthophosphates. The following example illustrates the condensed polyphosphate group (inorganic).

(a) Alkalinity of the bleach should not be higher than pH 10, since above than pH 10 their stabilising action decreases.

(b) Temperature of the bleach should not be above 60°C, higher temperature reduces the stabilising action.

(c) This should be used in conjunction with ammonia and not caustic soda.

There are several disadvantages in using polyphosphates as stabilisers. At temperature above than 80°C and in a strongly alkaline medium above pH 11, polyphosphates exhibit low-stability. Particular problems includes sensitivity to hydrolysis in hot, strongly alkaline bleach baths, and the lack of selectivity for heavy metal ions in the presence of alkaline earth ions.

3.2.7   Organic stabilisers

They are generally based on sequestering agents, protein dehydration products, or certain surfactants. There are products which acts as stabilisers only or which combines with other properties such as detergency and softening.

3.2.7.1   *Some organic stabilisers (sequestrants) used in peroxide bleaching*

Amino carboxylic acids

The aminopolycarboxylic acids form water-soluble complexes in combination with alkaline earth and heavy metal ions. These compounds have an increased stability at high temperatures, as well as high selectivity for heavy metal ions. However, at high temperatures (>90°C) salts of type Na_4-EDTA aminocarboxylic acids exert a fibre-protective action if oxygen is present, peroxide or atmospheric oxygen. Important among this group are as follows:

- Nitrilotriacetic acid (NTA)
- Ethylenediaminetetraacetic acid (EDTA)
- Diethylenetriaminepentaacetic acid (DTPA).

Polyhydroxy carboxylic acids

These compounds includes citric acid, oxalic acid and gluconic acid. The advantages of these are exceptional sequestering properties, in other words they are able to maintain a solution of metal ions even in the presence of precipitation agents where acid/lye concentrations are very limited. They also exhibit high selectivity for individual metal ions. The main disadvantage is that the effect of this type of compound is strongly dependent on the pH, and for this reason, only limited use is possible.

Phosphonic acids

The salts of phosphonic acids are known as phosphonates, and are derived from the following group of phosphonic acids:

$$-\overset{|}{\underset{|}{C}}-PO_3H_2$$

The important ones in this group are as follows:
1-Hydroxy ethylidene-1,1-diphosphonic acid (HEDP) (see below).

$$H_2O_3P-\overset{\overset{\displaystyle OH}{|}}{\underset{\underset{\displaystyle CH_3}{|}}{C}}-PO_3H_2$$

Ethylenediaminetetra (methylenephosphomic acid) (EDTMP) (see below).

$$H_2O_3P-H_2C \diagdown N-CH_2-CH_2-N \diagup^{CH_2-PO_3H_2}_{CH_2-PO_3H_2}$$

$$H_2O_3P-H_2C \diagup$$

Diethylenetriaminepenta (methylenephosphonic acid) (DETPMP) (see below).

$$(NaO)_2P-CH_2 \diagdown N-CH_2-CH_2-N-CH_2-CH_2-N \diagup^{CH_2-P(ONa)_2}$$

with $O=P(ONa)_2$ on the central nitrogen via CH_2, and terminal groups $(NaO)_2P-CH_2$ and $CH_2-P(ONa)_2$ bearing $=O$.

These stabilisers are versatile, they have a high level of thermal stability, stable whether the liquor is strongly alkaline or oxidising, and they act selectively on heavy metal ions. With regard to detergent and dispersing efficiency, they exhibit average to good builder properties. They also act in the same way as sequestering agents, which are able to prevent or at least delay precipitation or crystallisation. They can also be used as water softeners, exchangers and corrosion agents.

Even though the stabiliser controls the release of oxidation power of peroxide, there are many other parameters affecting the peroxide bleach liquor.

Effect of pH

pH has got strong effect on hydrogen peroxide. At very low pH like 1–3 peroxide is more stable, but as the pH goes higher the stability reduces and at high pH like 11.5–13 it is least stable. The bleaching takes place around 10.5 due to accumulation of perhydroxyl ions in the bleaching bath. At neutral or weak alkaline media, hydrogen peroxide does not produce any whitening effect and may cause degradation of cellulose.

Effect of temperature

The release of perhydroxyl ion, which is responsible for the bleaching power of hydrogen peroxide is affected by the temperature greatly. A lower temperature (below 80°C) the evolution of perhydroxyl ion is very slow, thereby the bleaching action is also poor. Hence the bleaching has to be done at higher temperatures like 90–100°C. In case of pressurised vessels the bleaching can be done much faster at higher temperatures than boiling up to 120°C.

Effect of concentration of peroxide in the bleaching bath

High concentration in the bath can affect the strength of cotton and may damage the fibre. General bleaching concentration varies from 1–4% owf per the liquor ratio and in modern machines with lower MLR(Material Liquor Ratio) the optimum concentration may be 1–2% owf.

Effect of time

Time required for bleaching is generally inversely proportional to the temperature of bleaching bath, even though the total time required for a particular bleach will depend on the whiteness required, type of fibre, liquor ratio, pH of the bath, presence and type of stabiliser, the witness at the beginning of the bleach other than temperature.

Advantages of peroxide over hypochlorite bleaching

1. Hypochlorite is a much stronger bleaching agent than peroxide and hence there are higher chances of fibre damage.
2. Peroxide can be used as bleaching agent almost all fibres including protein fibres, whereas hypochlorite bleach uses a high alkaline bath and hence not suitable for protein fibres like silk, wool, etc., which are highly affected by alkalinity.
3. Peroxide can be used along with scouring bath, thus making it one process wherever possible, thereby reducing cost, time, saving energy, etc. This is not possible with hypochlorite as it is not stable at higher temperature.
4. The whiteness produced is generally non-yellowing as against the hypochlorite bleach which is having possibility of yellowing.
5. Peroxide bleach can be used for bleaching yarn dyed material in most cases, but most of the dyes, especially reactive are mostly destroyed by hypochlorite. An exception may be vat dyes, which also may change in tone.
6. Peroxide bleaching is more environment friendly than hypochlorite, which gives unpleasant odours and health hazards, due to the formation of highly toxic chlorinated organic by-products (AOX -A term for adsorbable organic halogen compounds) during the bleaching process.
7. The peroxide is less corrosive than hypochlorite which needs special equipments like wooden vessels or lead lined or rubber coated equipments for its treatments.
8. Hypochlorite needs an antichlor treatment to completely remove the residual chlorine, which if present on the cloth can affect the further process. Whereas the removal of peroxide, which is also when left on the fabric can affect the following process can be removed very easily.
9. Sodium hypochlorite does not produce completely satisfactory white in spite of many advantages. For a good white a peroxide bleach is followed by a hypochlorite bleach.

10. Peroxide is a much safer bleach than hypochlorite as the latter produces a slight damage to cellulose fabrics.
11. Hypochlorite is not used for synthetic fibres also, as it effects greater damage to these fibres mainly due to the high alkalinity of the bath and hence peroxide bleach is always a better bet.
12. Peroxide bleach always gives a better handle to the fabric, compared to hypochlorite bleach which gives a harsh handle to the fabric (cotton).
13. Handling of hypochlorite's is more difficult than peroxide and needs many precautions compared to hydrogen peroxide.
14. Sodium hypochlorite which is used mostly for bleaching has to be used immediately after it is made. This calls for a hypochlorite making unit in the vicinity which is also laborious and risk involved whereas hydrogen peroxide is available of the shelf and stable.

Disadvantages

1. Hypochlorite bleaching is cheaper than peroxide bleaching.
2. Peroxide has to be stabilised with silicates which can cause harsh feeling and uneven dyeing. Silicates also create deposits on machinery and even on bowls. Of late organic stabilisers are available which can avoid all these issues.
3. Peroxide do not give acceptable whites in certain cases like acrylics. It has limitations to remove pigments available in the natural fibres.
4. Peroxide in higher strength can affect the skin and give irritation.
5. Pin holes during bleaching is a major problem of hydrogen peroxide bleaching which is caused by catalytic action of metal ions present in the bleaching bath or fabrics.

3.2.8   Persulphates

The following persulphates are used in textile processing

Table 3.13. Properties of Persulphates

Type of persulphate	Ammonia	Potassium	Sodium
Formula	$(NH_4)_2S_2O_8$	$K_2S_2O_8$	$Na_2S_2O_8$
Molecular weight	228.2	270.03	238.2

Type of persulphate	Ammonia	Potassium	Sodium
Active oxygen content, min%	6.9	5.8	6.5
Persulphate content of commercial product %	min. 98	min. 98	min. 96
Solubility g/100 g of solution at 10°C	49	3	46
20°C	54	5.5	54
30°C	59	8.8	58

3.2.9   Peracetic acid

Formula $CH_3CO.O.OH$. Molecular weight–76

Table 3.14. Properties of per acetic acid

Properties	Details
Concentration of peracetic acid	15%
Appearance	Almost colourless liquid
Concentration w/w%	36/40 as CH_3COOH
Hydrogen peroxide content	3.5–4.5 as H_2O_2
Specific gravity at 20°C	1.135
Flash point	40°C approximately
Odour	Pungent (15%)
Solubility	Soluble in water and polar organic solvents
pH	2

Peracetic acid can be produced by the action of hydrogen peroxide and acetic acid in the presence of a mineral acid like sulphuric acid.

$$H_2O_2 + CH_3COOH \xrightarrow{H+} CH_3.CO.O.OH + H_2O$$
$$\text{Peracetic acid}$$

It can also be produced by mixing 1 part of hydrogen peroxide and 6 parts of acetic anhydride in the presence of caustic soda or EDTA for about 4 h at room temperature which gives 80% peracetic acid. Excess of acetic anhydride can cause the formation of diacetyle peroxide which explosive and hence should be strictly avoided.

$$CH3 - \underset{\underset{\underset{CH3 - \overset{|}{C} = O}{O}}{|}}{C} = O \quad + H2O2 \quad \overset{Cat.}{\underset{NaOH}{\rightleftharpoons}} \quad CH3.CO.O.OH + CH3COOH$$

Peracetic acid can be used for bleaching of nylon, viscose rayon, cellulose acetate and even cotton. Peracetic acid is commercially available for textile bleaching in 5, 15 and 40% solutions which is known as "equilibrium peracid".

The major use of peracetic acid in as an effective bleaching agent in the household detergents and in the laundry industry. To reduce the danger in the on-site production of peracetic acid from acetic anhydride/hydrogen peroxide, activators can be used in household detergents to generate peracetic acid in situ. The commonly used activators are sodium perborate and acetylated O or N compounds such as tetraacetyl ethylenediamine (TAED). The perborate activators are assumed to act via the stages of peracetic acids. Perborate hydrolyses in aqueous solution and hydrogen peroxide is produced. Reactive

$$Na_4(H_4B_2O_8) + 4H_2O \rightarrow 2Na_2[B(OH)_4] + 2H_2O_2$$

$$H_2O_2 + OH^- \rightarrow HO_2^- + H_2O$$

$$X\text{-}Ac + HO_2\text{-} \rightarrow X^- + AcOOH$$

anion is produced from hydrogen peroxide under weakly alkaline medium and activator reacts. The TAED/H_2O_2 bleaching mechanism is as follows:

$$\begin{matrix} CH_2\text{-}N(COCH_3)_2 \\ | \\ CH_2\text{-}N(COCH_3)_2 \\ TAED \end{matrix} \quad + 2HO_2^- \quad \rightleftharpoons \quad \begin{matrix} CH_2\text{-}NH\ COCH_3 \\ | \\ C\ H_2\text{-}NH\ COCH_3\ O \\ DAED \end{matrix} \quad \begin{matrix} + 2CH_2CO\text{-}O_2^- \\ \\ \text{Peracetic acid anion} \end{matrix}$$

$$CH_2CO\text{-}O_2^- + H_2O_2 \xrightarrow{pH\ 8\text{-}9} CH_3COOOH + OH^-$$

$$CH_3CO\text{-}OOH + CH_3\text{-}CO\text{-}O_2^- \rightarrow CH_3COOH + CH_3COO^- + 2(O)$$

The peroxide anion reacts with TAED to form DAED (diacetylethylenediamine) and peracetic acid anions. At pH 8–9 the peracetic acid anion is in equilibrium with the free peracetic acid. This equilibrium peracetic acid oxidises its own per anion to form active oxygen which is active bleaching agent. Quite a number of substances utilised in textile chemical processing are known to pose various environmental problems. Chlorite bleaching is recommended to be avoided by the environmentalists as residual chlorite in wastewater and chloro derivatives of organic impurity in cellulosic natural fibres lead to the problem of absorbable organic halogen.

Peracetic acid has a very high oxidation potential and hence is extremely reactive.

In general, all organic proxy acids have an acid reaction and their stability decreases with increased pH and temperature. The reaction products of peracetic acid are hydrogen peroxide, oxygen, water, and acetic acid. In presence of heavy metal and ions such as copper, manganese or iron this acid undergoes catalytic decomposition. Peracetic acid bleaching is an environmentally friendly process, a good replacement for conventional bleaching processes. Bleaching can be done in neutral or mildly alkaline conditions with an acid (10%) concentration of 30 gpl. For materials that undergo sequent dyeing, the bleaching process can be done in about 30 min at 50°C, while for an OBA treatment; it goes on for about an hour or so at 60–70°C. The quality of the bleached material is in line with that of H_2O_2 bleaching. Peracetic acid can be catalysed to bleach cotton fibres at temperature as low as 30°C by incorporating 2,2 bipyridine and sodium lauryl sulphate in the bleach solution. The 2,2 bipyridine is sorbed from solution and forms the tris 2,2 bipyridine ferrous complex (trischelates) with ferrous ions in the fibres, and it is the trichelate associated with the fibres that catalyses bleaching. Sodium lauryl sulphate reduces peracetic acid decomposition in the bath by ionic association with trichelates.

With all the advantages, peracetic acid still has not been successful as a commercial bleaching agent. It can be used as a bleaching agent for cotton linen etc. Its working conditions are–strength 1.5–2.5 g/l, pH 6–7, temperature 50–80°C and time 20–60 min. Sequestering agent may be added to avoid the catalytic action of iron, copper, etc. Peracetic acid bleach, followed by a peroxide bleach gives a good whiteness.

Recipe for bleaching nylon

Quantity	Unit	Bath additions
0.75	%	Sequestering agent
0.75	%	Sodium bicarbonate
2	%	Caustic soda
5	%	Peracetic acid

pH 6–7.5.
Temperature 80°C.
Time 30–45 min.
Recipe for bleaching cellulose acetate
Liquor circulating machines

Quantity	Unit	Bath additions
300	g	Peracetic acid (30–40%)
50	g	Sodium hexameta phosphate
100	g	Wetting agent
99	l	Water

Bleaching is done in the above liquor after adjusting the pH to 5–6 with NaOH at a temperature of 66°C for 1 h. Rinse hot and cold.

Recipe for bleaching acrylic

Liquor circulating machine

Quantity	OWF	Bath additions
1	%	Sodium hydroxide
1	%	Sodium bicarbonate
1	%	Sequestering agent
0.01	%	Wetting agent
5	%	Peracetic acid (pre-diluted)

Bleaching bat to be made by adding chemicals in the same order. Enter the goods run cold for 15 min and then raise the temperature to 75°C, hold for 20 min cool and drain. Rinse hot, rinse cold.

Advantages

1. Peracetic acid can be used in the existing machines without any problem.
2. It is environmental friendly since the ultimate resultant products are only acetic acid and oxygen.
3. Can be used for bleaching the yarn dyed (vat) fabrics.
4. Versatile–used for cotton, linen, nylon, acrylic, acetate, etc.
5. Whiteness achieved by peracetic acid bleaching is superior with least fibre damage.

Disadvantages

1. Handling peracetic acid in concentrated form (30–40%) is difficult. It is explosive, with pungent odour, highly corrosive and burns the skin. It has to be transported with special care and protected tankers.
2. Cost wise not attractive and can be used only in special cases.

3. Peracetic acid is not as stable as stabilised hydrogen peroxide, especially in presence of metal ions. Peroxide is still the most used bleaching agent in this category.
4. Mainly used only in home laundry products where the peracetic acid is produced in situ.

3.2.10   Potassium permanganate

Potassium permanganate is powerful oxidizing agent. Potassium permanganate is the potassium salt of permanganic acid. Permanganic acid is extremely unstable which decomposes into nascent oxygen and manganese dioxide. Potassium permanganate exercises bleaching action in alkaline as well as in acidic medium. In alkaline medium rate and external release of oxygen is much less and hence it has very limited use as oxidising agent under these conditions.

3.2.11   Ozone bleaching

Ozone is one of the strongest oxidising agents and has strong tendency to react with almost any organic substance as well as with later. The reaction proceeds via several intermediates such as peroxides, epoxies and perhydroxyl and hydroxyl radicals. Some of these intermediates are likely to contribute to the bleaching effect in some manner. The formation of these intermediates is highly dependent on the process conditions.

3.3   Reductive bleach

- Bleaching agent: Derivate of sulphurous acids
  Sodium di-thionite ($Na_2S_2O_4$)
  Stabilised sodium di-thionite
  Sodium bi-sulphite.

Usually stabilised sodium di-thionite (e.g., Clarit PS Ciba, Blankit IN BASF) is used.

- Application only efficient as pre- or subsequent bleach
- Low importance for cellulose fibres
- Use for PA and wool
- No full white possible.

Table 3.15. Process for reductive bleach with stabilised Hydros

Process	Concentration ($Na_2S_2O_4$ Stab. sodium di thionite)	Temperature	Time
Discontinuous	2–6 g/l	60–90°C	0.5–3 h

pH 6–8.

Calculation of the amount to be used in reinforcing liquors for wet-on-wet treatment

The outgoing residual moisture content (RMo) of the goods must be greater than the ingoing residual moisture content (RMi).

RMo should be greater than RMi (RMo > RMi).

The difference should not be less than 12%

RMo – RMi = minimum 12%.

The outgoing residual moisture content (calculated on the dry weight of the goods) must not exceed the point at which dripping from the material takes place.

Necessary amount of the each product in the reinforcing liquor is calculated with the formula

$$K2 = (K1 \times 100)/(a\ RMo–a\ RMi)$$

Where

K1 = Amount in the basic recipe.
K2 = Amount of reinforcing liquor.
RMo = Outgoing residual moisture content.
RMi = In-going residual moisture content.
a = Exchange factor.

The smaller the difference RMo – RMi, the greater must be the amount of the product in the reinforcing liquor. In the rapid bleaching method with any chemical the liquor can be reinforced to eight times the concentration of the basic recipe.

The exchange factor takes into account the percentage of liquor taken-up or given-off by the ingoing goods depending on the type of fabric and its fibre content, the factor is 0.7–0.8 (70–80%).

A = (Liquor exchange in %)/100.

Example

Fabric 100% cotton 340 g/m.
Speed of bleaching range 43 m/m.
Operating system.

Recipe

Quantity	Unit	Chemicals
8	ml/l	Detergent
2	ml/l	Stabiliser

30	ml/l	Sodium silicate 38°Bé
40	ml/l	Caustic soda 36°Bé
40	ml/l	Hydrogen peroxide 35% by weight

Reinforcing liquor of eight times greater concentration metered to the range as two separate component.

Component A

A mixture of hydrogen peroxide 35% by weight and stabiliser liquid in the ratio 1:20.

Component B

Sodium silicate 38°Bé (71.5°Tw) $30 \times 8 = 240$ ml/l.Caustic soda 36°Bé (66.4°Tw) $40 \times 8 = 320$ ml/l.Wetting agent $8 \times 8 = 64$ ml/l.

Calculation

$$K2 = (8 \times 100)/ 70\text{–}56) = 800/14 = 57 \text{ g/l.}$$

aRMo – aRMi = 70–56% = 14% = 7 times greater concentration in the reinforcing liquor.

In 1 min $340 \times 43 = 14.62$ kg of fabric is passed through the range.

Component A

Consumption of hydrogen peroxide 35% by weight = 40 ml/l = $14.62 \times 40 = 584.8$ ml/min.

Or hydrogen peroxide may be metered at the rate of 0.58 l/min.

Component B

Sodium silicate 38°Bé + caustic soda 36°Bé + Wetting agent

Meter as reinforcing liquor at the rate of 14.62: 5 = 2.9 l/min, with further water to maintain the liquor level.

Multiply the calculated amount by the exchange factor a (0.7 in this example).

Wet-on-wet calculations

1. *Effective pick-up in wet-on-wet method*

   [(Pick-up at exit–pick-up at inlet) + (Pick-up at inlet × Exchange factor)]/100.

   Example Pick-up at inlet = 50%

   Pick-up at outlet = 95%

Exchange factor = 0.7

Effective pick-up = [(95–50) (50 × 0.7)]/100 = 0.80.

2. *Initial concentration in ml/l*

   Initial concentration in ml/l = (ml/kg)/(effective pick-up).

   Example 35/0.8 = 43.75 ml/kg.

3. *Initial volume in the trough (l)*

   = (Volume of trough × ml/l)/1000

   Example: Volume = (90 × 43.75)/1000 = 3.9 l

4. *Effective pick-up in %*

   = {Chemical on weight of fabric (ml/kg)} × 100/Concentration in the bath (ml/l)

   Or % exit pick-up–% Entry pick-up + (% Entry pick-up × Exchange factor).

5. *Exchange factor*

   Exchange factor = (Effective pick-up %–Exit pick-up % + Entry pick-up %)/Entry pick-up.

Calculation of bath concentration (More accurate method)

Concentration of chemicals in original and feed liquors, build-up of chemicals and bath dilutions.

Let

x = Chemicals or impurities in ml/min ($CvRKt.F_A.10^{-2}$) or ($Ckg . Kt . F_A . 10^{-2}$).

$C0$ = Concentration of chemicals in ml/l (F) at time 0.

$C1$ = Concentration of chemicals in ml/l (F) at time t.

Cv = Concentration of chemicals in the residual liquor on the goods before padding (ml/l).

Ckg = Concentration of chemicals on the goods after padding.

D = Rate of flow in l/min for wet goods.

$D = Kt . F_A . 10^{-2} +W+N$, for dry goods F = W + N.

dr = Density of R in g/ml.

df = Density of liquor (F) in g/ml.

F = Liquor in litres.

F_A = Liquor exchange rate % of R.

$F\%$ = Liquor in kg/kg of goods %.

Fn = Liquor on goods after padding l/kg. If the density dv, df is not sufficiently close to 1. Then,

$Fn = (F\% /df) + R([1 - (F_A /100)] . [1 - (dv/df)]$

Kt = Kilogram material per min.

N = Feed liquor in litres per min.

R = Residual or initial moisture content in l/kg.

t = Time in min.

W = Water inflow in litre per min.

(a) Feed liquor (V) maintaining the concentration in the pad liquor

    (1) Liquor density about 1

        Strength of feed liquor $(V) = \{C1[Fn\text{-}R + R . (F_A/100)] - [Cv . R . (F_A/100)]\}/ C1(Fn\text{-}R)$

    (2) Liquor density greater or smaller than 1

        $Fn = (F\% /df) + R [1\text{-} (1\text{-}F_A/100) . [1 - (dv/df)]$ l/kg

        $V = \{C1[Fn - R + R (F_A/100) - Cv . R . (F_A/100)]\} / [C1 (Fn\text{-}R)]$

Example of calculation

If a fabric with residual moisture of 60% has to be mercerised with 52°Tw (30°Bé) caustic soda (equivalent to 296.8 g/l) and if the increase in the weight of fabric (existing) is 130%, what will be the feed liquor strength (neglecting the shift in concentration of the liquor on the fabric due to the absorbed NaOH).

    $F_A = 100\%$ R = 0.6 l, dv = 1, df = 1.26, F% = 1.3% and c1 = 296 g/l

    $Fn = (1.3/1.26) + 0.6 (1\text{-}100/100) . 1 - (1/1.26) = 1.03$

    $V = [296.8 (1.03 - 0.6 + 0.6)]/[296.8 (1.03 - 0.6] = 2.04$

The feed liquor has to be 2.4 times stronger than the pad liquor.

(b) Concentration of chemicals (C1) in the pad liquor and strengthening factor (VC1) for the feed liquor preparation.

    If the fabric should contain a given amount of chemical Ckg after padding:

    To calculate the strengthening factor VC1 for feed preparation:

    $C1 = Ckg - Cv .R [1 - (F_A/100)]$ ml/l

    $VC1 = (Ckg - Cv . R) / [C1 (F_A - R)]$

Example

After desizing and scouring a fabric is to be bleached, wet, with 50 ml H_2O_2 35% w/w The equipment available for bleaching consists of squeeze rollers and a saturator. The rate of liquor exchange is about 90%. The moisture content of goods entering is 60% and on exit 70%. What is the required concentration of original and feed liquors.

$$Ckg = 50 \text{ ml } H_2O_2 \ 35\% \text{ /kg of fabric}$$
$$F_A = 90\%, Fn = 0.7 \text{ l/kg}, R = 0.6 \text{ l/kg}$$

Since Cv = 0

$$C1 = 50 / (0.7-0.6 + 0.6 \times 0.9) = 78.1$$
$$VC1= 50 / 78.1 (0.7 - 0.6) = 6.4.$$

Besides the other chemicals required for bleaching, the bleach bath also contains 78.1 ml/l H_2O_2 35% w/w. The feed liquor is 6.4 times stronger than the original liquor. Hence it contains 500 ml/l H_2O_2 35% w/w.

Pad applications of chemicals

In fabric preparation chemicals are padded on dry fabric (wet on dry) or on fabric which are not dried in between (wet-on-wet). The latter process is generally followed to avoid an extra cost of dyeing. The saturator should, therefore, have an adequate volume to provide a sufficient contact time between fabric and saturated liquor.

To obtain consistent chemical application, the following conditions should be met.
1. Uniform nip pressure
2. Uniform initial moisture content when padding especially wet-on-wet
3. Constant saturated liquor level
4. Constant fabric speed
5. Constant temperature
6. Good liquor penetration and interchange.

Mangle expression

The amount of liquor taken-up by any fabric is often described as mangle expression. This can be termed as pick-up also.

Pick-up (E) = {(Wet weight of fabric after padding – Air dry weight)100} / Air dry weight.

Wet-on-wet application

For an effective wet-on-wet application the entry expression should be always less than the exit expression (if it is the other way the padding liquor gets diluted as padding takes place).

A minimum of 10% difference is considered as minimum. In practice 15–20% difference is kept using a high efficient nip in front of the saturator.

The actual pick-up of chemicals called effective pick-up is the difference between the entry and exit expression (Eo-Ei) and the amount by which the incoming water is exchanged for pad liquor, this may be written as follows:

$$\text{Effective pick-up} = (Eo – Ei) + (Ei \times F)$$

Where F is the interchange factor or the fraction by which in coming water is exchanged for saturated solution. The value of F can only be determined by finding the concentration of chemical of the fabric after padding and relating this to mangle expressions and concentration of chemicals in the pad liquor. It is usually about 0.70–0.80 and these values can be used for starting up and then adjusted when F has actually been measured.

Example

A cotton cloth is padded through a pad liquor containing 50 g/l H_2O_2 (50% w/w). Exit and entry expressions are 95% and 85%, respectively. The peroxide concentration after padding is 3.3 % H_2O_2 (50% w/w) o.w.f., what is F?

Effective pick-up = (% o.w.f. × 100) / % pick-up

= (3.3 × 1000) / 50 = 66%.

Using equation for effective pick-up = (Eo-Ei) + (Ei + F).

= (95 – 85) + (85 × F) = 66

Hence F = 56/85 = 0.66.

Concentration of chemicals for continuous processing in long baths

In immersion bleaching (Long bath) allowance has to be made in the feed liquor for decreasing content of different chemicals in the recipe. The loss of chemicals in the bath will depend on substantivity, stability of that particular chemical at the running temperature, reactivity, etc.

After impregnating with the liquor in the dry or wet state the fabric runs continuously through the reactor and the washing range. Suitable impregnating devices are the padding mangle open-width soaper and spray assemblies. The bleach bath in the reactor and kept at a constant level by the liquor brought in from the impregnating device by the material. The concentration of the impregnating bath and the respective feed liquor are calculated using following formulas.

1. Impregnating bath

   Cu = [{(F.Co.K/Kt) + Fn . Co + L} – Cv. R (1-F_A/100)] / [Fn-R + R (F_A/100)]

   L = Alkali uptake of material in ml in liquor × °Bé per kg of the material.

   K = Decomposition constant of peroxide. Determined in the laboratory, this value is less than 1, and dependent on the composition and temperature of the bath.

   Note: In calculation to determine the concentration of

   (a) Caustic soda, K = 0

   (b) Hydrogen peroxide, L = 0

   (c) Substances which are non-substantive or decompose L and K = 0.

2. Feed for impregnating bath
    Cu(N) = Cu (Fn-R+R. F_A/100) / (Fn –R) ml/kg.
    These calculations are intended for a constant output in kg of goods /
    min. Fabric speed can be varied to compensate for differences in
    weight.
    The liquor ratio is preferably 20:1.

Fabric speed, fabric weight and output

Example

Treatment liquor = 3000 l
Liquor ratio = 20:1
Treatment time = 20 min (approximately)
Output = 7 kg/min
Fabric weight = 90–300 g/m
Fabric speed = 23–80 m/min.

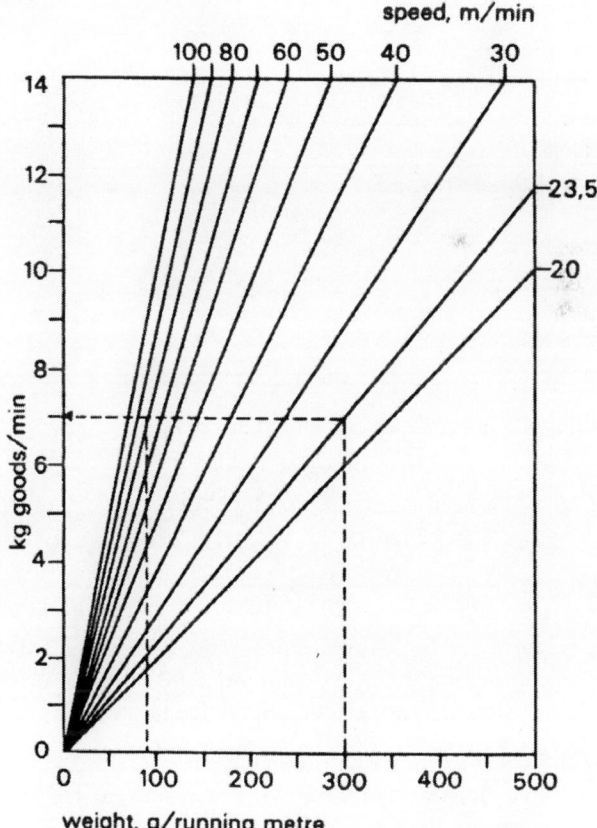

Figure 3.2. Wet desized fabric ready for bleaching

Bleaching conditions

Machine: Immersion accumulator (240 kg capacity).
Ratio 20:1
Temperature: 90°C
Bleaching Time: 30 minMaterial: Cotton
Fabric weight: 0.100 kg/ running metre
Fabric speed: 80 m/min

$F = 4800$ l
$F_A = 70\%$ during impregnation
$Fn = 0.8$ l/kg
$Kt = 8$ kg/min
$K = 8.96 \times 10^{-3}$
$R = 0.4$ l/kg
$Co =$ ml/l chemicals in the bleach bath.

The bleach bath shown in the table below has given good results in bulk working. Calculate the amount of peroxide in the original and feed liquors.

Table 3.16. Composition of baths

Chemical	Bleach bath recipe	Impregnation bath	Feed liquor
Wetting agent ml/l	3	3.5	6
Stabiliser ml/l	7	8.2	14
Sodium silicate 72°Tw ml/l	5	5.9	10
Caustic soda 66°Tw ml/l	10	22.8	39.8
Hydrogen peroxide 35% ml/l	20	181.6	308.8

Column 2 and 3 are already calculated answers.

We give below the calculation for peroxide.

Amount of peroxide in the impregnating bath

$$Cu = \{[(F\ Co\ .\ K)/Kt] - Cv.\ R\ (1 - F_A/100)\} / (Fn - R + R(F_A/100)$$

$$= \left\{\frac{(4800 \times 20 \times 8.96 \times 10^{-3})}{0.8} \times 20\right\} / (0.8 - 0.4 + 0.4 \times 0.7) = 181.6.$$

The impregnation bath should contain 181.6 ml/l of hydrogen peroxide. To calculate the amount of peroxide in the feed liquor

$$Cu(N) = Cu\ (Fn - R + R(FA/100)] / (Fn - R)$$

$$= 181.6\ (0.8 - 0.4 + 0.4 \times 0.7) / (0.8 - 0.4)$$

$$= 308.8.$$

The feed liquor should contain 308.8 ml/l hydrogen peroxide.

Build-up of chemicals in the bath as a function of time.

In rinsing and washing operations and in slop padding, the chemicals (X) introduced into the bath (F) by the material build up in accordance with following formula.

$$C1(t) = (C0. F - [(F.X/D) \exp(- D . t /F) + F.X /D] . 1/F$$

A graph is given to save the trouble of complicated calculations shown in the figure are plots of the e-functions, decreasing, for 100% initial difference between the chemical content of the fabric and bath with the "time constant" T as unit, and the decrease in the bath contamination the decrease in the bath contamination as a function of the rate of water flow in l per kg material (D/Kt).

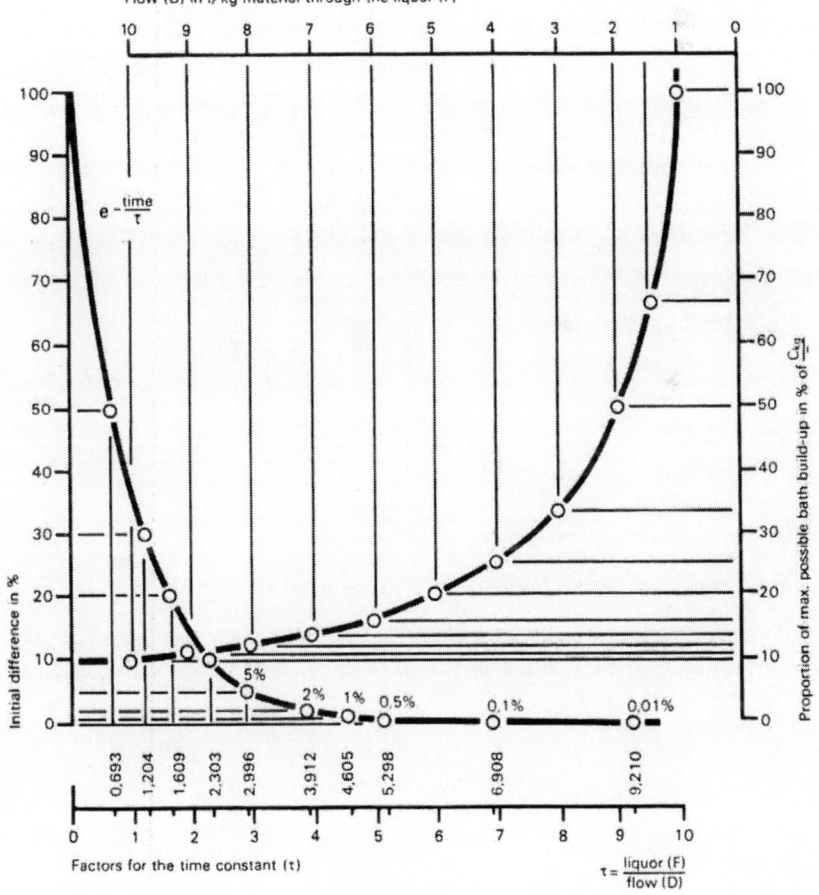

Figure 3.3.

Before a textile material impregnated with chemicals runs through an un-contaminated liquor (F), the difference between them in content of chemicals is 100%. As the material is fed in, the liquor will be progressively changed with the exchangeable chemicals and impurities in accordance with the function $e^{-time/T}$. After 9.21 T (min) the build-up of the chemicals in the liquor will be 99.99%. The initial difference between the chemicals carried and concentration C1 is still 0.01% for a maximum flow of 1 l water per kg of material (D/Kt) through the bath (F). The bath concentration amounts to

$$C1 = X/D \text{ ml/l}$$

As the rate flow D increases the load in the bath decreases to a fraction of the maximum possible build-up.

Example

After treating with caustic soda a cotton fabric of 200 gm/running metre and containing 630 ml caustic soda 66°Tw (36°Be)/kg material is rinsed at a speed of 50 m/min. The rinsing unit contains 650 l liquor and fresh water is run in at the rate of 20 ml/min in counter current. The fabric on entry has a moisture content of 90% and the liquor exchange rate is 95%. How long it will take for the build-up of caustic soda in the rinsing liquor to reach 99.99% and how high will the caustic soda concentration per litre rinsing liquor then be.

Use the graph

$X = 5985$ ml/min $(Ckg \times Kt \times F_A \times 10^{-2})$
$Ckg = 630$ ml
$D = 28.6$ l/min
$F = 650$ l
$F_A = 95\%$
$Kt = 10$ kg/min
$N = 0$
$R = 0.9$ l/kg of goods (caustic soda content is neglected).

The time constant T, derived from F/D i.e. 650: 28.6 comes to 22.7. The build-up of caustic soda in the bath will have reached 99.99% after 9.21 T = 9.21 × 22.7 = 209.3 min.

The build-up in the bath works out at

$$C1 = x/D = 5985/28.6 = 209.3 \text{ ml/l}.$$

At the rate of flow of about 2.9 l water/kg material this is equivalent to about 35% of the possible maximum.

To ensure constant conditions, bath are often replenished. In other words chemicals are added to the treatment bath in amounts corresponding to the maximum build-up caused by the passage through the bath of textile material contained containing chemicals. The amount of chemicals required for replenishing C1 (max) in ml/l is calculated from the formula mentioned earlier

$$C1 = x / D \ ml/.$$

3.4   Bleaching methods with peroxides

3.4.1   Peroxide requirements for various processes of cotton bleaching

Table 3.17.

Process	Concentration (50% H_2O_2)	Temperature	Time
Discontinuous	1–6 ml/l	80–95°C	40–60 min
		98–115°C	15–40 min
Pad-batch	20–40 ml/l	20–30°C	8–24 h
Immersion	10–20 ml/l	80–85°C	15–60 min
Pad-steam			
Short time	20–40 ml/l	100–102°C	1–5 min
Medium time	15–30 ml/l	100–102°C	6–24 min
Long time	10–15 ml/l	85–102°C	25–180 min

3.4.2   Cold peroxide bleach

Cold bleaching is often carried out in the textile industry because it offers the following interesting advantages:

- Simple, inexpensive installation.
- Reduced consumption of energy, reaction at room temperature.
- Direct application on grey fabric straight after singeing.
- Adequate degree of whiteness and absorbency for various dyeing processes.

3.4.3   Cold peroxide bleach–Pad roll–Cotton, Polyester/ cotton wovens

Impregnation   Batching   Washing off

Figure 3.4. Process

Guideline Recipe

Semi-continuous/Pad-batch

Quantity	Unit	Chemicals
Material		Cotton, PE/Cotton wovens
6–10	g/l	Detergent liquid or scouring agent
6–10	g/l	Dispersing agent
6–10	g/l	White spirit
1–2	g/l	Stabiliser
15	g/l	Sodium silicate 38°Bé
6–12	g/l	Caustic soda
30–40	g/l	Hydrogen peroxide 35% by weight

Pad at 20–30°C and 100% pick-up, Pile-up or batch–Store for 14–16 h at room temperature–Wash-off with 1–2 g/l sequestering agent and 1 ml/l caustic soda 36°Bé.

Notes:

Mix the dispersing agent and white spirit and add the mixture to the water while stirring.

1. Use only adding rollers with synthetic colouring.
2. For wet-on-wet treatment the padding liquor should be about 5–8 times stronger.
3. Equivalent quantities for other NaOH and H_2O_2 concentration please see the Table given in this book elsewhere.

Cold pad batch recipe (Alternate)

Quantity	Unit	Bath additions
0.1	g/l	Magnesium sulphate crystals
0.1	g/l	DTPA (40%)
2–5	g/l	Wetting agent
0–15	g/l	Organic stabiliser
10–15	g/l	Sodium silicate 79°Tw
10–12	g/l	Sodium hydroxide 100%
40–60	g/l	Hydrogen peroxide 35%

Cold pad batch recipe (Alternate)

Quantity	Unit	Bath additions
6–10	ml/kg	Suitable wetting agent
25–35	g/kg	NaOH 100%
25–40	ml/kg	Hydrogen peroxide 50%

5–8	ml/kg	Acid cracking agent
0–5	g/kg	Persulphate (Preferably sodium)

Impregnate cold (20–35°C) with liquor pick-up of 90–100% dwell for 16–24 h at room temperature, followed by continuous boil-off.

Notes

1. For already pre-treated fabrics (e.g., desized, alkaline scoured, mercerised materials) it is recommended to reduce amount of caustic soda to 25 g/kg.
2. An addition of persulphate is especially recommended for PVA-containing sizes.
3. System offers very good performance in the presence of iron catalysts.Particularly suitable for sensitive goods like viscose and modal fibres, or for blends with polyamides, however, for these blends do not use persulphate.

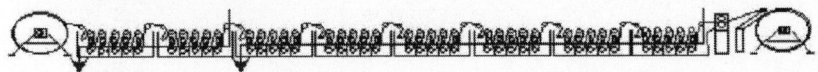

3.4.4   Continuous desizing and cold pad batch bleaching (CPB)

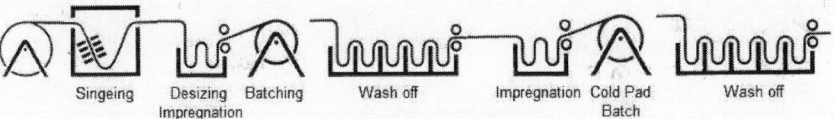

Singeing   Desizing Batching   Wash off   Impregnation Cold Pad   Wash off
           Impregnation   Batch

Figure 3.5. Process

Enzymatic shock desizing eliminates sizes which are CPB bleaching sensitive to alkali (PVA) and makes it possible to greatly reduce the amounts of bleaching agents which ensures a higher final quality.

Recipe

Enzymatic shock desizing	Cold pad batch bleaching	Unit	Recipe
4	4	ml/l	Wetting/Scouring agent e.g., Sandozin MRZ liquid
1		ml/l	Sequestering agent
1		ml/l	Suitable desizing agent

	6–8	ml/l	Peroxide stabiliser
	60	ml/l	NaOH 36°Bé
	40	ml/l	H_2O_2 (50%)
70–80°C	20°C		Bath temperature
	16–24 h		Batching time

Notes:

Stabiliser: Too low concentration does not stabilise H_2O_2 sufficiently, too high a concentration blocks the bleaching reaction.

Wetting/Scouring agent: The higher the concentration, the better will be the efficiency of extraction.

Sodium silicate 36°Bé: Without sodium silicate a lower degree of bleaching must be expected as well as visible vegetable residues.

Caustic soda: Too low concentration does not activate the H_2O_2 sufficiently and the degree of whiteness is lower; with too high concentration there is a risk of lowering the degree of polymerisation (DP).

Peroxide: Above 80 ml/l the degree of whiteness does not increase significantly.

Sodium Persulphate: This oxidiser improves the desizing effect and the degree of whiteness.

The optimum batching time is ca. 24 h. At the end of the reaction, spotting with titanyl chloride must remain yellowish which indicates good stabilisation of the residual hydrogen peroxide.

3.4.5   Process calculations

3.4.5.1   Process technology: Wet-on-wet

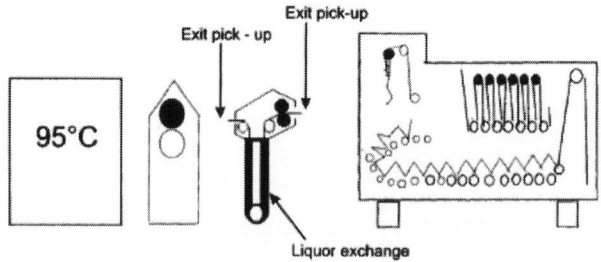

Figure 3.6. Process

Requirements for good quality treatment

- Uniform nip pressure
- Constant chemical concentration on the fabric (ml/kg)
- Uniform initial moisture content when padding wet-on-wet
- Constant saturator liquor level
- Constant fabric speed
- Constant temperature of the saturator
- Good liquor penetration and liquor interchange.

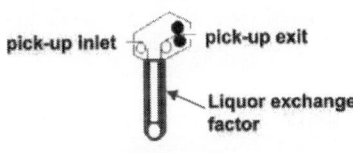

pick-up inlet   pick-up exit

Liquor exchange
factor

Figure 3.7.

Final calculation of the above example for impacta and dosage details.

Table 3.18. Process parameters

Entry pick-up	55%	Exchange factor	0.7	Fabric width	1.8 m
Exit pick-up	100%	Effective pick-up	83.50%	Fabric weight	170 g/m2
		Volume of impacta	90 L	Fabric weight	306 g
				Speed	60 m/min

Table 3.19. Dosage details

Chemicals	ml/kg	ml/l	Initial fill-up	Dosing of chemicals ml/min
NaOH 48°Bé	25	29.94	2.69	612
H2O2 50%	40	47.9	4.31	979.2
Peroxide stabiliser	10	11.98	1.08	244.8
Scouring agent	4	4.79	0.43	97.9
Total amount			8.51	933.9
Water	Automatic dosing		81.49	6328.1

Example

Pick-up inlet: 50%

Pick-up outlet: 95%

Exchange factor: 0.7.

Effective pick-up

Effective pick-up

$$= \frac{(\text{pick up exit} - \text{pick up inlet}) + (\text{pick up inlet} \times \text{exchange factor})}{100}$$

$$= \frac{(95\% - 50\%) + (50\% \times 0.7)}{100} = 0.8$$

Initial concentration in ml/l

$$= \frac{\text{ml / kg}}{\text{effective pick-up}}$$

$$= \frac{35\text{ml/kg}}{0.8} = 43.75\text{ml/l}$$

Initial volume of chemicals in the trough

$$\text{Volume in (l)} \; \frac{\text{Volume of impacta} \times \text{ml/l}}{1000} = \frac{90.1 \times 43.75\text{ml/l}}{1000} = 3.91$$

$$\text{Effective pick-up in \%} = \frac{\text{chemical on weight of fabric (ml/kg)}}{\text{concentration in impacta or bath (ml/l)}} \times 100\%$$

Or

= (% Exit pick-up - % Entry pick-up) + (% Entry pick-up × Exchange factor).

Example

Pick-up inlet % = 55%

Pick-up exit % = 100%

Exchange factor = 0.7.

Effective pick-up % = (110 - 55) + (55 × 0.7) = 83.5%.

Exchange factor

Exchange factor

$$= \frac{\text{Effective pick-up\%} - \text{Exit pick-up \%} + \text{Entry pick-up\%}}{\text{Entry pick-up\%}} =$$

Parameters influencing exchange factor

Temperature	An increase in temperature of the Impacta (up to 600°C) leads in to a higher exchange factor
Fabric weight	With heavier fabric the exchange value decrease
Speed	With the increase in speed of the machine the exchange value decrease
Fabric construction	Exchange value is related to the construction of the fabric (i.e., reed and pick) the denser the fabric lesser the exchange factor
Yarn quality	Type of yarn has direct influence on the exchange factor, open end yarn pick-up more liquor than ring frame yarn, so open yarn will have more exchange value
Type of fibre	In case of PE & PA and other hydrophobic fibre have less affinity for water
Level of saturator	Level of the saturator has direct influence on the exchange factor. Hence the level is always kept constant to eliminate any error for constant saturator concentration

3.4.6   Discontinuous methods

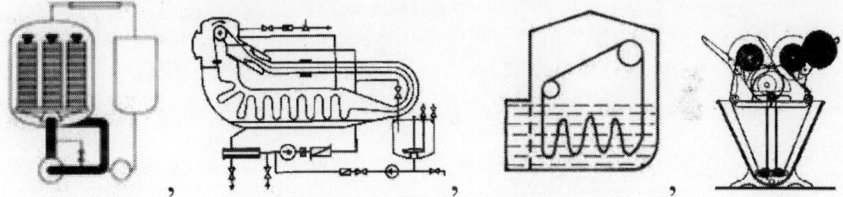

Figure 3.8. Discontinuous machines

Guideline Recipes

Form	Loose fibres, Tow, Yarn, Woven and knit goods		
Substrate	Co, PES/Co, PA/Co, CV/Co, Fl/Co, Fl, PES/Fl, Ju, CV, PES/CV Desized or scoured or pre-bleached goods		
Machine	Package MC, Jet, Overflow	Winch	Jig
MLR	6:1–15:1	8:1–20:1	2:1–5:1
Guide recipes	One bath causticising and bleaching		
Wetting agent ml/l		0.5–1.0	1–2

Low foaming wetting agent ml/l	0.5–1		
Stabilizer for peroxide ml/l	0.25–5.0	0.25–5.0	0.5–1.0
Caustic soda solid g/l	1–2	1–2	2–3
Hydrogen peroxide 35% ml/l	5–8	4–6	5–10
Working method			
Treatment temperature °C	90–130	90	90
Time min	30–60	60–90	60–120

Notes:

The stabiliser amount normally depends on the amount of peroxide used and as per manufacturers' recommendations.

1. An addition of urea 1 g/l will give a better white.
2. Demineralisation before bleaching helps in the final whiteness.
3. It is possible to include OWA.

3.4.7   Bleaching knit goods on overflow machines (batch process)–Peroxide

After boiling-off the cotton is hydrophilic and capable of taking up dyes but still has a pronounced brownish self-colour. In this state it is only suitable for very dark dyeings. For pale and brilliant shades, however, a bleaching treatment is absolutely necessary.

The purposes of bleaching are as follows:

- Complete removal of seed husks.
- High degree of whiteness for white goods.
- Uniform, stable basic white for goods which are to be dyed.
- To avoid catalytic damage to the goods, it must be free from heavy metal ions.

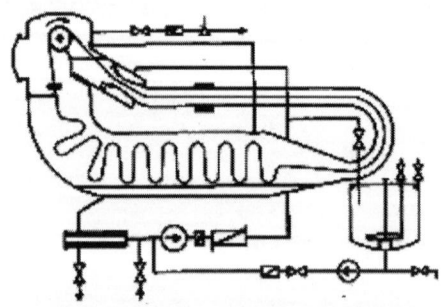

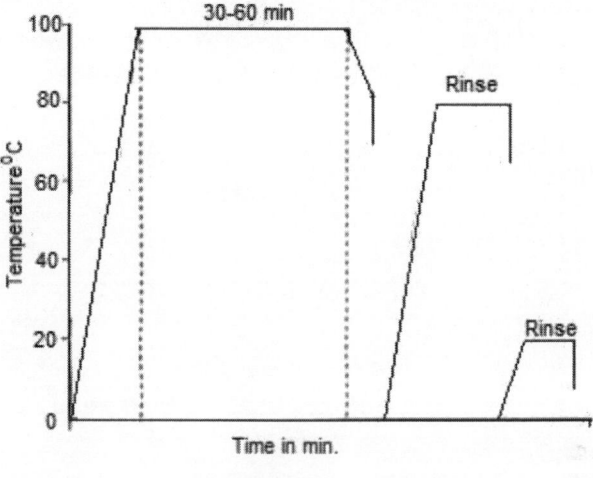

Figure 3.9.

Guideline recipe

Quantity	Unit	Chemicals
0.25–0.5	g/l	Wetting cum scouring agent
1–2	g/l	Lubricant
0.25–0.50	g/l	Stabiliser
0.5–1.0	g/l	Caustic soda 36°Bé
2.0–4.0	ml/l	Hydrogen peroxide 35% by weight

Chemicals are added in the starting bath

3.4.8   Combined boil-off and bleaching

In keeping with the trend to simplified process control, energy and water saving, the boiling-off and bleaching can be combined and carried out in one bath. The attainable effects are not quite good as those with separate processing and the consumption of chemicals is somewhat higher. The combination method is nevertheless sufficient for many qualities.

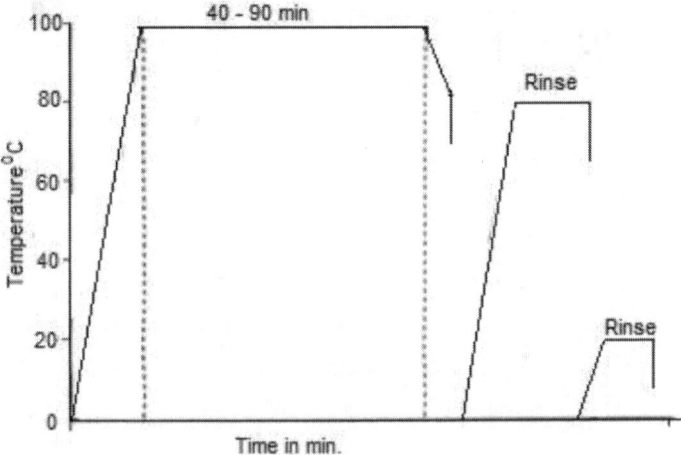

Figure 3.10. Process for combined boil-offAnd bleaching

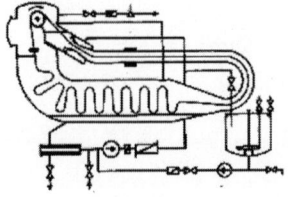

Figure 3.11.

Guideline recipe

Quantity	Unit	Chemicals
0.4–0.8	g/l	wetting cum scouring agent
1–2	g/l	Lubricant
0.5–0.75	g/l	Stabiliser
1.0–2.0	g/l	Caustic soda 36°Bé
3.0–5.0	ml/l	Hydrogen peroxide 35% by weight

Chemicals are added in to the starting bath in the same order and process followed as given in the bleaching graph above

3.4.9   Continuous cold peroxide bleach

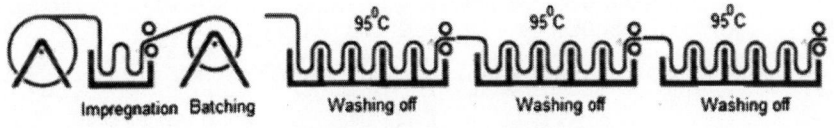

Impregnation Batching   Washing off   Washing off   Washing off

Figure 3.12. Process flow

Guideline recipe for woven and knit goods

Substrate	CO, PES/CO, PA/CO, FL/CO, FL, PES/FL, JU, CV, PES/CV			
Machine/process	Pad-/batch			
Material	Grey goods		Desized and scoured	
Type of treatment	Silicate +	No silicate	silicate +	No silicate
Guide recipes				
Water hardness	6.25–12.5	6.25–12.5	6.25–12.5	6.25–12.5
Suitable wetting and scouring agent ml/l	6–8	6–8	1–3	1–3
Stabilise ml/l		4–6		3–4
Sodium silicate 38°Bé ml/l	8–20		5–8	
Caustic soda solid g/l	10–40	10–140	6–12	6–12
Hydrogen peroxide 35% ml/l	40–60	40–60	30–40	30–40
Silicate scavenger ml/l	6–12		6–12	
Sodium persulphate g/l	4–7	4–7		

Working method
Liquor temperature 20–30°C.
Pick-up 100%.
(Wet-on-wet add-on 25–30% Metering of feed liquor required).
Batching 6–24 h at room temperature (Short batching time if post steaming is planned).

3.4.10   Hot peroxide bleaches

3.4.10.1   Continuous methods

Recipes for continuous methods for cotton materials

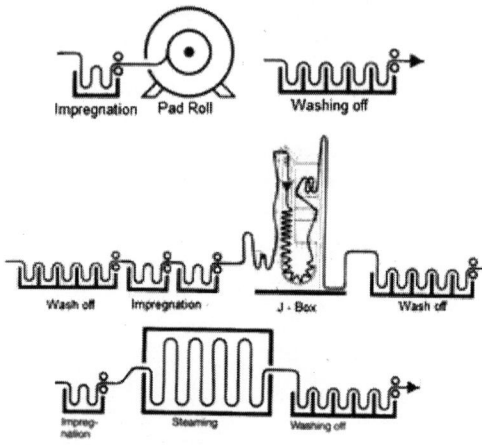

Figure 3.13. Pad-roll, J-Box, Pad-steam

Guideline recipes

Piece goods	Pad-roll, J box	Pad-steam
Wetting or scouring agent ml/l	1–2	4–4
Stabiliser ml/l	0.5–1	2
Sodium silicate 38°Bé ml/l	5–10	10–30
Caustic soda, solid g/l	2–5	10–15
Hydrogen peroxide 35% bt wt. ml/l	10–15	20–40
Impregnation °C	20–30	20–30
Pick-up %	80–100	80–100
Chamber, J-box temperature °C	90	
Time min.	60–90	
Steaming temperature °C		103–105, 142
Time min		1–3 min. 1 min.

Notes

1. If goods are impregnated wet-on-wet, the liquor uptake must be exactly determined and the feeding liquor reinforced accordingly by exchange method.

2. For other concentration of NaOH and H_2O_2 see the tables.
3. Stabilise quantity can vary as per manufacturers' recommendations. Mostly depends on the peroxide concentrations.
4. If there is restrictions in the use of silicate, the silicate quantity can be replaced by further quantity of stabiliser. However silicate is one of the best stabiliser per peroxide.

3.4.11   Continuous hydrogen peroxide bleach

(a) One stage desizing, scour boiling/bleaching

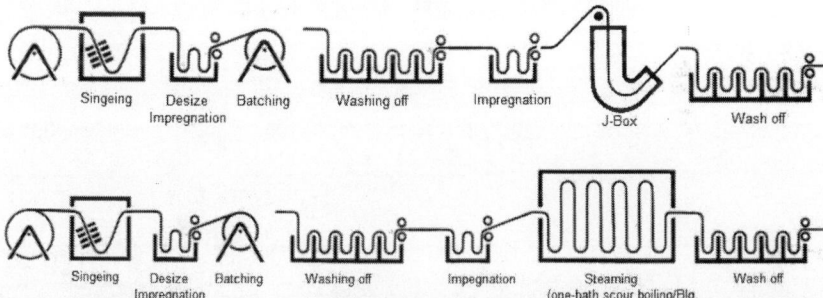

Figure 3.14.

Guideline recipes

Material	Cotton, polyester/cotton–Woven grey goods					
Machine	J-Box, U-ox, L-Box, Pad-Roll,			Pad-steam NT/HT, Combi steamer, conveyer, etc.		
Dwell time	Medium to long			Short to medium		
	Silicate +		Silicate free	Silicate +		Silicate free
Guide recipes						
Water hardness °	6.25–12.5	6.25–12.5	6.25–12.5	6.25–12.5	6.25–12.5	6.25–12.5
Wetting and detergent ml/l	8–15	3–8	3–8		3–8	3–8
Stabiliser ml/l		4–6	6–10		4–8	8–12
Sodium silicate 38°Bé ml/l	8–12	8–12		12–20	12–20	
Caustic soda solid g/l	10–20	10–20	10–20	20–30	20–30	20–30

H$_2$O$_2$ 35% ml/l	30–50	30–50	30–50	30–60	30–60	30–60
Sodium persulphate g/l	1–3	1–3	1–3	3–5	3–5	3–5
Working method						
Saturator-impregnate						
Liquid Temperature °C	20–30	20–30	20–30	20–30	20–30	20–30
Pick-up %	100	100	100	100	100	100
Steaming-dwelling						
NT temperature °C, Time min	90–100, 20–120		90–100, 20–120	100–106, 2–20		100–106
HT temperature °C, Time min					130–140, 1–2	130–140, 1–2

Washing-off–As hot as possible, possibly with soda ash and a detergent.

(b) One stage scour boiling / bleaching

For cellulosic, including bast fibres and their blends with synthetics

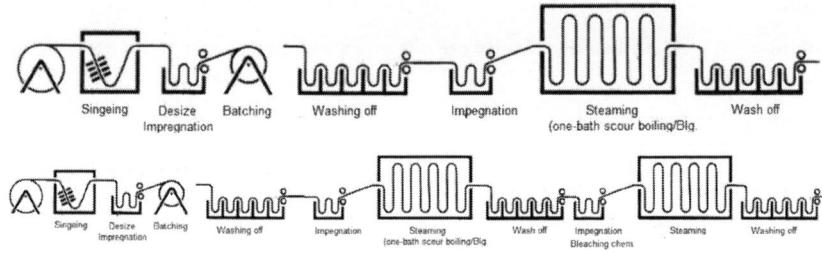

Figure 3.15.

Table 3.20.

Material	Cotton, polyester/cotton PA/CO, CV/CO, FI/CO, FI, PES/SI, JU, CV, PES/CV, etc.–Woven and knits grey goods	
Machine	Pad-steam NT/HT, Combi steamer, Conveyer, etc.	Pad-steam NT/HT, Combi steamer, Conveyer, etc.

Stage of the material	Desized			Desized and scour boiled/ pre-bleached		
	Silicate +		Silicate free	Silicate +		Silicate free
Guide recipes						
Water hardness °	6.25–12.5	6.25–12.5	6.25–12.5	6.25–12.5	6.25–12.5	6.25–12.5
Wetting and dispersing agent ml/l	5–10			3–6		
High alkali stable wetting agent ml/l		3–6	3–6		1–3	1–3
Stabiliser suitable for silicate-peroxide bleaching ml/l		4–6			3–4	
Stabiliser for silicate free cont. bleach			6–10			5–8
Caustic soda solid g/l	10–20	10–20	10–20	5–12	5–12	5–12
H_2O_2 35% ml/l	10–60	10–60	10–60	10–40	10–40	10–40
Working method						
Saturator-impregnate						
Liquid temperature °C	20–30	20–30	20–30	20–30	20–30	20–30
Pick-up %	100	100	100	100	100	100
Steaming-dwelling						
Temperature NT °C	85–100		85–100	85–100		85–100
Temperature HT °C		130–140	130–140		130–140	130–140
Dwell time min. NT	2–14		2–14	2–14		2–14

Washing-off–As hot as possible, possibly with alkali shock (with soda ash and detergent) neutralise.

Notes

1. When polyamide blends are treated by this method please add protective colloid to prevent fibre damage during the process.

2. With wet-on-wet impregnation pick-up will be 20–30%, metering of feed liquor is necessary and make changes in the recipe accordingly.

(c) One stage scouring boiling/bleaching (Alternate)

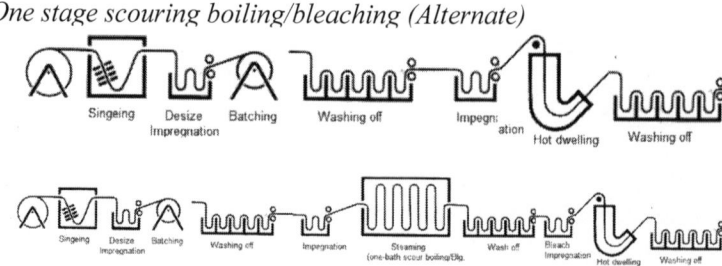

Figure 3.16. One bath scouring/Scour - Bleaching

Guideline recipes

Material	Cellulosics including bast fibres and their blends: CO, PES/CO, CV/CO, FI/CO, FI, PES/FI, JU, CV/PES, CV					
Machine	J box, L box, U box, Pad-roll			J box, L box, U box, Pad-roll		
Stage of the material →	Desized			Desized and scour boiled/pre-bleached		
Type of treatment→	Silicate +		Silicate free	Silicate +		Silicate free
Guide recipes						
Water hardness°	6.25–12.5	6.25–12.5	6.25–12.5	6.25–12.5	6.25–12.5	6.25–12.5
Anionic, low-foaming scouring dispersing agent with high alkaline stability ml/l	5–10			3–5		
Weakly anionic, low-foaming scouring dispersing agent with less alkaline stability ml/l		3–6	3–6		1–3	1–3
Stabiliser silicate recipe ml/l		4–6			3–4	
Stabiliser for silicate free cont. bleach			6–10			5–8

Sodium silicate 35°Bé ml/l	5–10	5–10		5–8	5–8	
Caustic soda solid g/l	5–20	5–20	5–20	2–5	2–5	2–5
H_2O_2 35% ml/l	10–50	10–50	10–50	10–30	10–30	10–30
Working method						
Saturator-impregnate						
Liquid temperature °C	20–30	20–30	20–30	20–30	20–30	20–30
Pick-up %	100	100	100	100	100	100
Steaming-dwelling						
Temperature °C	85–100	85–100	85–100	85–100	85–100	85–100
Time min	20–120	20–120	20–120	20–120	20–120	20–120

Washing-off–as hot as possible, possibly with alkali shock.

Notes

1. There are chances of silicon deposits on the machine. It is better to include silicate scavenger in the recipes.
2. When polyamide blends are treated by this method add protective colloid to prevent fibre damage during the process.

(d) Total immersion bleaching

In a different method of bleach, the goods passes through a fully flooded bleach bath. The bath volume varies considerably depending on the machine system and size may be up to 5000 l. In case of woven fabrics it is better to desize and scour boil before going for immersion bleach so that the peroxide stability is not impaired by the entrained impurities. It is ideal method for knits in rope form.

Not very commonly used these days

The advantages of total immersion bleach are:

1. Very good bleach effect.
2. No deformation of fabric structure reducing the danger of creasing to a minimum.
3. Fully continuous method with good production capacity, depending on the speed.

4. Combination possibilities e.g., scour boiling, sodium hypochlorite bleach, hydrogen peroxide bleach.
5. Minimum cost as the bleach bath is replaced only once in a week or even later.
6. Versatile use of forms e.g., woven fabric open width, tubular knit fabrics, terry towelling, etc.

But care has to be taken for the following factors to ensure the quality of bleach.
1. Regular control of the concentration bleaching agent and NaOH.
2. Metering pumps for continuous feeding of the chemicals.
3. Stability of feed liquor in case of all-in addition.

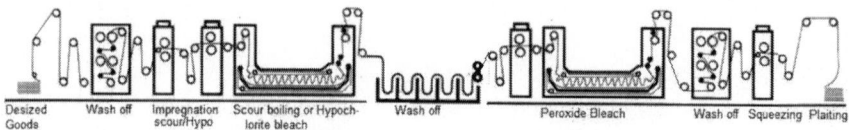

Desized Goods   Wash off   Impregnation scour/Hypo   Scour boiling or Hypochlorite bleach   Wash off   Peroxide Bleach   Wash off   Squeezing   Plaiting

Figure 3.17. Immersion Bleaching

Guide recipes

Form	Woven and knit goods, terry towelling
Substrate	CO, PES/CO,CO/PA, CV/CO,CV/PES, CV
Machine	Total immersion system
Type of treatment	Silicate free or low-silicate
Stage of the material	Desized/scour boiled/pre-bleached with hypochlorite
Guide recipes	
Low-foaming wetting, scouring, dispersing, degreasing agent stable to NaOH ml/l	2–4
Stabiliser suitable for immersion system	2–6*
Sodium silicate 35°Bé ml/l	0–2
Caustic soda solid g/l	10–30
H_2O_2 35% ml/l	10–50
Working method	
Steaming-dwelling	
Temperature °C	60–90
Time min	10–40

a. Example: Stabiliser AKM liquid, (Clarient)
Washing-off–As hot as possible, neutralise.

3.4.12   Bleaching of knitted goods

Full white bleaching of knitted goods in soft flow (circulating machine)

Bleaching can be done in open width or tubular form. As explained in scouring section, the bleaching can be done at atmospheric or at higher temperature (110°C).

Normal bleaching procedure is given below.

Notes

1. Precautions has to be taken to avoid pin hole damages. This can be done by adding chelating agents. Or by giving a demineralisation treatment before bleaching operation.

2. If necessary lubricating agents can be added, especially and low GSM fabrics and qualities which are prone to develop rope marks and crease marks. Addition of lubricants also helps to reduce weight loss.

3. For higher temperature bleaching the whole procedure and additions of chemicals are same except that the temperature is raised above 100°C and holding time is reduced. This helps in reducing cycle time.

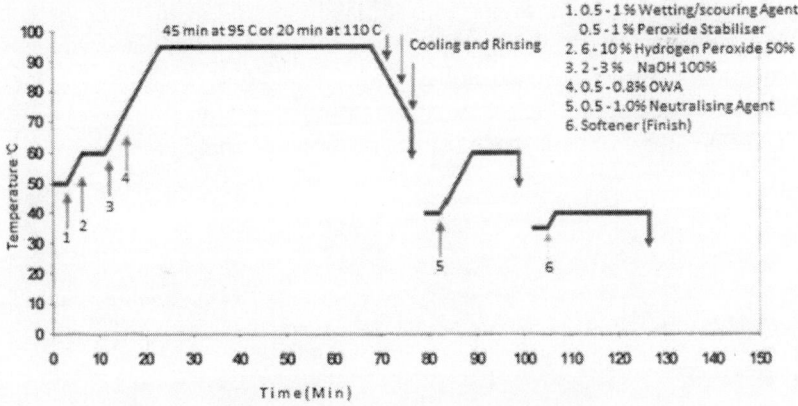

Figure 3.18. Normal full bleach process

4. It is always better to do a demineralisation program before or after or in situ to achieve a better whiteness. Given below a graph showing the whiteness improvement by demineralisation.

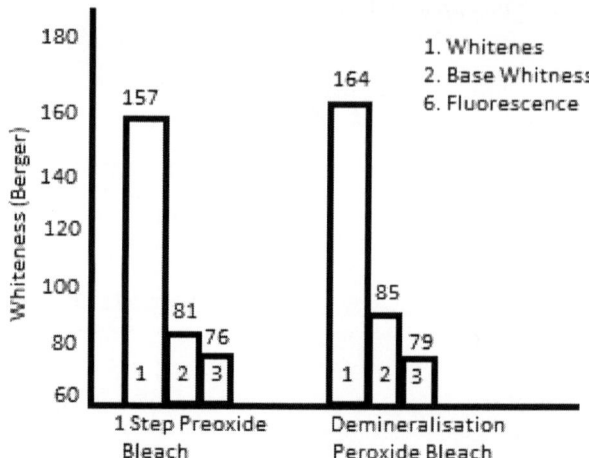

Figure 3.19. Difference in whiteness between one step and two step process

Procedure for demineralisation/bleaching one bath two steps

In case of good grey qualities the process demineralisation peroxide bleach can be carried out in a one bath two-step process.

In this method a demineralisation process is done in the beginning of the process and continued with bleaching process. The process is continued like a one-step bleaching.

All the process is done in one bath except for neutralisation and finishing. Even though there is saving in water and time in this method, there are other disadvantages in these process. Continuous demand of better whiteness by customers, do not go hand in hand with the combined processes. Washing-off after each process is always very important in achieving better whites. In such cases the two bath process gives better whiteness.

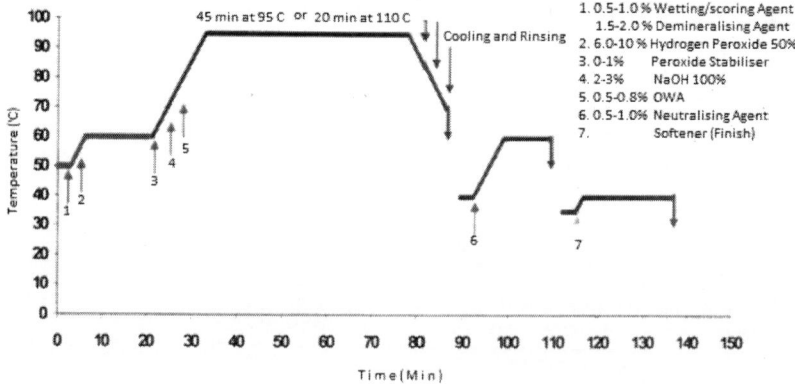

Figure 3.20. One bath two step process–Bleaching/demineralisation

Demineralisation/bleaching–Two bath two step process

The demineralisation process is separated and after the treatment the bath is drained and new bath is taken for the bleaching process. Wetting agent has to be added as the grey fabric has to be wetted out and penetration of the demineralising agent has to penetrate in side for better leaching of the heavy metals.

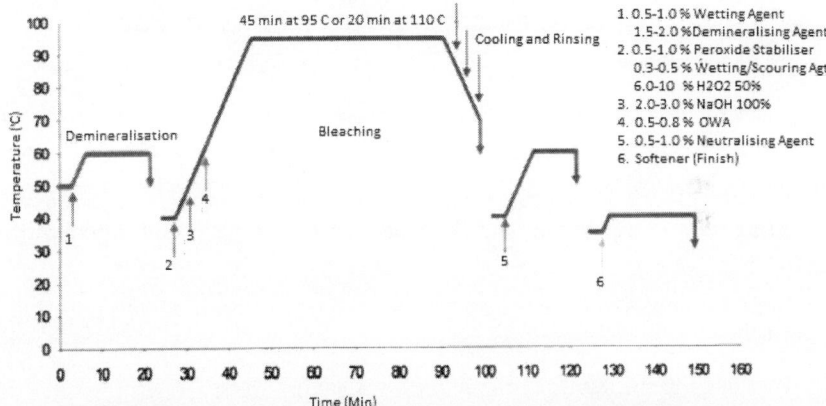

Figure 3.21. Two step–Two bath process demineralisation/bleaching

The second part of the process is a normal bleaching process. The requirement of the wetting scouring agent is reduced as the fabric has been wetted out in the demineralisation process.

Notes

1. In all the above process lubricating agent has to be added wherever necessary.

2. If there is a demand for higher whiteness it is better to follow the two bath two step process.

3. When there are many batches for bleaching the bath can be reused and water consumption can be reduced. Rinsing and process liquors are pumped into the tanks and used again for subsequent processes. That means that the second rinsing bath can be used again as first rinsing bath in the following process.

4. At temperatures up to 110°C the water quantities for cooling proq cesses are rather small and have only little influence on the costs.

5. It is a headache for a processor to achieve same whiteness for fabric from different sources. The application of higher temperatures is a possibility to compensate differences of origin. However one has to take into consideration the strong degreasing treatment at a higher temperature and therefore the need of softener.

6. For the best possible elimination of impurities the rinsing processes should be one at a high liquor ratio.

3.4.13   Continuous bleaching machineries for knitted goods

1. "Steep master" system–Kusters, Germany

Example, with pre-padding, drum chamber, post steaming system, washing compartments.

Material: Pre-desized and/or scour boiled woven and open-width knits.

Pre-treatment stages: Possibilities for peroxide bleaching and one stage scour boiling/bleaching soaping for printed goods.

Process flow: Example (60–80 m/min speed).

Wet-on-wet padding with peroxide bleach liquor.

Drum chamber = Total immersion bleach bath, 20–25 min at 60°C.

Post-steaming 1–2 min at 10–102°C.

Open-width washing.

For post-steaming, it is needless to say that the goods should leave the drum chamber with residual peroxide.

2. "Startrans+" System–Goller, Germany

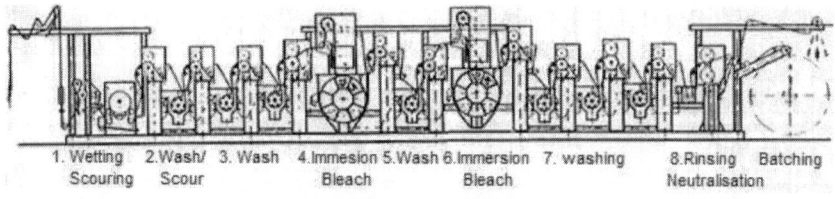

1. Wetting   2.Wash/   3. Wash   4.Immesion 5.Wash 6.Immersion   7. washing   8.Rinsing   Batching
     Scouring   Scour   Bleach   Bleach   Neutralisation

Figure 3.22. Goller System

Examples

Material	Grey 100% Terry Towelling
	Mercerised 100% Cotton knitted

Recipes common for both

No.	Operation	Recipe and Parameters
1.	Wetting out/Scuoring	20^0C
2.	Washing desizing	90^0C
		1 g/l Non-ionic detergent
		1 g/l Sodium polyphosphate
3.	Washing/Demineralising	90^0C
		1 g/l Non-ionic Detergent
		1 g/l Demineralising agent
4. & 6.	Total Immersion Bleach	Bath set at 80^0C with
		2-8 ml/l Stabiliser
		4-5 ml/l Caustic Soda 50%
		15 ml/l Hydrogen Peroxide 35%
		2 ml/l Optical Brightener liq.
		Feed Cold @ about 80-100 litres/hour
		22 ml/l Stabiliser
		13 ml/l Caustic Soda 50%
		110ml/l Hydrogen Peroxide 35%
		3.5 ml/l Optical Brightener liq.
5.	Intermediate washing	90^0C
7.	Washing	90^0C, 80^0C, 60^0C
8.	Rinsing	20^0C

Example: With scouring compartments, pre-padding, dwell chamber system (2 units in some cases) washing-off compartments.

Materials: Woven, knits (open width), terry towelling.

Pre-treatment stages: Possibilities for desizing, scour boiling and bleaching.

Process sequence: Example (20–30 m/min).

1. Scouring.
2. Wet-on-wet padding with peroxide bleach liquor.
3. Dwell chamber–Total immersion bleach bath having two units (2 × 10 min).
4. Open-width washing.

3. "Galaxy System"–Bruckner, Germany

With sodium hypochlorite/hydrogen peroxide combination bleach.

Material: Tubular knits.

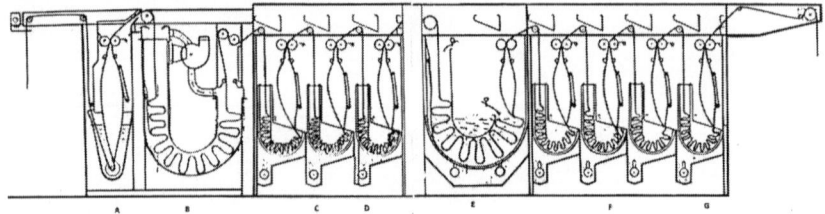

Figure 3.23. Bruckner System

A- Impregnation Compartment for sodium hypochlorite bleach
B- Cold dwelling Cpmpartment
C- Cold rinsing compartments 3 nos
D- Impregnation compartment for hydrogen peroxide
E- Total immersion compartment for peroxide bleach
F/G - Hot rinsing compartmetns 3 nos
G- Cold rinsing compartment

Process flow: Example (30–50 min)

1. Padding with sodium hypochlorite bleach liquor.
2. Dwell chamber (J-box) cold 20–30 min.
3. Cold rinse
4. Cold wet-on-wet impregnation with peroxide bleach liquor.
5. Dwell chamber (J box) hot, 20–45 min at 85°C (total immersion).
6. Washing-off.

The tubular goods pass through the impregnation and "Tubolavar+" units in a balloon state.

Metering chemicals as single component.

4. Jemco (USA)

Material: Tubular knits.

Process flow: Example (40–60 min).

Normal version with two units one for bleaching one for washing. More units can be joined to increase capacity. Processing in rope form is feed to the farthest end of the kiers the bottom of which is slopped towards the front of the machine. The kier is divided into different compartments and filled with bleaching liquor. As the machine runs the rope is moved towards the front due to the slope and from the front of the machine the rope is passed through a tube where the bleaching liquor is forced through pumps which take the rope to the farthest end of the next compartments. The fabric is passed through

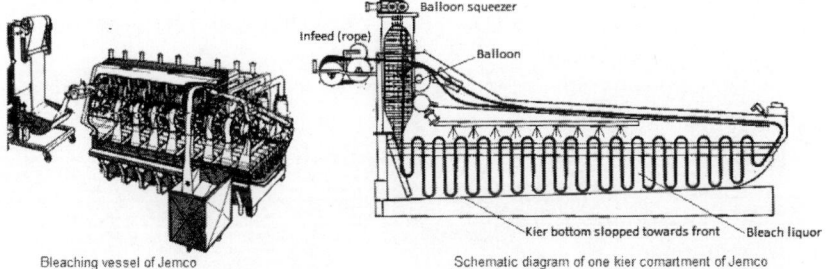

Balloon squeezer

Infeed (rope)

Balloon

Kier bottom slopped towards front   Bleach liquor

Bleaching vessel of Jemco

Schematic diagram of one kier comartment of Jemco

Figure 3.24. Jemco System

many compartments to give enough time for bleaching. At the beginning the bleaching liquor is prepared and taken to the kiers and heated to the required temperature. As the bleaching of the liquor continues further replenishing of chemicals (predominantly peroxide) is done after checking the strength of the same in the kier liquor. The bath can be continuously used for many batches thus reducing the cost of chemicals. After passing through the kiers compartments once the required bleach level is achieved the fabric in rope form passes through a balloon padder with detwisting device and squeezed and passed through another kier of almost same design, for washing, neutralising and if necessary passed through OBA and again ballooned and squeezed.

The salient features of the machines are:

1. Rope processing without rope squeezing means no stretching or rope marks.
2. Precise control of bleaching chemicals either with traditional batch or new direct dozing with recipe capability.
3. Automatic, precise kier and multi-compartment washer with temperature control.
4. Kier and jet washer each have individually controlled, self-adjusting compartments that stay full, even with changes in fabric width and weight.
5. Low abrasion, no pilling and minimal weight loss.
6. Fabric torque eliminated by patented DE twister and automatic balloon inflation.
7. Direct costs of less than half that of beck or winch bleaching and significantly lower than low-liquor Jet dye machinery.
8. Lower water usage–10 l/kg of fabric approximately.
9. Lower steam usage–less water means less steam, approximately 0.7 kg steam/kg fabric.

10. Lowest direct labour–one operator runs kier and washer, including softener application.
11. Lower chemical usage than any batch system.
12. Chemicals are more efficiently used, resulting in kier draining only about once a month for 24/7 operation.
13. Direct dosing of concentrated peroxide, caustic, auxiliaries and optical permits recipe optimizing for a wide range of fabrics.
14. Large volume kier provides more dwell time for a lower required chemical concentration.
15. Balloon squeeze padding system reduces chemical carry-over to a minimum.
16. Full-bleach white, prepare for print, prepare for dye, softener application knits (jersey, rib, interlock, pique), fleece and woven in rope form.
17. Less floor space and capital cost per kg than dye machines.

Jemco is one of machine with an economical bleaching system. Several model sizes are available with production capacity up to 900 kg/h and can be used for 100% cotton, and polyester/cotton blends.

Process

In a practical running procedure, chemicals are made in an overhead tank and drawn into bleaching kier. Volume of the bleaching kier is made up and heated to 90⁻95°C and the fabric is run into the vessel. The fabric is threaded through each tube after collecting enough material in the compartment and brought to the squeezing rolls. Before the squeezing, the tubular fabric is ballooned and detwisted and fed to the squeezer without creases and fed to the washing kier, here the fabric is washed and the last compartment the fabric passed through the finishing solution with additional OBA (optional) and again ballooned, squeezed and plaited into trollies.

Guide recipes

Example: Overhead tank capacity 5000 l, Bleaching kier capacity–10,000 l.

Table 3.21. Guideline recipe

Capacity of OH tank 5000 l	Concentration in bleaching vessel capacity 10,000 l)	Bath additions
Quantity in kg	(g/l)	
1.6	0.04	Detreating agent
20	2	Wetting agent

40	4	Caustic soda flakes
25	2.5	Sodium metasilicate
250	25	Hydrogen peroxide
30	3	Peroxide stabiliser
30	3	Optical brightening agent

As the fabric runs the volume is in the bleaching vessel is made-up continuously and strength of chemicals are checked and replenished wherever required. The bath is used continuously till it starts affecting the whiteness of the fabric.

3.4.14   Bleaching with hypochlorites

The sodium hypochlorite bleach is employed.

- As a bleach for complete removal of cotton seed husks
- As a bleach to increase the degree of brilliance of pale to medium shades
- As a bleach for white grounds in printing
- As a pre-bleach for full white goods.

The advantages of the sodium hypochlorite bleach are its economy, its applicability temperature (20–30°C) and the relatively rapid bleaching effect under the usual working conditions. Sodium hypochlorite is a strong oxidising agent. In order to protect the cellulose during bleaching operation (loss of DP) it is essential to control the concentration of bleaching liquor, the pH, the treatment time and temperature. The bleaching effect obtained with sodium hypochlorite is not permanent.

After bleaching the goods are rinsed in cold water and then given an antichlor treatment to remove completely all chlorine residues in the cellulosic fibres which could lead to later damage. Sodium bisulphite is generally applied in an acetic acid medium as the antichlor agent. Sodium hypochlorite treatment is followed by a hydrogen peroxide bleach. The antichlor treatment with sodium bisulphite can be omitted provided there are no long delays between bleaching stages.

Bleaching with hypochlorite is best under following conditions

1. Concentration   2–5 g/l of available chlorine.
2. Temperature   Room temperature.

3. Time   1.5 – 2 h.
4. M:L ratio   1:1–1:10.
5. Alkalinity   10–20 (% of available chlorine).
6. pH   10.5–11.0.

Hypochlorous acid (HOCl): Hypochlorous acid is an active, unstable compound known only in the water solutions. This acid by virtue of its ready decomposition with liberation of oxygen, finds limited use as an active oxidising and bleaching agent.

$$HOCl \rightarrow HCl + O$$

Hypochlorites: Hypochlorite solutions, when properly prepared exhibit greater stability than the parent acid, although not as active, are generally applicable for bleaching in place of the acid. Stability of hypochlorites (usually sodium hypochlorites and calcium hypochlorites) depends on the following factors.

(a) Hypochlorite concentration

(b) Concentration of certain catalysts

(c) Alkalinity or pH of the solution

(d) Exposure to light.

Effect of concentration and temperature

Table 3.22. Half-life of hypochlorite solutions

Available Cl_2	Half-life days at			
	100°C	60°C	25°C	15°C
200	0.016	0.6	44	
100	0.079	3.5	220	175
50	0.25	13	790	800
25	0.63	28	1800	5000
5	2.5	100	6000	

The practical bleaching temperature is 400C for 60 min

Effect of catalysts

The most powerful catalysts promoting the decomposition of NaOCl are cobalt, nickel, copper.

Table 3.23. Catalytic effect of Cu and Ni on high con- centration bleach liquor (200 g/l available chlorine and 12 g/l excess NaOH).

Nickel ppm	Copper ppm	Half-life days
	0.12	53.5
0.11	0.22	50
0.15	0.39	45
0.2	0.6	40
0.28	0.85	35
0.35	1.14	30
0.53	2.15	20
1.07		10
1.8		7

Effect of metal cations

Transition metal cations (Cu^{++}, Fe^{+++} etc.)can catalyse degradation of cellulose by hpochlorite. To minimize this use stainless steel equipments for bleaching, and avoid any rust or metal pieces in the fabric/fibre (it is better to give a deminerisation treatment before hypochlorite bleaching)

Effect of alkalinity and pH

It is observed that greatest irregularities occur at the points where there is neither excess NaOH nor excess $NaHCO_3$. Small amounts of either apparently tend to change the properties of the solution drastically in the region of pH values 10.0–10.6.

NaOCl is a mild base. At above pH 10 (>10) it dissociates in water as follows

$$NaOCl \xrightarrow[\text{pH>10}]{H_2O} Na^+ + OCl^-$$

But at pH 5-8.5 the hypochlorite ion reacts with water to form hypochlorous acid

$$OCl^- + H_2O \underset{}{\overset{pH = 5\text{-}8.5}{\rightleftharpoons}} HOCl + OH^-$$

If base (NaOH) is added to the solution the equation shifts to the left, HOCl concentration is reduced and hence the bleaching is reduced. As seen in the equation above pH>10 only OCl- is the major species in the solution

If acid (say HCl) is added the equilibrium shiftes to the right. HOCl concentration is increased. Rapid bleaching occurs. Chances of degradation of the fibre is more. At pH 5-8.5 HOCl is the major specis in the solution

On addition of extra acid (HCl), chlorine gas is generated, HOCl concentration is reduced. Bleaching becomes slow.

$$HOCl + HCl \rightleftharpoons Cl_2 + H_2O$$

At pH<5 Cl2 is the major species in the solution.

Table 3.24. Effect of pH on HOCL concentration

pH	% of pH existing as HOCl
10.0	0.30
9.0	3.0
8.0	21.0
7.5	50.0
7.0	73.0
6.5	91.0
6.0	96.0
5.0	99.7

pH for bleaching with hypochlorite is usually kept at 9-10, where the HOCl concentration is low. As the HOCl is consumed in bleaching it is replenished by the equilibrium (see equation) and thus fibre damage is also minimized. pH of the bleaching solution is dropped during the operation, it is always better to buffer the solution by adding Na2CO3.

Table 3.25. Stability of hypochlorite at different alkali concentrations

Available Cl_2 g/l (at start)	NaOH g/l	Na_2CO_3 g/l	$NaHCO_3$ g/l	Half-life days
212	14.6	6.6	0	44
	0	45.3	0	19
	0	45.3	excess	0.3
113	7.5	3.4	0	200
	0	23.2	0.84	20
	0	23.2	excess	0.5
57	3.7	1.7	0	800
	0	11.5	0.59	200

	0	11.5	57.7	0.6
29	1.9	0.84	0	2000
	0	5.8	0.25	600
	0	5.8	28.8	3.2
5	0.17	0.16	0	9500
	0	0.6	0.01	800
	0	0.6	5.9	200

3.4.15   Available chlorine and active chlorine

Active chlorine means the chlorine capable of taking part in an oxidising reaction or in other words the oxidising power of one atom of elemental chlorine.

Available chlorine may be considered simply by a measure of the oxidising power and may be determined on solutions containing no hypochlorite.

Active chlorine is exactly half of the available chlorine.

Expressions which denote available chlorine content. There are three systems of expressing the available chlorine content (a) Nominal percentage (b) Grams per litre of available chlorine (c) Percent by weight of either available chlorine or sodium hypochlorite.

Trade percent = g per litre/10.

Trade percent = % available chlorine by weight × Specific gravity.

Grams per litre= trade % × 10.

Grams per litre= % available chlorine by weight × Specific gravity.

Percent available chlorine by weight = Trade percent / Specific gravity.

Percent available chlorine by weight = g per litre / (10) × Specific gravity.

Preparation of sodium hypochlorite

The most stable form of sodium hypochlorite is prepared by chlorinating a caustic soda solution in correct proportion to maintain a small residual caustic soda and alkalinity in the finished product.

$$Cl_2 + 2NaOH = NaOCl + NaCl + H_2O.$$

$$70.914 + 80.01 = 74.454 + 58.454 \; 18.06.$$

On the basis of molecular (Chemically pure materials 1 lb of chlorine + 1.13 pounds of caustic soda (NaOH) will produce 1.05 lb of sodium hypochlorite and 0.83 pound of sodium chlorite.

The following table gives the general requirement of caustic soda, chlorine required for the preparation of a protective % of available chlorine.

Table 3.26. Properties of different percentage of Hypochlorite solutions

Trade %	Chlorine gpl	Sodium hypochlorite % by weight	Pounds/1000 gallon bleach liquor		Excess caustic (NaOH)		Specific gravity of finished hypochlorite 20°C
			Chlorine	NaOH 98% gpl	App. % by wt		
1	10	1.03	83.4	121.4	3	0.29	1.02
1.5	15	1.53	125.1	172	3.3	0.32	1.03
2	20	2.03	166.8	221.6	3.5	0.34	1.03
2.5	25	2.52	208.5	272.1	3.8	0.36	1.04
3	30	3	250.2	321.8	4	0.38	1.05
3.5	35	3.47	291.9	372.2	4.3	0.41	1.06
4	40	3.94	333.6	421.9	4.5	0.42	1.07
4.1	41	4	341.9	432.3	4.6	0.43	1.07
4.5	45	4.41	375.3	472.4	4.8	0.45	1.07
5	50	4.87	417	522.1	5	0.46	1.08
5.1	51	5	428.7	536.3	5.1	0.47	1.08
5.4	54	5.25	452	564	5.2	0.48	1.09
5.5	55	5.32	458.7	572.6	5.3	0.49	1.09
5.7	57	5.5	476.2	593.5	5.4	0.5	1.09
6	60	5.76	500.4	621.8	5.5	0.5	1.09
6.3	63	6	522.9	649	5.6	0.52	1.1
6.5	65	6.21	542.1	672.4	5.8	0.53	1.1
7	70	6.64	583.8	722.4	6	0.54	1.11
7.5	75	7.07	625.5	722.9	6.3	0.57	1.11
8	80	7.5	667.2	822.6	6.5	0.58	1.12
8.5	85	7.93	708.9	873	6.8	0.6	1.13
9	90	8.34	750.6	922.7	7	0.62	1.13
9.5	95	8.76	792.3	973.2	7.3	0.64	1.14

10	100	9.16	834	1022.9	7.5	0.65	1.15
10.5	105	9.56	875.7	1073.4	7.8	0.68	1.15
11	110	10.05	917.4	1123	8	0.69	1.16
11.5	115	10.36	954.1	1173.5	8.3	0.71	1.17
12	120	10.76	1000.8	1223.2	8.5	0.73	1.17
12.5	125	11.14	1042.5	1273.7	8.8	0.75	1.18
13	130	11.53	1084.2	1323	9	0.76	1.18
13.5	135	11.91	1125.9	1373.8	9.3	0.78	1.19
13.6	136	12	1134.2	1384.2	9.4	0.79	1.19
14	140	12.28	1167.6	1423.5	9.5	0.79	1.2
14.5	145	12.69	1209.3	1474	9.8	0.81	1.2
15	150	13.03	1251	1523.7	10	0.83	1.21
15.5	155	13.38	1292.7	1574.2	10.3	0.85	1.22
16	160	13.75	1334.4	1623.9	10.5	0.86	1.22
16.5	165	14.11	1376.1	1674.3	10.8	0.88	1.23
17	170	14.45	1417.8	1724	11	0.89	1.24
17.5	175	14.81	1459.5	1774.5	11.3	0.91	1.24
17.8	178	15	1484.5	1804.1	11.4	0.92	1.24
18	180	15.14	1501.2	1824.2	11.5	0.92	1.25
18.5	185	15.49	1542.9	1874.6	11.8	0.94	1.25
19	190	15.83	1584.6	1924.3	12	0.95	1.26
19.5	195	16.17	1626.3	1973.9	12.2	0.96	1.27
20	200	16.5	1668	2024.5	12.5	0.98	1.27

3.4.15.1   Formation of hypochlorite

From caustic soda (NaOH)

$$2NaOH + Cl_2 = NaOCl + NaCl + H_2O$$

From soda ash

$$2Na_2CO_3 + Cl_2 + H_2O = NaOCl + NaCl + 2NaHCO_3$$

$$212.008 \qquad 70.914 \quad 18.016 \quad 74.454 \quad 58.454 \qquad 168.030$$

3.4.15.2 Formation of calcium hypochlorite [Ca (OCl)$_2$]

From hydrated lime [Ca(OH)$_2$]

$$2Ca(OH)_2 + 2Cl_2 = Ca(OCl)_2 + CaCl_2 + 2H_2O$$
$$148.192 \quad 141.828 \quad 142.994 \quad 110.994 \quad 36.032$$

From quicklime (CaO)

$$2CaO + 2H_2O + 2Cl_2 = Ca(OCl)_2 + CaCl_2 + 2H_2O$$
$$112.16 \quad 36.032 \quad 141.828 \quad 142.994 \quad 110.994 \quad 36.032$$

Estimation of bleaching chemicals

Alkali

To 100 ml of distilled water add exactly 10 ml of the alkaline solution followed by a maximum of 5 drops of phenolphthalein indicator. Titrate with 0.1 N sulphuric acid until the sample turns colourless from red/pink.

Calculation

No. of ml of 0.1 N sulphuric acid consumed × 1.060 = g/l of soda (calcined).
No. of ml of 0.1 N sulphuric acid consumed × 0.400 = g/l of NaOH (100%).
No. of ml of 0.1 N sulphuric acid consumed × 0.909 = g/l of NaOH (38°Be).

3.4.15.3 Recipe for sodium hypochlorite bleach

(a) Discontinuous process

A very good whiteness can be achieved by hypochlorite bleach. This process is almost obsolete. But is very economical and can be followed for cotton, cotton/polyester, cotton/viscose, polyester/viscose. When a combined hypochlorite and peroxide bleach is employed, the results are found to be excellent.

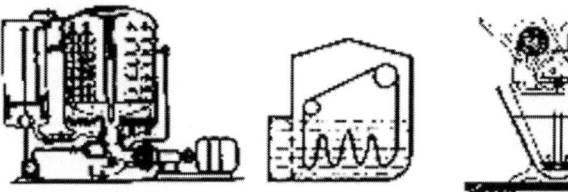

Figure 3.25. Yarn dyeingMc, Winch, Jig

Guideline recipe

Method/chemicals	Yarn 10:1–20:1	Piece goods winch 20:1, Jig 5:1
Wet out the goods with suitable wetting agent	1 g/l	1 g/l
Add available chlorine	1–2 g/l	1–2 g/l

Add caustic soda to adjust pH	10.5–11	10.5–11

Bleach for 1 h. at 20–30°C

Drop the bath

Rinse or bleach with peroxide
without rinsing as per recipe given
elsewhere in this book

Note:

For lower liquor ratios available chlorine should be higher. A combination of chlorine and peroxide gives the best whiteness especially with knitted goods.

(b) Semi-continuous method

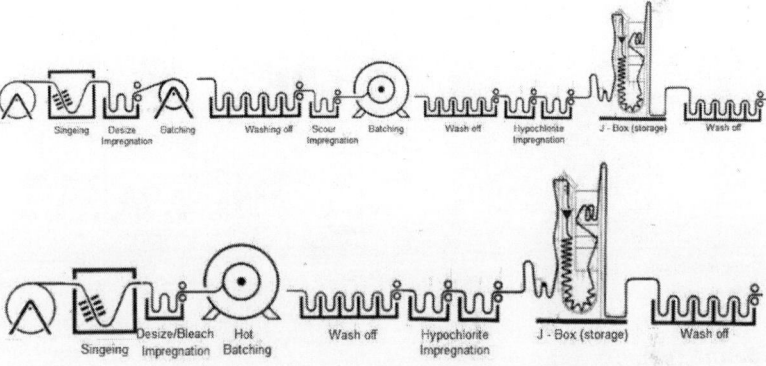

Figure 3.26. Semi-continuous method

– Impregnate at 20°C with available chlorine 3–5 g/l and a suitable wetting agent (1–2 g/l to which add caustic soda to get a pH of 10.5).

– Either batch or store in J box for 1–2 h at room temperature.

– Rinse.

General pad-batch recipe

Quantity	Unit	Chemicals
6–10	ml/l	Scouring agent
1–2	ml/l	Stabiliser
10	g/l	Sodium silicate 38°Bé
6–12	g/l	Caustic soda solid
30–40	g/l	Hydrogen peroxide 35%

Note

1. The bath is set as per order of addition given.
2. Method–pad goods at 20–30°C at 100% pick-up–Pile-up or batch–store 14–15 h–wash of in a soaper.
3. This method is suitable for cotton and cotton/polyester.
4. For a hot peroxide bleach (especially for knitted goods) the following procedure can be followed. Set-up the pad liquor as above with the half the amount of caustic soda (i.e., 3 g/l) and 10 g/l sodium silicate then–pad at 20–30°C with 120–130% pick-up–store for 30–50 min steam continuously for 10–20 min at 100°C. Rinse hot and then cold.

3.4.15.4   Hypochlorite bleach–other materials

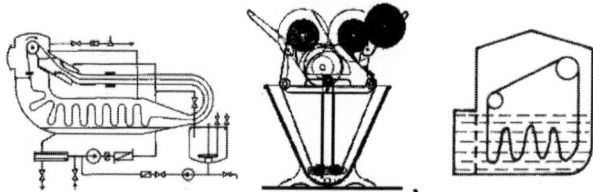

Figure 3.27. Circulating Mc, Jig, Winch

Guideline recipes

Form	Yarn, woven and knit goods		
Substrate	Co, PES/Co, CV/Co, FI/Co, FI, PES/FI, CV, PES/CV grey goods, desized or scoured goods		
Machine	All circulation dyeing machines	Jig	Winch
MLR	6:1–15:1	2:1–5:1	8:1–20:1
Guide recipes	One bath causticising and bleaching		
Wetting agent ml/l		1–3	1–2
Low-foaming wetting agent ml/l	1–2		
Available chlorine	1–2	2–4	1–2
Dilute caustic soda for pH	10.5	10.5	10.5

Working method

Wet out 10 min at room temperature		2 ends	
Add hypochlorite			
Add diluted NaOH to adjust pH 10.5	20–30	20–30	20–30
Bleach at temperature	20–25	20–25	20–25
Time	60–90	60–90	60–90
Drain and rinse cold			
Antichlor treatment with sodium bisulphite in acetic acid medium or peroxide bleach			

Alternative with sodium bisulphite: 2–4 g/I sodium bisulphite powder
pH 4 with acetic acid
treat for 20 min at 40°C
rinse cold.

Continuous method–Yarn, woven PES/CO, CV/CO, FI/CO, PES/FI, CV, PES/CV

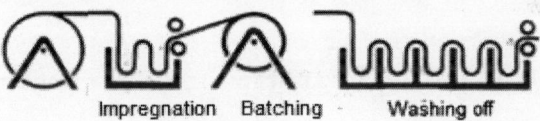

Impregnation   Batching   Washing off

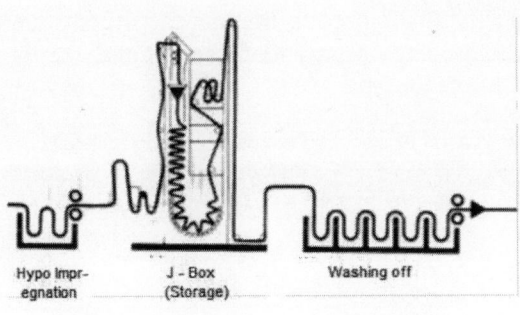

Hypo Impr-   J - Box   Washing off
egnation   (Storage)

Figure 3.28. Pad-Batch, J-box systems

Guideline recipes

Process	Piece goods–Pad-batch	U box, L box, J box, conveyer total immersion system
Impregnate at 20°C with available chlorine	3–5 g/l	2–6 g/l
Solvent based detergent and scouring agent	1–2 g/l	2–5 g/l
Caustic soda solution for	pH 10.5	10.5–11
Store for 1–2 h at room temperature		
Rinse		
Working method		
Saturate: Impregnate		
Temperature °C	20–25	20–25
Pick-up %	100	100
Batching temperature °C	20–30	20–30
Time	2–3 h	10–60 min
Washing-off	Cold	Cold

Note:

1. After hypochlorite bleaching the fabric has to be given an antichlor treatment with sodium bisulphate and acetic acid or peroxide bleach with sodium sulphite.
2. 2–4 g/l sodium bisulphite powder.
3. pH 4 adjusted with acetic acid.
4. Treat in cold in a washing machine with 2–3 rinsing compartment.

3.4.15.5   Classification of cellulose according to Cu fluidity values

Cuprammonium fluidity is used to find out the degradation of cellulose especially during bleaching or other processes. Higher the fluidity greater the degradation of the cellulose.

Table 3.27. Classification of bleached cotton according to fluidity

Fluidity (rhes)	Classification
1–4	Very mildly prepared cotton, excellent bleaching
5–8	Satisfactory bleaching
8–9	On the boundary lines of safe bleaching
above 9	Badly over bleached, degraded cellulose (tendered)

Table 3.28. Barium activity numbers of processed fabric (BTRA)

Fabric	Barium activity number
Poplin	107–157
Cambric	105–158
Voile	120–156
Satin	120–125
Drill	104–142
Polyester/cotton	111–155

Recipes for peroxide bleach after hypochlorite bleach

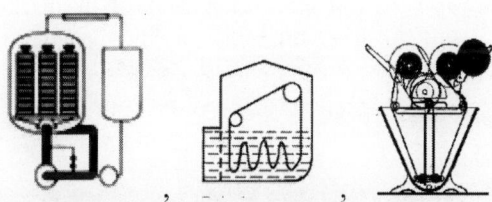

Figure 3.29. Yarn dyeing machine, winch, jig

Guideline recipes

Discontinuous process	Yarn machines 10:1–20:1	Piece goods	
		Winch 20:1	Jig 5:1
Stabiliser ml/l	0.25 - 0.5	0.25	0.5
Caustic soda 36°Bé ml/l	1-2	1	2
Hydrogen peroxide 35% by weight ml/l	3-6	3	6
Process			
Bleach for 1 h at 90°C			
Rinse hot			
Rinse cold			

Requirements of a good pre-treatment
- Good absorbency 1–4 TEGEWA drop test
- No residual size > 6 on TEGEWA scale (starch size)

- Whiteness Berger Dyeing/printing 70 = -80, Full white 80–88
- No residual Co seeds
- No fibre damage < 3 damage factor
- Low conductivity Especially for printing < 100 µS/cm
  (a) pH fabric 6.5–7.0.

3.4.16   The preparation of viscose and blends of viscose

Viscose is important mainly due to its comfort, and it is being used more and more for fashion wear and casuals and viscose is emerging as the fibre of choice. Today, viscose is considered the best alternative to cotton among man-made fibres and consumption is fast growing. It is a fully biodegradable textile fibre made from wood pulp. The viscose fibre (cellulosic fibre) is a regenerated cellulose fibre. Fibre composed of regenerated cellulose, in which substituents have replaced not more than 15% of the hydrogen of the hydroxyl groups. About 85% of the total viscose fibre production is produced as staple fibres and about 15% as filaments. A comparison of comfort, aesthetic and utility performance characteristics between cotton, viscose and polyester is given below.

Table 3.29. Comparative performance of cotton, viscose and polyester fabric

	Comparative rating		
Parameters	**Cotton**	**Viscose**	**Polyester**
Comfort			
Moisture regain	Good	Very good	Poor
Thermal protection	Good	Very good	Poor
Air permeability	Very good	Good	Poor
Softness	Good	Very good	Poor
Smoothness	Poor	Good	Very good
Static dissipation	Good	Very good	Poor
Aesthetic			
Drape	Good	Very good	Poor
Lustre	Poor	Very good	Very good
Crease recovery	Poor	Poor	Very good
Uniformity	Poor	Very good	Good

Utility performance

Anti-pilling	Good	Very good	Poor
Wash and Wear	Good	Poor	Very good

Even though it is cellulosic, the preparation or processing is different from cotton due to its poor wet strength. Most the process available in the cotton process is principally suitable for viscose material but we have to make changes in the recipes to take care of the wet strength of the material.

Given below the strength characteristics of regenerated cellulose fibres in comparison to cotton.

Table 3.30. Comparison of fibre properties of regenerated cellulosic fibres and cotton

Fibre properties of reg. cellulose	Modal	Polynosic	Viscose	Cotton
Fineness (dtex)	1.7	1.7	1.7	1.7
Dry breaking resistance (cN/tex)	34	38	25	28
Dry breaking elongation (%)	14	11	17	10
Wet breaking resistance (cN/tex)	20	26	14	32
Wet breaking elongation (%)	15	12	21	11
Humidity absorption (%)	12.5	10–14	13	8
Swelling (%)	75	60–70	90	90
Fibrillation tendency*	1	3	1	2

In case of cotton, of the total water (liquid) on the fibre (1) 70% may be adhesively bound between the individual fibre (capillary water) and cohesively bound to the surface of the fabric, (2) 25% may be the water of swelling in the amorphous region, (3) 5% water swelling linked by H-bonds. But surprisingly it is (1) 25, (2) 65, (3) 15, respectively. Whereas first one can mostly be removed by squeezing (mechanical methods) and second one can be removed by drying but the third one can be removed by over drying only.

The water retention value (swelling index) of viscose is very much higher than that of the cotton. In aqueous liquors, viscose fibres tend to swell more strongly than modal fibres or cotton. This swelling process happens very quickly and is almost complete after 10 s at the lower temperature range. Fabrics become much stiffer when wet because the fibres are so swollen. In their swollen state, viscose fibres can become set to a certain extent. This is called hydro plasticity.

Viscose has lower tenacity in both wet and conditioned state than cotton. Viscose has lower tenacity in both wet and conditioned state than cotton–more care is necessary to prevent fabric breakages and tears in wet processing.

How manufacturing and physical properties influence processing of viscose

Different ratio of crystalline and amorphous regions compared to cotton	Viscose has higher dye affinity than cotton
Viscose has irregular serrated skin and cross-section	Inferior diffusion and penetration. More kinetic energy needed. Hot reactive dyes
Viscose loses tenacity when wet	More care needed to avoid damage
Wet swelling increases with temperature	Very important in package dyeing. Liquor circulation should mainly be IN to OUT. OUT to IN should be < 30 s
Swelling of fibres makes wet fabrics stiff. Swelling and heat can set creases	Use longer L.R. than for cotton. Keep liquors above 50°C, cool at maximum 1°C per min. Use suitable anti-crease lubricants.
Viscose may contain residues of sulphur	Mild peroxide bleach may be necessary to remove sulphur
Viscose is creamy white in colour, naturally clean. Fabrics free from natural fats and waxes, motes and seeds	Little preparation required. Bleaching chemicals can be reduced. Always give SOME preparation to avoid carryover of sulphur
Dyes have higher substantivity and faster fixation	Use "Migration" dyeing techniques (at up to 110°C). Add salt after dye

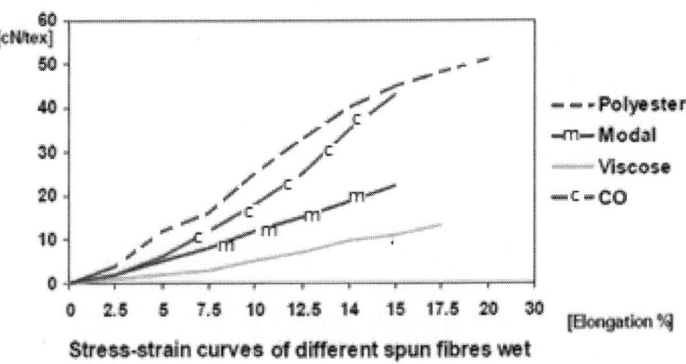

Stress-strain curves of different spun fibres wet

Figure 3.30.

Viscose and modal fibres are supplied in a pure state and with a higher degree of whiteness than cotton. Bleaching is only required for a full white or pastel shades. Viscose/cotton blends require bleaching baths with a reduced chemical content.

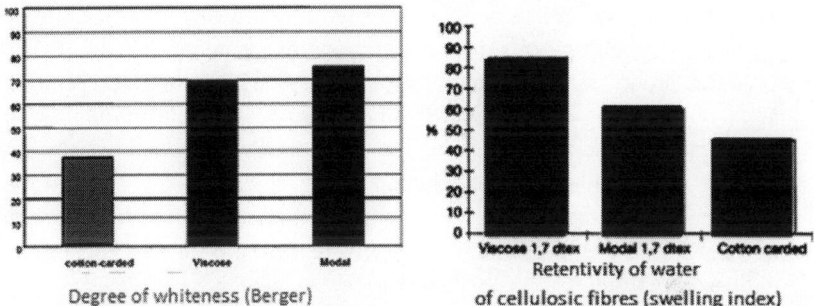

Degree of whiteness (Berger)   Retentivity of water of cellulosic fibres (swelling index)

Figure 3.31. Whiteness and water retentivity of cotton and viscose

Scouring and bleaching need to be mild in nature. The fabrics should always be scoured, and never put straight into dye because it is important to remove any residual sulphur to prevent dye reduction. Spinning lubricants used on viscose tend to yellow with heat, and so should be removed for best whites and bright pastels. The liquor ratio may need to be a little higher than for cotton, because of the higher water retention but also because of the high swelling. (Reduce jet capacity.) Break outs are always an issue on viscose, particularly where there is lengthways tension on the fabric. Best avoided by sewing the fabric at an angle, not straight across. Use multiple "zig-zag" stitching for additional security. Fabrics will tend to stiffen in tight constructions so lubrication is important.

There should be as little tension on the fabric as possible, because of the low wet modulus it will stretch easily and dimensional stability may never be achieved. Usually means slower turnaround times and dyeing machines with winches to help the fabric into the jet, or the cigar-type machines with the fabric movement into the jet being downwards (like the Gaston County Futura). The more "soft flow" the dyeing machine–the better.

Better base white than cotton, therefore bleaching mainly required only for full white or pastel shades.

Typical pre-treatment routes

Grey state → Washing/desizing → Dyeing/printing → Finishing

Grey state → Washing/desizing → Causticising →Dyeing/printing → Finishing

Grey state → Causticising → Washing → Dyeing/printing → Finishing

Grey state → Washing/desizing → Causticising →Bleaching →Dyeing/printing → Finishing

Grey state → Visco-Combi-Batch (Ciba) → Washing → Dyeing/printing → Finishing

Woven fabrics

Generally, warp sizes consist mainly of water-soluble polymers but they are often blended with partly insoluble starch components and fatty or oily size additives.

Removal of spinning oils and preparations may happen during desizing and washing processes with effective surfactants, which should be compatible with the desizing recipe. Other impurities: Besides the washing/emulsifying surfactants the use of multi-purpose complexing agents is helpful to remove inorganic or metallic impurities.

Identification of the size present is important prior to deciding on a suitable desizing procedure, however, many factories will opt to use their standard cotton desize treatment:

Example: Applying in the quench box of a singeing machine and pad.

Recipe

Quantity	Unit	Chemicals
2–4	ml/l	Amylase enzyme
1.0–2.0	ml/l	Wetting agent

Impregnate at 60°C (as per enzyme requirement); pick-up 80–100%; batch minimum 4–24 h–normally overnight; hot wash-off.

3.4.16.1   Demineralising

Demineralising is not a must for viscose fabric because as such it is clean fibre. However, for better white demineralising can be done at room temperature.

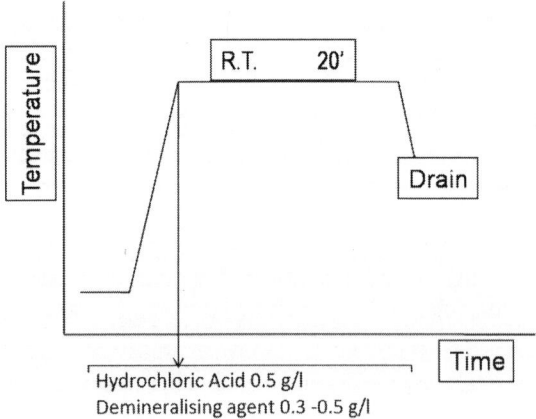

Figure 3.32. Demineralisation process

The process can be carried out at room temperature for 20 min with hydrochloric acid (if the machine metallurgy allows the same) and a demineralising agent.

Discontinuous scouring/desizing by surfactants
Woven viscose fabric with water soluble starches and for knit goods

Table 3.31. Guide Recipe

Quantity	Unit	Chemicals
1–3	ml/l	Wetting and detergent
0.5–2.0	ml/l	Chelating agent for heavy metal ions
0–4.0	g/l	Soda ash

Treat for 20–40 min at 60–80°C liquor ration 5:1–20:1 and hot rinse pH desizing bath should be maintained depending on the sizing material used.

Relax/scouring
Viscose rayon fibres, unlike natural cellulosic fibres, are free from natural fats and waxes, motes and seeds, and the scouring process, therefore, need not be as severe as for cotton, and can be based on soda ash or tetra sodium pyrophosphate recipes rather than caustic soda. A typical scouring recipe on a 5 box continuous open-width washing range would be as follows:

Box 1

Quantity	Unit	Chemicals
3	g/l	Scouring agent
2	g/l	Soda ash
1	g/l	Wetting agent or detergent

Box 2

Quantity	Unit	Chemicals
1	g/l	Wetting agent or detergent at 95°C

Box 3

Quantity	Unit	Chemicals
1	g/l	Wetting agent or detergent at 95°C

Box 4
Water only at 70°C

Box 5

Water only–Hot

Knitted fabric–on jet or overflow machine

Table 3.32. Guide Recipe

Quantity	Unit	Additions
2–4	g/l	Low foaming jet scouring agent
0.5–1.0	g/l	Wetting agent
0.5–2.0	g/l	Defoamer (if necessary)
1–3	g/l	Soda ash (pH about 9)

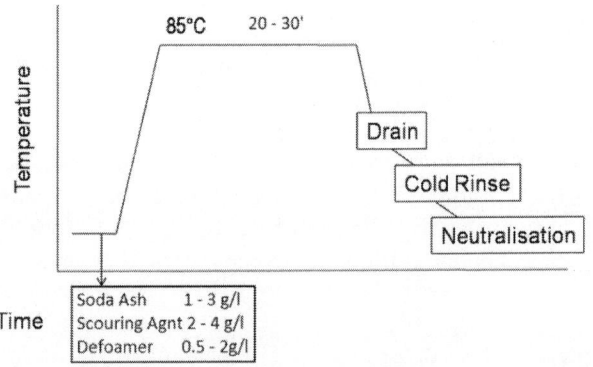

Figure 3.33. Scouring process

Run 20–30 min at 60–90°C
Cold rinse, neutralise.

Peroxide bleach on jet or overflow machine

Recipe for full white

Quantity	Unit	Additions
1–3	ml/l	Peroxide stabiliser (as recommended)
2–4	g/l	Wetting cum scouring agent
0.5–2.0	g/l	Defoamer (if necessary)
2–6	ml/l	Hydrogen peroxide 35%
0.3–0.7	%	Optical whitener

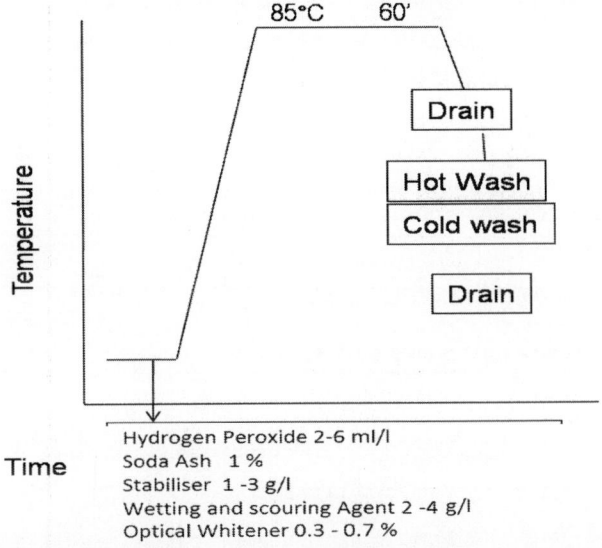

Hydrogen Peroxide 2-6 ml/l
Soda Ash   1 %
Stabiliser 1 -3 g/l
Wetting and scouring Agent 2 -4 g/l
Optical Whitener 0.3 - 0.7 %

Figure 3.34. Full white process

Load the fabric , run for 5 min, add chemicals, auxiliaries and optical whitener, run at room temperature at 5 min and raise the temperature to 85°C, and hold at this temperature for 60 min, cool and drain. Neutralise with core neutralising agent, hot wash followed by cold wash.

Notes on soft flow processing

1. Remove oils and waxes in preparation.
2. No twists in loading–load machine below capacity.
3. Sewing–sew more than once–at an angle.
4. Abrasion–avoid slippage–always run at 50°C or higher.
5. Cooling creases–cool at 1°C per minute.
6. Danger of sulphur residues and reduction.
7. Tension in soft flow–problems with shrinkage.
8. Migration of unfixed dye while waiting for drying.
9. Fabric drying out in patches–watermarks (physical/optical effect). Keep fabric wet right up to the dryer.

Pad-steam peroxide bleach (PSPB)

Quantity	Unit	Chemicals
4.0–6.0	ml/kg	Suitable peroxide stabiliser
4	g/kg	NaOH 100%

Quantity	Unit	Chemicals
6.0–15	ml/kg	H_2O_2 50%
1.0–2.0	ml/kg	Low-foaming washing and rewetting agent

Process: Impregnate cold.
3–12 min steaming (saturated conditions).
Hot wash-off.

Discontinuous sodium chlorite bleach

Recipe for viscose knit goods and woven goods

Quantity	Unit	Chemicals
1.0–2.0	ml/l	Wetting and detergent
1.0–2.0	ml/l	Sodium chlorite 80%
1.0–2.0	g/l	Buffering salt (e.g., PK2 from Degussa-Hulis)
2.0–3.0	g/l	Sodium nitrate (Corrosion Inhibitor)
1.0–2.0	ml/l	Formic acid 80% pH about 4.5

Process: Treat 30–60 min at 80–85°C.
Liquor ratio 5:1–20:1.
Rinse, antichlor.
Hot rinse.

Discontinuous reductive bleach

Recipe for viscose knit goods and woven goods

Quantity	Unit	Chemicals
3.0–5.0	g/l	Stabilized sodium dithionate
0.5–1.0	m/l	Wetting and detergent

Process: Treat for 30–60 min at 75–85°C.
Liquor ratio 5:1–20:1.
Wash-off at 40°C with addition of 0.5–1.0 ml/l H_2O_2 35%.
Warm rinse.

For viscose knit goods and pre-desized woven goods

There is a high-risk of forming permanent creases in rope dyeing machines because viscose fibre swells and fabrics become stiff. Fabric will therefore not run well at temperatures below 50°C. Fabric must always be run at temperatures of 50°C or higher. Cool at a gradient of not more than 1°C per minute to prevent hydro plastic creasing.

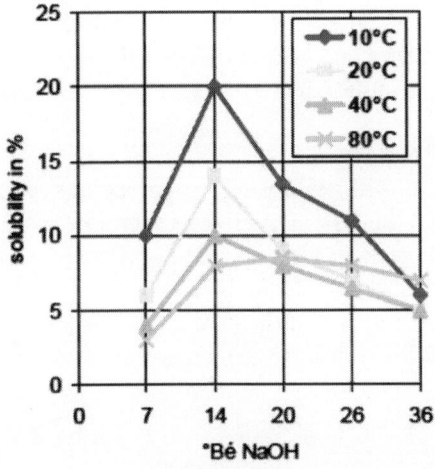

Figure 3.35.

As the solubility of regenerated cellulose in caustic soda decreases markedly with increasing temperature, as shown in the above graph it is advisable to rinse viscose in hot water to avoid damage.

Problems faced in processing viscose fabrics

Harsh handle of finished article: The water retention value (swelling index) of viscose is very much higher than that of cotton. In aqueous liquors, viscose fibres tend to swell more strongly than modal fibres or cotton. This swelling process happens very quickly and is almost complete after 10 s at the lower temperature range.

Fabrics become much more stiff when wet because the fibres are so swollen. In their swollen state, viscose fibres can be set to a certain extent. This is called hydro plasticity.

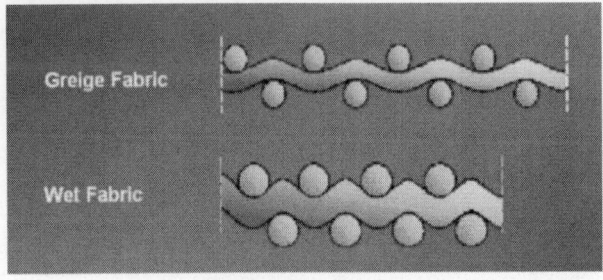

Figure 3.36.

Insufficient dimensional stability: To guarantee a good dimensional stability and crease/folds-free end article the following notice should be observed. The material should be treated under low-tension conditions, allow the material shrink as freely as possible (= eliminates tension in yarn and fabric) and avoid undesirable lustre and brilliancy by exclusion of roller-guidance under high tension and hard squeezing, especially in swollen state.

3.4.16.2   Causticising

Causticisation modifies the viscose fibre surface or skin to enable more rapid diffusion of dye into the fibre. Benefits are therefore more obvious in printing than in dyeing, where long diffusion times are employed, and with selected reactive dyes, yield gains of up to 50% are possible. Typical conditions for pure viscose are treatment with 6–8°Bé caustic soda at 25–30°C for at least 2 min followed by low-tension washing with boiling water to assist the rapid

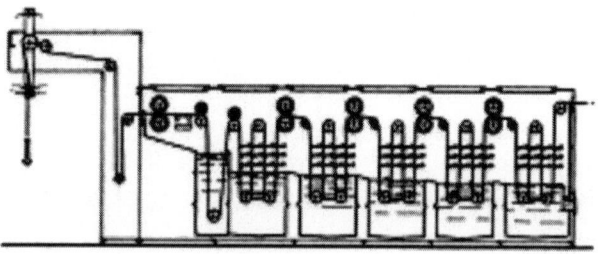

Figure 3.37. A typical causticising machine

removal of alkali liquor pick-up of about 120% should be achieved. The best after-washing device is probably a sieve drum continuous open-width range. Low-uniform processing tensions are essential for consistent results and good quality. Rinsing should be done as hot as possible to minimise swelling and ensure rapid and complete removal of caustic soda. An addition of 2–4 g/l soda ash in the initial wash boxes will promote removal of alkali, and help maintain a good fabric handle. Neutralisation with suitable acid is also advisable. Enhancement of dyeing and printing levelness by swelling the fibre with caustic treatment. Pad-batch processing is popular although dedicated continuous plant, employing a scary or conveyor for tensionless swelling and reaction, is preferable.

This step is preferably carried out by

- (a) Continuous pad dwell methods on conveyors or scarys
- (b) Pad-batch semi-continuous methods on A-frames
- (c) Jig
- (d) Discontinuous methods in rope form only to obtain special surface effects.

To get higher colour yield the causticising should be applied with 6–8°Bé NaOH.

To reduce excessive shrinkage and harsh handle, the treatment with pure caustic soda may be modified (any one):

1. KOH up to 7°Bé
2. Mixture of both
3. NaOH in mixture with common salt (50–150 g/l NaCl).

Process: Causticising bath: 40–50 g/l NaOH 100% containing 2–4 ml/l detergent.

– *Padding*
 Impregnate cold; liquor pick-up 80–120%.
 Dwell for – h at room temperature.
 Hot wash-off.
– *Or continuous/relaxing*
 at 60–80°C [SEP] Hot wash-off. (Treat 1-15 min)
 process (Ciba) (Visco Combi)

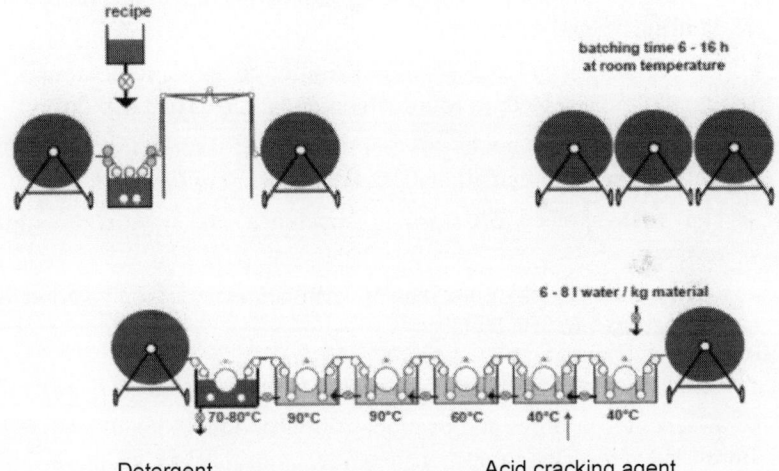

Figure 3.38. Visco-combi process

Recipe

Quantity	Unit	Chemicals
8.0–12.0	ml/kg	Peroxide stabiliser (Suitable)
40	g/kg	NaOH 100%
8.0–15.0	ml/kg	H_2O_2 35%

In order to get a stable recipe bath it is very important to add the chemicals in the correct order. A stirrer would be very helpful to get a homogeneous mix; a high concentration of NaOH tends to sink due to its specific Weight.

To make 1000 l solution

1. Take 800 l water
2. Add NaOH 36°Bé 100 l and then
3. 8–12 l of peroxide stabiliser
4. 4–6 l acid cracking agent are added one after another with stirring
5. At the last add 8–15 l H_2O_2 35%.

Add rest of the water to make-up 1000 l.

Notes:

1. It is advisable to install a dwelling zone of 10–15 s after the impregnation unit to allow the fibre to swell before being wound-up.
2. The tension during the batching-up should be kept as low as possible by using a separate central winding-up device.
3. The length of one batch should be limited to avoid, for example, the tearing of the selvedges (our recommendation ± 1000 m).
4. Rinsing is done as hot as possible with the tension as low as feasible while taking care that all NaOH is removed from the fabric.
5. The first choice for washing machines are open-width drum washers.
6. Open-width vertical or horizontal washing machines with a good tension control are also suitable.

Scouring of regenerated cellulose and/or modal and its blends with polyester

Regenerated cellulose fibres and/or modal fibres and their blends with polyester fibres are pre-treated under tension-free conditions if possible in order to take full advantage of the shrinkage tendency of the fibres. At the same time a more uniform and closed appearance of the goods is achieved which is important for the following processing operations and the final effect.

Scour boiling is carried out either on continuously arranged winches (rope form) or even better on an open-width scour boiling machine such as the Mezzera system (manufacturer: Mezzera, Italy) where the tendency to creasing is prevented (scour boiling on the winch. The goods are often causticised before scour boiling to obtain a more closed appearance of the goods, increased dye yield and improved basic crease recovery angle.

Causticising is carried out by the pad batch method followed by rinsing on a machine positioned before the scour boiling machine.

The pre-treatment possibilities are:

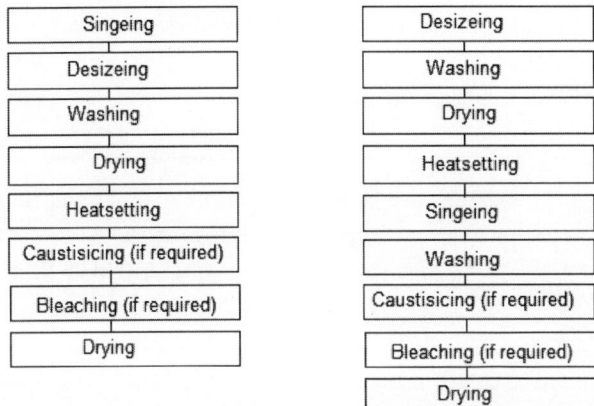

Figure 3.39. Process routes for Pretreatment

Guideline recipes

Form	Woven goods	
Substrate	CV, PES/CV	
Machine	Pad-batch, Mezzera OW	Mezzera OW
Method	With pre-causticising	Without causticising
Guide recipes		
Pad-batch		
-Saturator		
Caustic soda 36°Bé ml/l	4–6	
Suitable scouring agent ml/l	5–10	
- Impregnate		
Liquor temperature °C	20–30	
Pick-up %	90–100	
- Batch		
Temperature	Room temperature	
Time h	1–2	
Open width scour boiling	Intermediate rinse at 80°C	
Caustic soda g/l	3–6	3–6
Suitable scouring agent ml/l	2–5	2–5
-Working method		
Bath temperature °C	80–85	80–85
Time min	20	20
Open width washer		

Compartment 1 °C	80–85	80–85
Compartment 2 °C	75–80	75–80
Compartment 3 °C	50–60	50–60
Compartment 4 °C	30–40	30–40
	Neutralise if needed	Neutralise if needed

If the goods have been pre-causticised the smaller amount of caustic soda is sufficient for scour boiling. Depending on the degree of contamination the scour boiling bath is used for 1–2 months and the products reinforced from time to time.

3.4.17   Pre-treatment of Bamboo Fibres

Pre-treatment route

Singeing → Desizing → Scouring → Mercerisation→ Bleaching OR
Singeing → Desizing → Scouring and Bleaching → Mercerisation

3.4.17.1   Singeing

The surface of pure bamboo fabric or bamboo/cotton blended fabric may have a lot of hairiness which has to be removed by gassing (singeing). If singeing is not proper it may affect the dyeing appearance, solidity, evenness, etc.

Singeing parameters

Gassing speed: 80–100 m/m.

Burner temperature: 1100°C.

Note: Burner fire should be even. Gassing result should reach Grade 3–4.

3.4.17.2   Desizing

In bamboo warp sizing usually starch based sizing agents are used. Hence, high-activation, high-stability and selective enzymes which can target on starch sizing agent should be used which it should not damage bamboo fibre. We require that the enzyme will not be damaged by axunge or waxiness. This enzymes need to to be kept active in a certain range of pH 5–9 (or as recommended by enzyme manufacturer).

Table 3.33.

Quantity	Unit	Chemicals
2–5	ml/kg	Enzyme (e.g., Tinozym L40)
3–4	ml/kg	Wetting agent

Padding at 100% expression, at 25–90°C and store for 8–16 h (or as recomθ mended by enzyme manufacturer). Hot wash, cold wash and make ready for scouring.

3.4.17.3   Scouring and bleaching

Bamboo fibre's natural colour is light yellow and so whiteness degree of bamboo fibre is lower than cotton. For fabric woven with bamboo/cotton blended yarn, if scouring and bleaching does not meet requirement, the whiteness degree of cloth cover will be uneven. Because bamboo fibre is sensitive to both acid and alkali, volume of caustic soda must be minimum.

Guideline recipe for Scour/Bleach

Quantity	Unit	Chemicals
2–4	g/l	Peroxide stabiliser
1–3	g/l	Caustic soda
8–10	ml/l	Hydrogen peroxide 35%

Temperature, 90–98°C.
Time, 40–60 min.

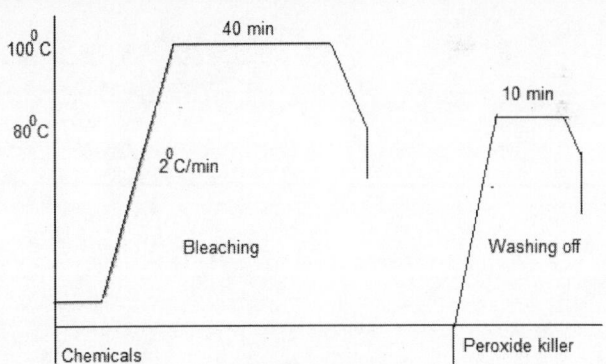

Figure 3.40.

3.4.17.4   Mercerisation

For bamboo fabrics and bamboo/cotton blended fabrics, mercerisation not only improves fibres' absorbance of dyestuffs, but also it can increase the brightness of the dyed fabrics due to plain surface and even lustre. It also improves the drape, shining, etc. Mercerisation can be done chainless or chain mercericers.

Mercerising parameters

Caustic soda: 210–240 g/l.

Temperature: room temperature.

Time: 40–60 s.

PART IV

4

Pretreatment of substrates – other natural fibres

4.1 Pre-treatment of linen/flax

About 1.5–2% of the world wide fibre production is linen. Linen is a bast fibre. Main cultivational areas are CIS, Poland, China, France, Czechia, Turkey, Rumania, Belgium, The Netherlands. Flax fibres vary in length from about 25–150 mm (1–6") and average 12–16 µm in diameter. There are two varieties: shorter tow fibres used for coarser fabrics and longer line fibres used for finer fabrics. Flax fibres can usually be identified by their "nodes" which add to the flexibility and texture of the fabric. The cross-section of the linen fibre is made-up of irregular polygonal shapes which contribute to the coarse texture of the fabric.

4.1.1 Features of linen

Colour of average linen is yellowish buff to grey in colour. Linen fibres have a characteristics silky lustre, much more pronounced than that of untreated cottons. It is stronger than cotton, and its tensile strength increases when it is wet. Compared to cotton, linen is less elastic, so they crease and wrinkle easily. Heat conductivity of linen is better than cotton, so carries heat away from the body faster. Hence, garments made of linen feel cooler than those of comparable weight in cotton. Linen can absorb up to 20% of its weight in moisture before it feels damp and easily releases moisture to the air, leaving the wearer feeling fresh and cool. The more it is washed, the softer and brighter it becomes. Its brightness comes from the nodes of the linen fibre, which reflect light. Linen helps to protect the skin from UV rays. They are natural anti-bacterial health fabrics. With repeated washings, the fabric attains new sheen and airiness which makes it even more skin-friendly. Linen is naturally anti-static hence repels dirt.

4.1.1.1   Cross-section of a flax stem

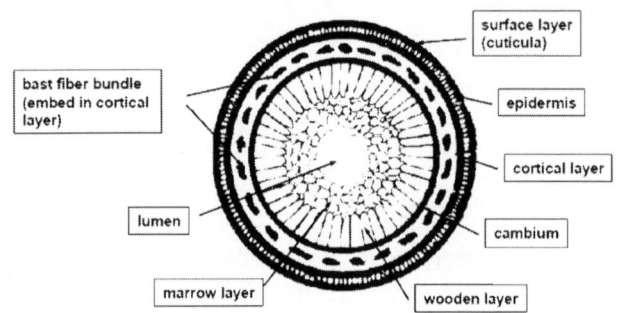

Figure 4.1.

Production of linen fibre

Following steps are involved

Plucking–Harvesting the plant (mechanically)

Rippling–Removal of seeds and leafs by an iron comb (mechanically)

Retting–Destruction of vegetable glue for separating layers (dew or water retting; chemically/biologically)

Breaking–Breaking of wooden parts (mechanically)

Scutching–Removal of wooden parts (mechanically)

Hackling–Sorting out, cleaning and dividing of the bast fibre bundles (mechanically).

With cellulose representing more than 90% of the total fibre composition, cotton is a relatively pure raw product in contrast to linen, which contains only around 60% cellulose. Many more impurities need to be removed from linen.

Table 4.1. Comparison of cotton and linen composition

Composition	Cotton	Linen/flax
Cellulose	90–95%	60–65%
Pectins and hemi-celluloses	1–5%	~ 20%
Lignins	Nil	~ 3%
Waxes	~ 0.60%	~ 1%
Water soluble parts	~ 2.50%	~ 12%

Main objectives of pre-treatment of linen are removal of–sizes, incrustations, alkaline earth and heavy metal ions. It should achieve high degree of whiteness, good hydrophilicity along with high fibre protection. At the same time it should be economical, reproducible and environmental-friendly.

Possible pre-treatment steps

1. Enzymatic desizing/cracking
2. Alkaline cracking
3. Oxidative cracking
4. Acid cracking
5. Peroxide bleach (± silicate)
6. MEGA bleach
7. Hypochlorite/chlorite bleach
8. Mercerising
9. Ammonia treatment.

Pre-treatment of linen woven goods and its blends with other cellulose fibres is carried out in rope (kiers or rope bleaching machines) or open-width process on the jig, in pad roll or on continuous pre-treatment ranges.

Blends with polyester should be pre-treated in open-width only. A pad-batch cold peroxide bleach is also suitable, ideally in combination with a subsequent alkaline boiling step–e.g., a continuous boil-off on an open-width washing range–or a peroxide bleach under hot conditions.

Cotton/linen blends are often sized with starch, which can be removed efficiently by an enzymatic desizing/cracking step.

Acid scouring

It is advantageous to start the pre-treatment sequence with an acid cracking process (depending on size on the fabric and machinery constellation). Depending on origin, linen fibres can be heavily contaminated with alkaline earth and heavy metal ions (e.g., Ca and Fe). Without an acid cracking process, there is a danger of: forming deposits and precipitations on fabric and machinery parts, filtration/incrustation of alkaline earth salts in yarn packages and poor bleaching results in combination with higher fibre damage.

An acid cracking process can be carried out in pad batch, in continuous (e.g., washing range) or in discontinuous applications.

Recipe continuous scouring methods (100% linen)

Chemicals	Unit	Pad-batch	Continuous Long Time	Continuous Short Time
Scouring agent	ml/kg	3–6	2–6	1–2
Sequestering agent	ml/kg	4–8	6–12	2–4
Impregnation temperature	°C	20–40	20–40	
Dwelling temperature	°C	Room temperature	40–60	40–60
Dwelling time	min		1–3	0.50–2
	h	2–6		

Recipe discontinuous scouring methods (100% linen).

Chemical	Unit	Winch	Package mac/Jet	Jig
Scouring agent	ml/l	1–2	1–2	
Solvent based souring and wetting agent (low foaming)	ml/l			2–4
Sequestering agent	ml/l	1–3	1–3	2–4
Treatment temperature	°C	40–60	40–60	40–60
Treatment time	min	30–60	30–60	45–90

4.1.2   Enzymatic desizing/enzymatic scouring

In pre-treatment of linen woven goods, the classical desizing step is often not carried out separately, but within combination processes (e.g., enzymatic cracking, oxidative cracking, etc.). Starch sizes can be removed by several hot alkaline and / or alkaline/oxidative processes. Removing alkaline earth and heavy metal ions in the first step can help to avoid problems in subsequent processes.

Recipes for continuous enzymatic scouring (100% linen)

Chemicals	Unit	Pad - Batch	Pad - Roll	Pad - Steam
Suitable enzyme	ml/kg	2–6	2–4	2–6
Scouring and wetting agent	ml/kg	3–5	3–5	3–5
Invatex ED	ml/kg	2–4	2–4	2–4
Impregnation temperature	°C	20–60	20–90	60–90
Treatment temperature	°C	Room temperature	85–95	100–102
Treatment time	min			2–30
	h	4–12	1 – 2	

Recipes for discontinuous enzymatic scouring (100% linen)

Chemicals	Unit	Pad-batch	Pad-roll	Pad-steam
Suitable enzyme	ml/l	1–2	1–3	2–4
Scouring and wetting agent	ml/l	1–2	1–2	2–4
Invatex ED	ml/l	2–4	2–4	2–4
Processing temperature	°C	30–60	30–60	45–90

4.1.2.1   *Alkaline scouring*

With an alkaline cracking process, incrustations such as waxes, fats, pectin and lignin with its illuminating colour are removed from the fibres. An alkaline treatment with soda ash leaves the glue between the fibre bundles. This results in a harsher fabric handle.

By an alkaline treatment with caustic soda (up to 40 g/kg NaOH 100%) a much better cleaning effect can be achieved, which results in a softer handle and more intensive swelling of the fibres. Adding a complexing/ cracking agent helps to remove pectins and its alkaline earth and heavy metal salts.

Recipes for continuous alkaline scouring (100% Linen)

Chemicals	Unit	Pad-roll	Pad-steam
NaOH 100%	g/kg	20–30	25–40
Cracking agent with wetting, detergency, dispersing and complexing power	ml/kg	4–6	0–4
Alkaline cracking agent	ml/kg	4–8	2–4
Impregnation temperature	°C	20–60	20–80
Treatment temperature	°C	85–95	100–102
Treatment time	min		5–25
	h	1–2	

Recipe for Discontinuous alkaline scouring (100% Linen)

Chemicals	Unit	Winch	Package machine / Jet	Jig
NaOH 100% or soda ash	g/l	2–6	3 - 6	6–12
Suitable cracking cum wetting/ detergent agent	ml/l	1–3	1–3	2–5
Sequestrant	ml/kg	0.5–2	0.5–2	2–5
Treatment temperature	°C	90–98	90–98	90–98
Treatment time	min	30–60	30–60	45–90

In case of linen/synthetic blends the caustic quantity has to be reduced by 20–30%.

4.1.2.2   Oxidative desizing

Recipes for continuous oxidative scouring (100% Linen)

Chemicals	Unit	Pad-batch I	Pad-batch II	Pad-steam
Peroxide stabiliser	ml/kg	4–6	3–6	6–12
NaOH 100%	g/kg	30–40	30–40	20–50
H_2O_2 35%	ml/kg	20–40	20–40	3–5
Sodium persulphate	g/kg	3–6	3–6	
Impregnation temperature	°C	20–40	20–40	40–60
Treatment temperature	°C	Room temperature	Room temperature	100–102
Treatment time	Min			2–45
	h	6–25	6–5	

4.1.3   Bleaching

In pre-treatment of linen it is common to speak of a 1/2, 3/4 und 4/4 whiteness. A combined, multi-step bleaching process is necessary to obtain a 1/2 whiteness on grey yarn linen qualities. Oxidative bleaching can be done with sodium chlorite, sodium hypochlorite or hydrogen peroxide. Bleaching with sodium chlorite shows best discolouration of the lignin. At the same time the typical, harsh linen handle of the fabric is maintained.

Linen can be bleached with silicate free recipes or silicate recipes. In silicate free recipes there is no danger of forming Calcium silicate deposits and better removal of starch size and other impurities with more intensive swelling of fibres. Whereas silicate bleaching gives an even and higher degree of whiteness.

4.1.3.1   Continuous peroxide bleach (silicate)

Guide recipes for continuous silicate bleaching (100% Linen)

Chemical (silicate containing)	Unit	Pad-steam short-time	Pad-steam medium-time	Pad-steam long-time
Peroxide stabiliser	ml/kg	3 – 5	3–6	4–6
NaOH 100%	g/kg	2–8	2–8	2–6
Sodium silicate 36°Bé	ml/kg	2–8	3–10	4–12

H_2O_2 35%	g/kg	20–60	15–40	10–25
Suitable wetting and scouring agent		2–3	2–3	2–3
Impregnation temperature	°C	20–40	20–40	20–40
Treatment temperature	°C	100–102	100–102	100–102
Dwelling time	Min	1–5	5–25	25–180

Note: In case of Li/synthetic blends it is recommended to reduce NaOH concentration by 20–30%.

4.1.3.2   Continuous peroxide bleach (silicate free)

Guide recipes for continuous silicate free bleaching (100% Linen)

Chemicals (silicate free)	Unit	Pad-steam short-time	Pad-steam medium-time	Pad-steam long-time
Peroxide stabiliser	ml/kg	4–8	6–10	8–12
NaOH 100%	g/kg	6–12	6–10	4–6
H_2O_2 35%	ml/kg	20–60	15–40	10–25
Suitable wetting and scouring agent	ml/kg	2–3	2–3	2–3
Impregnation temperature	°C	20–40	20–40	20–40
Treatment temperature	°C	100–102	100–102	100–102
Dwelling time	Min	1–5	5–25	25–45

Note: In case of Li/SYN blends it is recommended to reduce NaOH concentration by 20–30%.

4.1.3.3   Discontinuous peroxide bleach (silicate)

Guide recipes for discontinuous peroxide bleach (silicate)

Chemicals (with silicate)	Unit	Winch	Package Mc/Jet	Jig
Peroxide stabiliser (suitable)	ml/l	0.5–1.5	0.5–1.5	1–3
NaOH 100%	g/l	0.5–1.5	0.5–2	1.5–4
Sodium silicate 38°Bé	ml/l	0–2	0–2	0–8
H_2O_2 35%	ml/l	4–10	5–15	10–30
Suitable wetting and scouring agent	ml/l	1–2	1–2	2–3

Treatment temperature	°C	90–98	90–98	90–98
Treatment time	min	45–60	45–60	60–20

Note: In case of Li/SYN blends it is recommended to reduce NaOH concentration by 20–30%.

4.1.3.4   Pad-batch cold peroxide bleach

Guideline recipes for pad-batch cold peroxide bleach (100% Linen)

Chemicals	Unit	Silicate free I	Silicate free II	With silicate
Peroxide stabiliser (suitable)	ml/kg	12 – 20	6–10	4–6
NaOH 100%	g/kg	25–35	25–35	8–15
Sodium silicate 38°Bé	ml/kg			6–12
H₂O₂ 35%	ml/kg	40–80	40–80	40–80
Suitable wetting and scouring agent	ml/kg	5–10	5–10	5–10
Sodium persulphate	g/kg	0–5	0–5	0–5
Impregnation temperature	°C	20–40	20–40	20–40
Dwelling temperature	°C	Room temperature	Room temperature	Room temperature
Dwelling time	h	6–24	6–24	6–24

Washing-off/Boiling-off

1. After a pad-batch cold peroxide bleach it can be continuous boiled-off. The procedure is shown in the diagram.

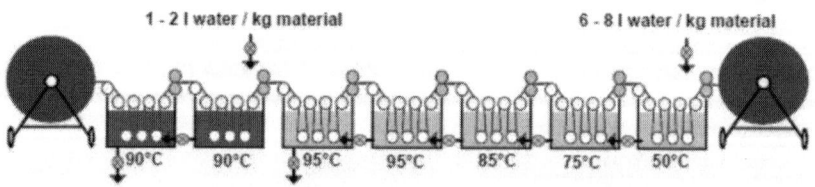

Figure 4.2.

First two wash boxes are sharpened with

10 g/kg NaOH 100%

3 g/kg Scouring agent

And the temperature is kept at 90°C. In all other wash boxes, the temperature is maintained at 95°C. Water addition is shown in the diagram.

2. Washing-off and neutralisation after a Pad-steam or Pad-batch cold peroxide bleach.

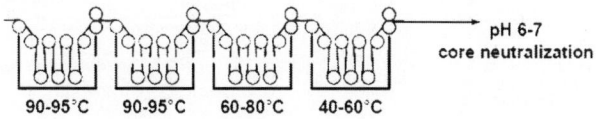

Figure 4.3. Washing-off after pad-steam

The neutralisation and washing should ensure thorough final cleaning of fabric and machinery, Complete removal of residual alkali and core neutralisation of the fibres and no formation of disturbing neutralisation salts.

4.1.3.5 Hypochlorite bleach
Recipe and process for hypochlorite bleach

Process	Concentration (NaOCl/active Cl$_2$)	Temperature	Time
Discontinuous	1–4 g/l	20–25°C	30–120 min
Pad-batch	2–6 g/kg	20–25°C	10–120 min
pH	9–11.5		
Addition of	Buffering salt (0.5–2 g/l) corrosion inhibitor NaNO$_3$ is necessary		
Antichlor treatment with	H$_2$O$_2$		

4.1.3.6 Chlorite bleach
Recipe and process for chlorite bleach

Process	Concentration (NaClO$_2$ 80%)	Temperature	Time
Discontinuous	1.5–5 g/l	70–85°C	1–3 h
Pad-roll/pad-steam	10–25 g/kg	80–100°C	20 min–4 h
pH	3.5–4 (With rormic or acetic or oxalic acid)		
Addition of	Buffering salt (0.5–2 g/l) Corrosion inhibitor NaNO$_3$ is necessary		

Results of bulk trials

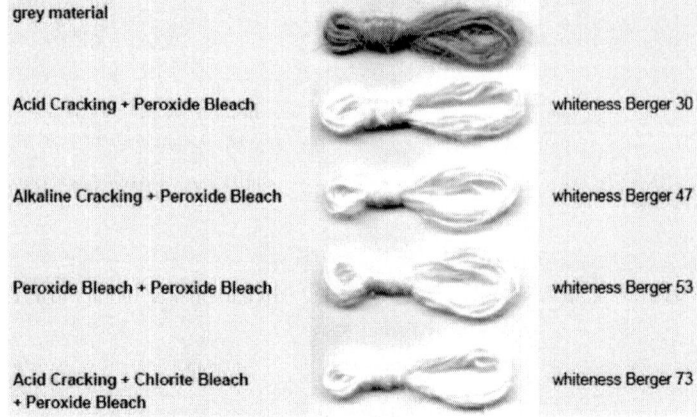

Figure 4.4. Appearance of linen afte bleach

Alkaline cracking (2 times) + acid scouring + chlorite bleach + peroxide bleach

Alkaline scouring + peroxide bleach

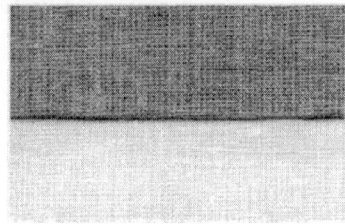

Pad-batch acid scouring + peroxide bleach on Jig

hydrophilicity:	3 sec
whiteness (Berger):	58
DP value:	2180
damage factor:	0.30

Pad-batch cold peroxide bleach

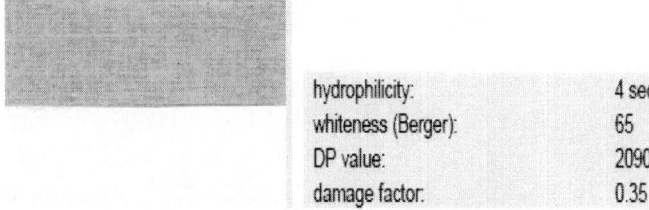

hydrophilicity:	4 sec
whiteness (Berger):	65
DP value:	2090
damage factor:	0.35

Pad-batch cold peroxide bleach + continuous boil-off

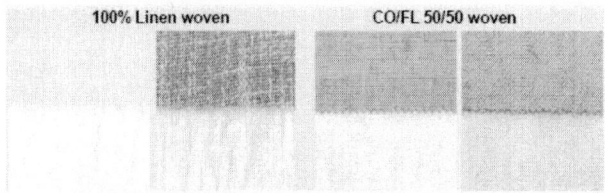

4.1.4   Quality tests for the efficiency of preparation (cotton)

4.1.4.1   Quality test for silicate on the fabric

Ash 5–10 g of the fabric in a platinum crucible. On cooling, mix with 5-6 times the amount of sodium potassium carbonate (made by mixing equal weights of Na_2CO_3 and K_2CO_3) and heat it until a clear melt is obtained. On cooling, dissolve the solidified melt in distilled water and add granular ammonium molybdate. Acidify with nitric acid (20%). In the presence of silicates an intense yellow colour or a yellow, crystalline precipitate is produced. To test if sample also contain phosphate, first dissolve the ash in 10 ml nitric acid (20%) filter and then mix the filtered residue with 5–6 times the amount of sodium potassium carbonate and continue as above.

4.1.4.2   Quality test for calcium on the fabric

About 5–10 g of fabric is weighed accurately and then ashed in a crucible. The ash is taken up in hydrochloric acid (10%) and ammonium chloride and ammonia (SG = 0.88) are added until the solution is alkaline. If a voluminous precipitate forms it should be filtered-off. The ammoniacal filtrate is acidified with acetic acid and calcium precipitated with oxalic acid as calcium oxalate. After filtration, precipitate is spotted onto a platinum wire and placed in a Bunsen flame. The presence of calcium is indicated by a characteristic red flame colour.

If a quantitative determination of calcium is required, then the filtered calcium oxalate is washed with a little distilled water and then shaken up in warm sulphuric acid (20%) and titrated against 0.1 N $KMnO_4$ solution.

$$1 \text{ ml } 0.1 \text{ N } KMnO_4 = 2.004 \text{ mg calcium}$$

Note:

The results from cotton and bast fibres are not always satisfactory as calcium and magnesium can be taken-up by the fibre during growth or from the liquor where they are present as hardness formers. The concentration of calcium and magnesium expressed as a percentage on weight of raw cotton, is usually in the following ranges:

Calcium 0.043–0.15%

Magnesium 0.046–0.11%

To correct for any calcium absorbed during wet processing a comparison should be made with grey fabric.

4.1.4.3   Qualitative test for iron on the fabric

Potassium thiocyanate test

Reagents:

Nitric acid 5% or 10% w/w analytical grade (A.R.).

Potassium thiocyanate solution 10% w/w made by dissolving 10 g potassium thiocyanate in 90 g distilled water.

Potassium bisulphate analytical grade (A.R.).

Hydrogen peroxide 35% w/w.

A small sample, 0.5–1.0 g of fibre (fabric, yarn) is placed on a watch glass or petri dish or porcelain tile. And then spotted with 1–2 drops of nitric acid (5%) and allowed to stand for 2–3 minutes so that any iron present is oxidized to ferric ions. To detect ferric ions, 2–4 drops of potassium thiocynate solution (10%) are added.

If no colour forms then no iron is present. A red colour indicates iron is present and the intensity of the colour indicates the amount. A pink colour indicates low-concentrations and a wine-red colour high concentrations.

In difficult cases or for dyed fabric, the test can be carried out on an ashed sample. For this, the fabric or yarn sample is ashed in a crucible over a Bunsen or in a muffle furnace and then the ash, after cooling, is dissolved in 10% nitric acid. If the ash is not soluble in nitric acid, the ashing is repeated and a little (about 1 g) potassium bisulphate is added to the crucible while it cools. The crucible is then reheated to red heat and after cooling, the ash is taken up in 10–15 ml distilled water. Care must be taken to ensure that the entire mass is fully dissolved, if necessary by warming. Add to this solution, 1 drop of hydrogen peroxide (35%) and then 1 drop of potassium thiocyanate solution. A pink or red colour indicates the presence of iron.

4.1.4.4   Potassium ferrocyanide test

A small sample of fabric on a watch glass, petri dish or porcelain tile is spotted with diluted hydrochloric acid (10%). The acidified area of the fabric is then spotted with dilute potassium ferrocyanide (1% solution freshly prepared). A dark blue colouration indicates the presence of iron.

4.1.4.5   Quality test for copper on the fabric

To test for copper, the sample should preferably be ashed, as their actions are not as sensitive as the iron test and a clear identification on the fibre, especially in small amounts is very difficult.

4.1.4.6   Copper tetramine test

The cooled ash is spotted with 5–10 drops of nitric acid (10%) or hydrochloric acid or by melting with potassium bisulphate (see iron test).

Add dilute ammonia (1:1 water: conc. ammonia) until the ash is alkaline. A blue colour shows the presence of copper. A white back ground, e.g., a sheet of paper, helps in observing the shade changes.

4.1.4.7   Sodium diethyldithiocarbamate test

Reagents:

Hydrochloric acid 20%: 800 g 25% HCl (A.R.) diluted with 200 ml of distilled water.

Ammonia solution 35%: 1:1 conc. ammonia (A.R.): water.

Potassium bisulphate (A.R.)

Citric acid solution 10% (A.R.)

Sodium diethyl dithiocarbamate solution 0.1% (A.R.)

Dissolve the ash as in the iron tests. Neutralise with ammonia solution (1:1 dilution) and bring to pH 6 with citric acid (10%) using 5 drops for each millilitre of test solution. Add further dilute ammonia (1:1) until it is again alkaline. Now add 1 drop sodium.

Diethyl dithiocarbamate solution (0.1%) to the solution. A yellow colour shows the presence of copper.

Note

The test may, however, be carried out on the fabric directly when localised, higher copper concentrations are suspected. The sample is first spotted with dilute nitric acid and then after neutralisation with ammonia, 1–2 drops of the diethyl dithio carbamate solution is spotted on to the same area.

4.1.4.8   Qualitative tests cellulosic damages

Various tests are available for detecting the presence of oxycellulose and hydrocellulose formed when cellulose is chemically damaged. As these materials are not clearly defined chemical compounds the intensity of reaction can vary considerably and to distinguish between oxycellulose and hydrocellulose is quite difficult. The tests can be influenced by impurities (size, wax, finish, etc.) and so the fabric tested should be as clean as possible. Typical reagents and reactions are shown below:

The alkaline silver nitrate test is most useful for detecting localised oxidative damage. The test is carried out as follows:

Prepare two solutions, (a) containing 8 g of silver nitrate in 100 ml of distilled water, and (b) containing 20 g of sodium thiosulphate and 20 g of caustic soda in 100 ml of distilled water. These solutions may be stored as stock solutions, if kept in dark bottles. The test solution must be prepared freshly from these stock solutions for each test, the test fabric should be weighed and about 20 ml of test solution prepared per 1 g of sample. Measure 20 ml distilled water into a beaker and add 2 ml of the solution (b). Add to this slowly, with constant mixing, 1 ml solution (a). It is most important that solution (a) is added to the diluted solution (b). Any precipitate which forms during the mixing should re-dissolved. When mixed, bring the test solution to the boil, immerse the fabric and continue to boil for 5 min with frequent stirring. Remove the sample and wash-off dark patches will be developed where damage had occurred.

Table 4.2. Tests result for cellulose damage after preparation

Test solution	Damaged cellulose
Fehling's solution	Brownish red colour
Nessler's reagent	Yellow then grey colour
Schiff's reagent	Red colour

Alkaline silver nitrate	Grey to black patches
Stannous chloride gold chloride	Purple colour

4.2   Pre-treatment and bleaching of wool and silk

4.2.1   Preparation of wool and woollen blends

4.2.1.1   Wool

The process up routes of woollen fabrics are based on the type of fabric, fabric weight, worsted yarn or normal spun yarn, structure of fabric, etc.

Given below processing routes of some of the fabrics:

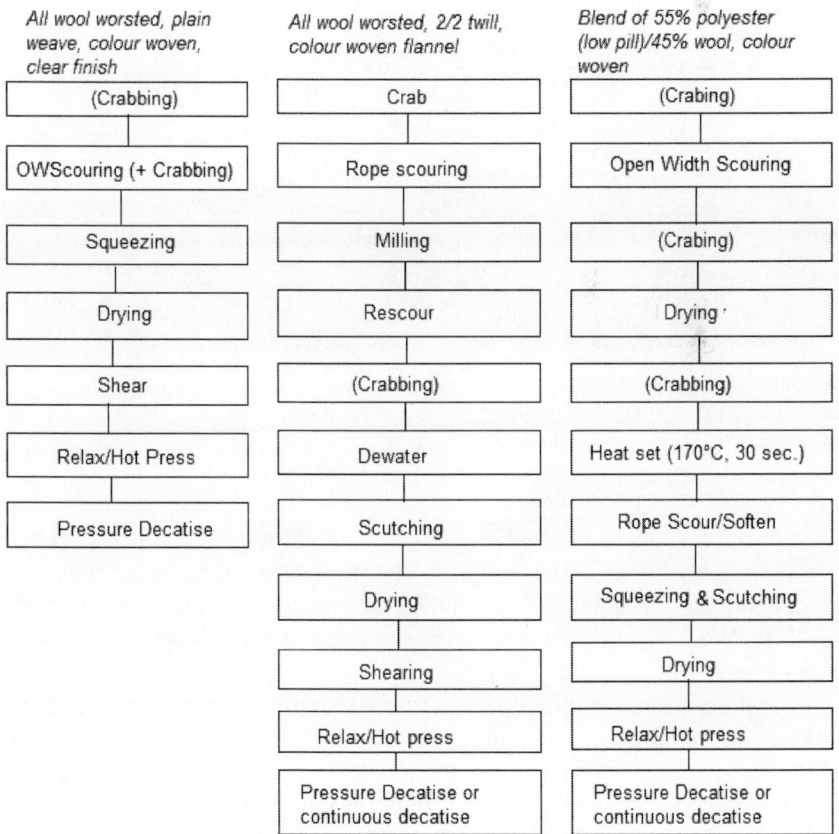

Figure 4.5. Different processing routes for woolen materials

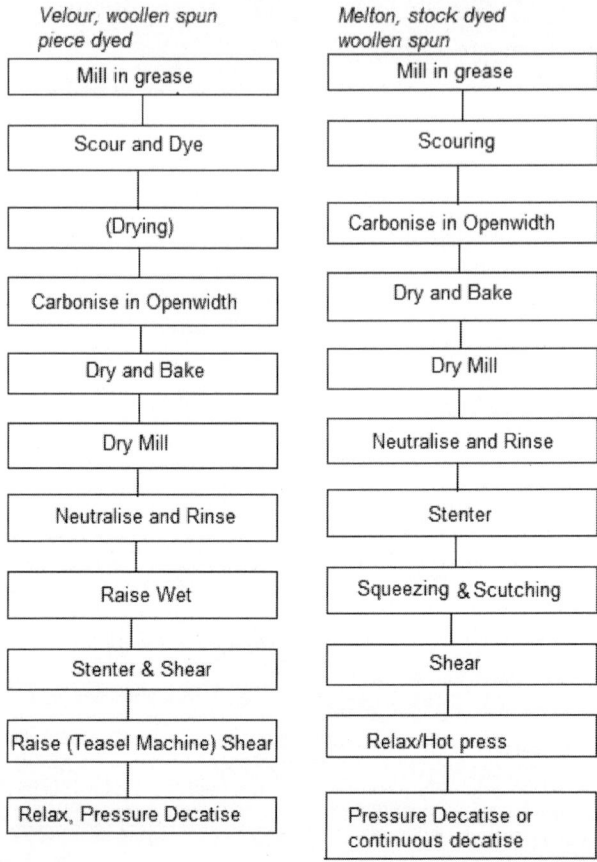

Figure 4.6. Different processing routes for woolen materials

Carbonising of wool

Generally, carbonising of wool is done in the fibre form to remove the foreign matter (mainly of cellulosic origin) from the fibre. It is further removed completely by the combing machine. But sometimes a different approach must be followed in the woollen cycle, where the quantity of contaminants requires a specific treatment with sulphuric acid, to avoid any possible problems during the dyeing process. Carbonising is also essential when the raw stock is mainly composed of rags or waste (dry carbonising with HCl gas at 80°C). In fact, with this type of material the carbonising process eliminates any vegetable residue in the staple after scouring, due to the good resistance of wool to the action of acids. Carbonising can be carried out also on staple fibres, yarns and fabrics. Washed and sometimes piece-dyed fabrics as well as grey fabrics can also be carbonised.

Procedure

Process and systems used: wetting vessels, squeezing cylinders, stenter, dry fulling unit.

Recipe for carbonizing wool

Quantity	Unit	Chemicals
4–6	%	Sulphuric acid
1–2	g/l	Acid stable wetting agent

Temperature 85–90°C.
Time 30-60 min.

The fibres are soaked with sulphuric (2.5–4°Bé) acid and wetting agent, squeezed by means of two cylinders and then dried in a stenter at 85–90°C, for 30–60 min. While drying the acid gets concentrated due to the evaporation of the water, as a result dehydrating and hydrolysing the cellulosic matter. Next, fibres/fabrics are carefully washed to remove completely any residual acidity, which could affect the fibre and subsequent operations. A series of washing processes also includes a neutralisation treatment with sodium acetate. During the fabric process, a dry beating process to remove the carbonised vegetable residues from the fabric texture precedes the washing phase. If the process is correctly carried out it yields faultless results. By the older method, the individual operations were carried out separately, often in different rooms. This procedure may lead to various faults which are mostly only apparent after dyeing. Carbonising after dyeing lessens the danger, but the colours are often altered in shade unless dyestuffs fast to carbonising have been used. The quicker the passage of the goods, the larger should be the addition of wetting agent.

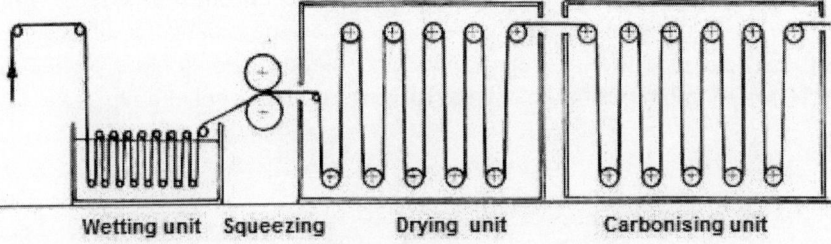

Wetting unit   Squeezing   Drying unit   Carbonising unit

Figure 4.7. Schematic diagram of a carbonising process unit

Notes:

1. On no account should the goods contain residues of soap or lime-soaps. The material should, if possible, be impregnated, carbonised, and neutralised in one continuous run.

2. If this is not possible and impregnated material has to lie for some time before carbonising, it should be plaited down and completely covered with a cloth impregnated with the carbonising liquor, as otherwise stains may be caused by direct sunlight falling on the goods, drops of water, draughts or uneven drying.

3. Stoppage of the machine during the operation and folds or creases in the cloth can produce streaks and fold marks.

4. After carbonising, the goods must be thoroughly washed and neutralised with a soda solution of 3–4°Tw (2–3°Be).

5. If in case the fabric without neutralising is taken for dyeing, it should be dyed at the commencement without acid in case of metal complex dyes where the dyeing is done in the presence of mineral acid.

6. Dyestuffs which require acetic acid to commence dyeing tend to give unlevelled results. Hence such cases it is essential that the goods be neutralised before dyeing.

7. It is necessary that the fabric which are carbonised in piece goods form should not have cotton selvedges (or any other yarn which can be damaged by mineral acid). In case a cotton selvedge is used it can be saved by brushing the selvedge with sodium silicate 42–52°Tw (25–30°Bé) or with a 20% soda solution thickened with chalk. This protective covering is removed in the subsequent washing operations.

Solvent carbonising process

Machine: Special solvent carbonising unit

An alternative method, first the fibre/fabric is soaked in a solvent and then treated with dilute sulphuric acid. The solvent, usually perchloroethylene, due to its low-surface.

Tension soaks the fibres faster and deeper than an aqueous solution. But the vegetable impurities being hydrophilic is not wetted out by the solvent. The solvent soaked fabric is further treated with a aqueous of sulphuric acid, the aqueous solution cannot remove the solvent from the wool and replace it. On

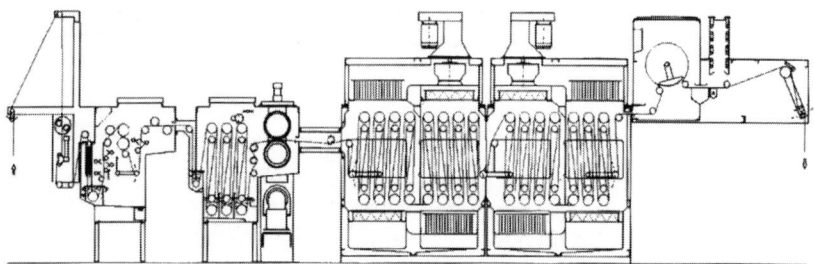

Figure 4.8. Diagram of a solvent carbonising unit

the contrary, the aqueous solution is absorbed by vegetable hydrophilic particles. In practice, with this system vegetable impurities absorb the acid solution selectively, and the acid carries out only a gentle action on wool. The advantage of this process is that it is more environmental friendly due to lower pollution and the possibility of by-passing the acid removal step (or if necessary, this step is considerably easier and faster).

Milling/fulling and washing of wool

Wool and woollen major blends normally undergoes a milling treatment, which is primarily meant for scouring and felting of wool. The conditions required for printing are temperature, moisture and pressure. In a traditional milling operation the fabric (carded or combed) is soaked in water in the presence of surfactants and run in rope form in a milling machine at around 40°C where the fabric is subjected to continuous pressure both warp and weft way, whereby it tends to felt, causing fabric shrinkage and a subsequent dynamic compacting. During the operation, the temperature and mechanical stress is carefully controlled to achieve the desired shrinkage degree. Sometime the fabric is sewn together to a tubular form to avoid irregular tensions on both selvedges. The percentage shrinkage in warp direction (lengthwise) is controlled from the beginning by means of markers positioned on the centre of the piece in warp direction, at one-metre distance and width wise shrinkage is assessed by measuring the width. In the milling machine a component called jaws consisting of vertical parallel steel plates (or older version machines this may be made of wood plate also), positioned in the front part of the machine that make the fabric shrink in the weft direction by squeezing the fabric.

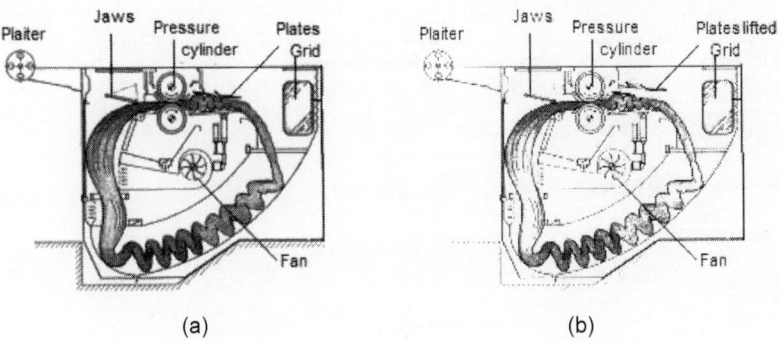

(a)   (b)

Figure 4.9. Milling machine (a) Milling step (b) Washing step

Pressure rolls are adjustable-pressure cylinders that make the fabric shrink in the weft direction and push the fabric inside the box. (Some machine manufacturers have adopted two upper cylinders for feeding the fabric, with a lower rubber cylinder with a rough surface.) A square section tube (Box) where the fabric is packed, slowed down by means of the adjustable plate. In this section the fabric shrinks in warp direction. A hinged plate on top of the box; it can be lowered by reducing progressively its section, slowing down the fabric. There are many variations in the machines by different manufacturers having facilities to improve milling/washing operations.

After the milling operation, the material must be washed to remove dirty water and the chemicals used. In newly designed machines the fabric milling process is often combined with a washing process (sometimes milling is combined with a quick washing process). Washing and partial felting is done by fast washing step, whereby the fabric, drenched with liquor, moves at a speed ranging between 400 and 600 m/min with the plate open, and runs into the grid. Beating, combined with high speed, causes a slight felting on the surface and the yarn swells, as a result hiding the comb marks.

Milling recipes

Recipe for milling wool, wool/PE, wool/PA, etc.

	Scouring stage I	Scouring stage II	Milling	Washing
Soda solution 3°Bé %	1%	0%		0.55%
Soap %	0.60%		0.60%	
Concentrated ammonia	0.50%	0.50%		
Non-ionic detergent		0.20%		0.20%
Time min	10 min	90 min	Variable	20 min

Grease milling of woollen material

Chemicals	Grease milling	First wash	Second wash	Third wash
Soda solution 2°Bé %	100%			
Soda Solution 3°Bé %		58%	17.00%	
Sodium metaphosphate			0.33%	
Non-ionic detergent	1%			0.75%
Treatment time		30 min	45 min	30 min
Temperature		Cold	40°C	40°C
Acetic acid				0.85%

Rapid milling

On combined milling and washing machines.

Recipe for Rapid milling

Quantity	Unit	Chemicals
0.5–0.75	%	Soda ash
1	%	Non-ionic detergent
80	%	Water

Heat the milling liquor to 50°C and distributed through perforated metal funnel mounted on the milling machines.

Crabbing

During the manufacture operation of fabric, the tension applied on the yarn remains within the woven or knitted structure. The fabric has to be relaxed before scouring to remove these strains to prevent distortion (e.g., "cross-footing") during subsequent processing. Crabbing is based on the effect of moisture and heat on the fabric. As stated before, steam, hot and boiling water modify the physico-mechanical properties and the linear dimensions of the fibre. As a result of hydrolysis, its strength is reduced depending on the temperature and duration of the treatment. Under the effect of steam or hot water, as well as the tension applied to the fabric, stipulate the changes occurring in the internal structure of separate fibres. At the same time internal stresses created in the fibres during preceding treatment are relieved and their "temporary" or "full" setting takes place. In addition to the conversion of wool into another modification, heat and moisture fix the disposition of separate fibres inside the fabric and on its surface.

Theory

The effect of steam on strained wool fibre may be in two ways–breaking of existing bond and formation of new bonds.

$$R - S - S - R + H_2O \longrightarrow R - SH + R - S - OH$$
Breaking of existing bonds

$$R - S - OH + R - NH_2 \longrightarrow R - S - NH - R + H_2O$$
Formation of new bonds

There are also many other different types of bonds existing in the wool fibre are given below. Different external parameters can change these existing bonds to achieve the release of stress and relaxation or setting (see below).

Cross links	Mobilising agents
$C = O$ ········· HN *Hydrogen bonds*	Water, steam, DMF, Urea etc.
$C = O$ $O = C$ $CH\text{-}CH_2 \text{ - S - S - } H_2C \text{ - } CH$ NH NH *Disulphide links*	Thioglycollic acids, reducing agents, Steam at high pressure or steam at atmospheric pressure for longer periods
$C = O$ $O = C$ $CH\text{- R1 - } COO \cdots NH_3 \text{ -R2- } CH$ NH NH *Electrostatic bonds* *R1, R2 - Hydrocarbon units*	Water or steam
$C = O$ $O = C$ $CH\text{- R3} \cdots \text{ R4- } CH$ NH NH *Hydrophobic bonds* *R3, R4 - Hydrocarbon side chains*	Organic solvents like alcohols, DMF etc.

Figure 4.10. Mobilising agents for changing the existing bonds in wool

In the above chart you can see that the water and steam is a common factor for mobilising most of the cross-links. Hydrogen bond breaking and reforming is an accepted and known subject. As we have seen in the structure of wool fibre (The substrates–Series I of the Handbook for Textile Processor by same author) wool is made of a matrix and micro-fibrils. The crystalline substance of the micro-fibrils is embedded in the matrix, the amorphous substance. The two morphological components, differ distinctly in sulphur content. During the kier decatising, it is assumed that this disulphide exchange, the splitting of disulphide bridges, takes place first in the matrix by means of water as a kind of preliminary stage. The so-called micro-fibril component is plastically deformable, i.e., these micro-fibrils are particularly stabilised by hydrogen bridges. Steam can actually now attack these hydrogen bridges, which were previously masked by a complicated system of disulphide bridges. The important step therefore is that the micro-fibrils are deformed: then by cooling, oxidation and the blowing through the air, this set micro-fibril condition is re-locked by the reorganised disulphide bridges in the matrix.

Crabbing influences subsequent dyeing and finishing processes and makes it possible to prevent many faults in fabrics. Faults are most often observed in worsted fabrics because of the special properties of the raw materials used in their manufacturing, their gaiting characteristics, and methods of treating. Moreover, a clean surface and accurate weaving pattern are of the greatest importance for these fabrics. Because of the complicated mechanical action to which separate fibres and subsequently yarns are subjected in the process of spinning and weaving, the internal stresses in different parts of worsted fabrics are usually not sufficiently uniform. This non-uniformity, which is usually not visible in grey cloth, becomes highly pronounced in subsequent treatment and particularly after dyeing, since fibres under stress are dyed more intensively than non-stretched fibres.

The non-uniformity of grey cloths is eliminated by crabbing, because all the internal stresses in separate threads and parts of fabric are eliminated, this being one of the major aims of crabbing. Crabbing fixes in place the fibres on the fabric surface so that in subsequent treatment they are able to withstand mechanical action and to retain the initial direction.

Fixation of the fibres inside the material and on its surface has one more aim. It impedes fabric shrinkage and formation of a surface coverage in subsequent processing and consequently contributes to the preservation of the structure of worsted fabrics and of the weaving pattern. Crabbing also assists in fabric cleaning as in the process of crabbing size and other soluble impurities are removed.

The crabbing operation is carried out in the presence of heat and moisture, during which the intermolecular bonds in wool are broken and then reformed in a more relaxed configuration. Relaxation and setting happens in the presence of moisture, temperature and pH. Crabbing is usually done by batch process, but nowadays continuous crabbing also is practiced.

Batch wise crabbing

The crabbing machine consists of a long rectangular bowl with a pipe for water supply and a hole for drainage. At the bottom, the bowl is provided with a steam coil for water heating. In the larger section of the bowl separated by partition and there is an even number of rollers (usually six rollers) which during operation must be immersed into water to half of their diameter. After filling the bowl with water and heating it to boiling point the fabric in open width form is batched on to a roller (cylinder) which is covered with a cotton wrapper and half immersed in hot or boiling water. Then this roller is

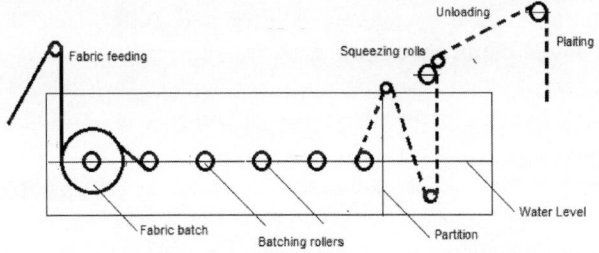

Figure 4.11. Diagramatic representation of Crabbing machine

connected to the brake to create tension and the fabric is reversed and rolled over the second roller, then over the following roller, and so on, this operation being continued until the present number of passages is achieved. The average speed of fabric motion is 30–40 m/min. This operation is carried out continuously, according to a certain sequence, i.e., as soon as the first roller is empty, a new lot of fabric is wound upon it. The fabric is next steamed and finally passed through the last partition where cold water is stored to stop the setting process. Then it passes through a couple of squeezing rolls, and plaited. The temperature and the level of the water in the bowls are automatically regulated. The duration of treatment may can be adjusted by modifying the number of pieces in the lot simultaneously wound on the roller or by changing the number of rollers through which the fabric passes depending on the style of the fabric to be treated. The number of rollers should be always even, as otherwise each of the ends of the fabric lot will get inside the roll and onto its surface an uneven number of times, so that the action of boiling water on various parts of the pieces will be non-uniform, and this will result in a different intensity of colouring. Usually crabbing in this machine lasts for 30–45 min. The optimum pH for crabbing has been shown to be around 6.

Notes

1. During crabbing the fabric should be wound without wrinkles, creases or misalignment, with a constant tension, as otherwise hard creases may be formed in the process which are very difficult to eliminate later on.
2. The present routine of crabbing is to be strictly adhered to, since the action of hot water for a longer time than necessary may cause deterioration of the fibre or reduction of its strength. This is also necessary to avoid variance in the dyed lot when two or more crabbed rolls are dyed together.
3. It is also necessary to take into account that alkali coming from the fabric gradually accumulates in crabbing bowls and may cause wool deterioration at boiling temperature.
4. Minding this, a systematic control and adjustment of the pH of the medium is necessary. Despite reversing the fabric direction, traditional batch crabbing machines remain associated with uneven treatment which, for piece dye qualities, can result in end-to-end or piece-to-piece variation. Improvements on traditional machines include the use of large crabbing cylinders and bigger treatment lots, minimising variation.
5. Absolute control of loading tensions, centring devices and anti-slipping compacting rollers, ensure uniform package build-up.

6. Despite reversing the fabric direction, traditional batch crabbing machines remain associated with uneven treatment which, for piece dye qualities, can result in end-to-end or piece-to-piece variation.

7. Improvements on traditional machines include the use of large crabbing cylinders and bigger treatment lots, minimising variation. Absolute control of loading tensions, centring devices and anti-slipping compacting rollers, ensure uniform package build-up.

4.2.1.2   Continuous crabbing

Now crabbing is usually conducted as a continuous process (Figure below), combining crabbing with washing. Even though continuous, the treatment time is generally quite short and the levels of set can be lower than those attainable with batch systems. Two basic types of continuous crabbing machine are available.

Cylinder type: After wetting out in hot water, the fabric in open-width form is passed around a large, rotating, heated cylinder. The fabric is pressed at high pressure against the heated cylinder by a specially engineered impermeable belt. Special seals resist escape of steam and entry of air at the edges of the belt. Fabric operating temperatures as high as 135–140°C are claimed and superheated steam is created in situ, setting the fabric. Setting is arrested by shock cooling. Chemical setting agents are sometimes added to the wetting tank to promote higher levels of set.

Superheated water machines: In this type of machine super-heated water and cylinder is having no pressure belt to maintain fabric/cylinder contact. Yarns are allowed to swell with minimal fabric compacting. The fabric enters and exits through barometric columns. Water temperatures are around 110°C, although a series of steam battery heaters situated around the main cylinder are claimed to elevate the fabric temperature during its contact time with the cylinder, promoting fabric setting.

4.2.1.3   Scouring wool and woollen blends (WO, WO/PES, WO/PA)

Wool scouring is mainly aimed at the removal of spinning lubricants and soil. Normally it can be done with a detergent, but if the wool is heavily stained sometimes solvent based scouring agents and storing and further washing in water. Scouring is more important where the wool is blended with fibres like cotton which needs much severe scouring treatment but cannot be done in the presence of wool, especially alkaline scouring. A weakly alkaline bath maintained at pH 8.5–9.0 with the addition of ammonia or soda ash along with an anionic and non-ionic detergents are used for scouring wool and wool/blend fabrics. Non-ionic detergents since the pH is not critical may not need an alkaline bath, but it is always better to keep the alkaline pH, especially in case of blends with cotton. A usual scouring treatment may be with 1% (owf) of the detergent at 40–50°C for 40–60 min for batch machine depending on the length of the batch.

A milling machine which is described above can be used as scouring machine in the batch process. The fabric is stitched into an endless loop after loading through the nip rollers which also acts as fabric transport throughout the process. Scouring chemicals are taken in the trough and run at stipulated temperature at a lower speed of about 200 m/min by which scouring times are reduced and fabrics are more effectively opened before the nip, reducing the tendency for processing marks especially creases. Machines with facilities to open the fabric in each turn of the fabric, baffle plates to facilitate milling operation.

4.2.1.4   Continuous scouring

Rope scouring of light weight fabric often causes creases on the fabric which will be hard to remove. Open-width continuous scouring (some machines with slight milling facility are available) machines are designed especially for such fabrics. In an eight compartment scouring range, the first three or four boxes are filled with weak alkaline solutions of detergents with a temperature of 40–60°C, which ensures desizing and cleaning of the fabric from dirt and contamination. In the last scouring boxes the fabric is washed with warm and cold water. The scouring boxes are provided with live and blind steam lines for heating up the liquors. Some machines have tangential jets of scouring and rinsing liquors, followed by squeeze rollers or suction slots. There are many machine designs like complete immersion systems with liquor interchanged by sucking large volumes of scouring solution through the fabric,

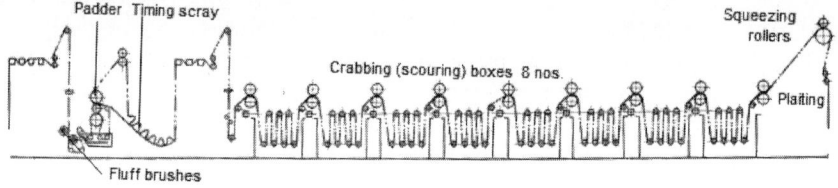

Schematic representation of continuous scouring machine

Figure 4.12.

suction drum type systems with varying flow direction etc. Driven guide rollers, relaxation zones or overfeeding onto drums ensure minimum processing tensions, important for lightweight fabrics and fabrics containing elastane filaments. Continuous open-width scouring machines are often used in-line with crabbing tanks to effect cleaning and setting of fabrics. (Some machines are provided with an enzyme impregnation and timing scary for the soaking and enzyme action, for fabric which enzyme desizing–see above figure.)

Recipe for scouring wool and woollen blends

Chemicals	Wool	Wool/PA, Wool/PE, Wool/PAC, Wool worsted
Non-ionic wetting/detergent g/l	0.3–1.5	0.5–1.5
Soda ash %	1–4	
Ammonia* ml/l	1–2	1–2
Formic acid ml/l		1–2
Temperature °C	40	30–40
Time min	30	20–30

*When mineral oil lubricants are used.

4.2.1.5   Discontinuous scouring and washing (alternate)

Guideline recipes

Form	Woollen yarn		Worsted yarn	
Substrate	Wo		Wo, Wo/PES, Wo/PAC, Wo/CA	
Type of treatment	Alkaline		Alkaline	Acid
Guide recipes	Olein lubricants	Mineral oil lubricants		
Non-ionic wetting agent ml/l	0.3–1.0	0.5–1.5	0.3–1.5	
Soda ash %	1.0–4.0			
Ammonia 25%		1.0–2.0	1.0–2.0	
Suitable scouring agent				0.3–1.0
Formic acid 85% ml/l				1.0–2.0
Working method				
Temperature °C	40	40	30–50	50
Treatment time min	30	30	20–30	20–30

4.2.1.6   Discontinuous bleaching

Guideline recipes for discontinuous bleaching

	Wool	Wool/Polyamide, Wool/Polyester, Wool/PAC, Wool worsted
Non-ionic detergent ml/l	0.3–1.5	0.3–1.5
Soda ash % OWF	1–4	1–2
Ammonia ml/l	1–2	1–2
Formic acid 85% ml/l		1–2
Temperature °C	40	30–30
Time min	30	20–30

Ammonia is used when mineral oil lubricants are used.
Liquor ratio 10:1–40:1.

Discontinuous hydrogen peroxide bleach for dyeing. of wool and its blend with synthetic fibres (WO, WO/PES, WO/PA, WO/PAC)

Guideline recipes: Bleach for dyeing

Materials	Loose fibres, slubbing, yarn woven and knit goods	
All normal Blg and dyeing M/cs (MLR 5:1–30:1)	Normal bleach	Rapid bleach
Guide recipes ml/l		
Stabiliser %	0.5–1.0	0.5–1.0
Trisodium phosphate crystals g/l	0.2	
Sodium tripolyphosphate g/l		0.4–0.5 (pH 7.5)
Ammonia 24% ml/l	0.75% (pH 9)	
Non-ionic detergent (e.g., Sandozin NIT)	0.5 ml/l	0.5 ml/l
Hydrogen peroxide 35% ml/l	10–20	10–20

Working method

Treatment temperature °C	50	70–80
Time h	5 (overnight)	1

4.2.1.7   *Discontinuous peroxide bleach for full bleach*

Material–Wool and its blends –WO, WO/PES, WO/PA, WO/PAC
After pre-bleaching with the previous recipe treat as follows:

Recipe for discontinuous peroxide bleach for full bleach

Materials	Loose fibres, slubbing, yarn, woven and knit goods
Machine	All dyeing and bleaching machines–MLR 5:1–30:1
Guide recipes	
Hydros g/l	2–5
Low foaming wetting agent ml/l	0.5–1
Working method	
Treatment temperature	
With optical brightener	60–70°C
Without optical brightener	60°C

Materials	Loose fibres, slubbing, yarn, woven and knit goods
Treatment time	60–90 min
Rinsing	
With optical brightener	Cold rinse only
Without optical brightener	At 40°C and cold, possibly with 0.5 ml/l H_2SO_4 conc. in cold for 15 min

4.2.1.8 Bleaching wool and its blends with synthetics

Recipe and process for bleaching wool and its blends with synthetics

Form	Loose fibres, tow, yarn, woven and knit goods	
Substrate	Wool, Wo/PES, Wo/PA, Wo/PAC scoured	
Machine	All bleaching and dyeing Mcs	
MLR	5:1–30:1	
Type of treatment	Normal bleach	Rapid bleach
Guide recipes		
Stabiliser ml/l	0.5–1	0.5–1.0
Tetrasodium pyrophosphate g/l	2	
Sodium tripolyphosphate g/l		0.4–0.5 (pH 7.5)
Ammonia 24% ml/l	0.75 (pH 9)	
Wetting agent (non-ionic) ml/l	0.5	0.5
Hydrogen peroxide 35% ml/l	10–20	10–20
Working method		
Temperature °C	50	70–80
Time h	5 (possibly overnight)	1
Rinse warm and cold		

Notes

1. Used bleach liquors can be reused up to 5 times when strengthened with 20–25% of the original amount of hydrogen peroxide 40–50% of the original amount of stabilizer and 100% of the original amount of sodium tripolyphosphate (pH 7.5).

Additional discontinuous reductive bleach for full white after the hydrogen peroxide bleach for wool and its blends with synthetic fibres

Recipe

Form	Loose fibres, tow, yarn, woven and knit goods
Substrate	Wool, Wo/PES, Wo/PA, Wo/PAC pre-bleached with hydrogen peroxide
Machine	All bleaching and dyeing Mcs
MLR	5:1–30:1
Type of treatment	Normal bleach
Guide recipes	
Arostit BLN gran	2–5 g/l
Low-foaming wetting agent	0.5–1 ml/l
Working method	
Treatment temperature	0.75 (pH 9)
Without OBA	60°C
With OBA	60–70°C
Treatment time	60–90 min
Rinse	
With OBA	Rinse cold only
Without OBA	at 40°C and cold with 0.5 ml/l H_2SO_4 cold for 15 min

4.3   Silk

Silk is the finest and most valuable natural fibre. All its beauty and fascinating properties are fully brought out by refined weaving/knitting constructions and wet-processing. There is a general tendency to use silk for a wider range of use. Previously confined to evening wear and luxury goods, nowadays it is increasingly found in summer and winter fashion collections e.g., blouses and shirts, sportswear, underwear, etc.

Main production of silk and silk fabrics comes from China. Other major producers are Japan, India, Russia, South Korea and Brazil. If one checks the history prior to 1957 Japan was the main producer and consumer of silk. Today still they remain the main consumer but not producer. They import silk for domestic consumption.

4.3.1   Structure of silk

Silk is one of the animal fibre and show similar chemical characteristics of the natural animal protein fibre, wool. True (mulberry) silk is spun into continuous threads about 2500 m long on average. These threads consist of two separate individual strands ("brines") made of fibroin which are covered and bonded together by a gum named "sericin".

4.3.2   Chemical composition of mulberry silk

Chemical composition of mulberry silk

Fibroin	70–80%
Sericin	20–25%
Waxes and fats	0.4–0.8%
Moisture	10–11%
Carbohydrates and starches	1.2–1.6%
Inorganic matter	0.7%
Pigments	0.2%

Fibroin and sericin both are protein or protein mixtures with different amino acid composition: fibroin contains mainly glycine, alanine, serine and lyrosine whereas sericin consists of serine. Since silk does not contain sulphuric containing amino acids and sulphate bonds as in wool, it is resistant to alkalis and to attack by moth and beetle larvae. Like wool silk is amphoteric and has an isoelectric point at pH 5.1. Swelling and reactivity are lowest in the electric region of pH 4–5. Silk has dyeing properties of its own arising from the absence of both scales and sulphur containing amino acids.

Tussah silk which is raised on oak and other leaves are the most important wild variety of silk. X-ray diagram indicate a lower degree of crystalline orientation than mulberry silk besides, major differences in macro- and micro-structure. In mulberry silk the fibroin strand has a rounded triangular cross-section whereas the tussah silk has a wedge shaped fibre cross-section. With its brownish to yellowish green colour and lower sericin content of 8–15% the Tussah silk has different chemical and physical characteristics than mulberry silk and hence its dyeing and finishing characteristics are also different.

Silk has a sheen because of its smooth surface and lacking of the scale structure of wool. Also silk is sensitive to mechanical loads, which reduce the sheen of the yarn and leave irreparable abrasion, chafe, or rub marks. For that reason silk must be handled very gently in all processing operations.

A typical silk preproduction flow chart is as follows:

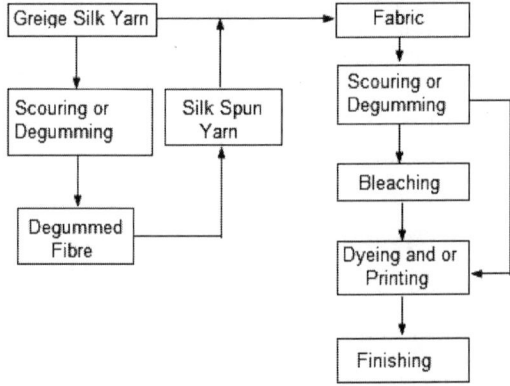

Figure 4.13. Process routes for Silk

4.3.3   Preparation

Silk as produced by the larvae and spun out from the cocoon has a yellowish to yellowish green colour, lacks the lustre and is harsh and straw like to touch. This is due the sericin cover of the actual silk fibre. The first step is to remove this and expose the actual silk fibre out to get its natural sheen and soft structure. Thus the first preparation step in case of silk is the removal of servicing which is not a part of the fibre. This process is termed as degumming. Degumming process is the removal of sericin from the fibroin. It is equivalent to the scouring process used for the purification of cotton and wool. The process is generally done by the silk producer. Commercially partly degummed silk (Souple) and fully degummed silk (cuit or bright) available in the market. Removing the gum reduces the weight of the mulberry silk to the extent of about 25% and of tussah silk by about 15% but provides the full luxurious sheen which is further developed by finishing process.

4.3.3.1   Cleaning and degumming

As sericin is insoluble in cold water it has to be removed by hydrolysis–which is probably easiest way–or by any other method. Alkali breaks down the long protein molecules into smaller fragments easily dispersed or solubilised in hot water. The hydrolysis of proteins can be carried out by treatment with acids, alkalis and enzymes. Acids are non-specific and tend to attack vigorously. Alkalis also attack both sericin and fibroin. As the rate of hydrolysis of sericin and fibroin is sufficiently differed to control the process in such a way as the fibroin is not affected. Thus, the process of degumming with soaps in the presence of mild alkali like soda ash is the most commonly followed process.

4.3.3.2   Degumming with soap solution

Since soap containing no alkali and soap having greater capacity to hydrolyse (e.g., Olive oil soap) is more effective in degumming points on the principle of action of soap in degumming. The alkali formed by the hydrolysis is supposed to form a chemical bond with sericin and produces soda salt. Further the swollen sericin is separated by the soap and dissolves in water due to the emulsification action of the soap.

Generally, 20–30% soap to the weight of silk to be degummed at MLR of 30–50 is sufficient for degumming. Degumming is usually done for 60–120 min to boil. Action of soap on silk is very mild and safe.

Soap containing no alkali is the main agent used for degumming. It hydrolyses the gum without forming a gel and is readily washed-off. The classical method of degumming mulberry silk.

30–40% neutral soap (o.w.g) in soft water.

Kept at boiling (90–95°C).

Time of treatment 2–4 h depending on the type of silk.

Recipe

Quantity	Unit	Bath additions
3–5	g/l	Neutral soap
0.5	g/l	Sodium hexameta phosphate

MLR 1:50

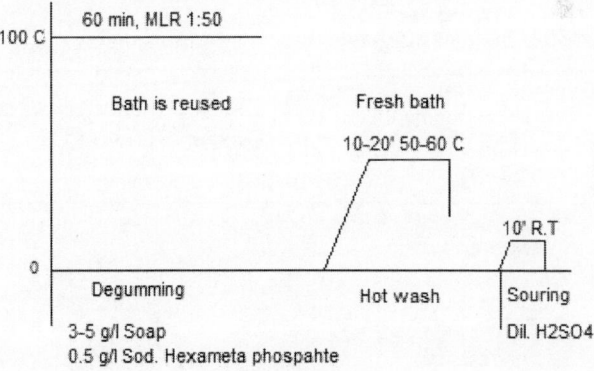

Figure 4.14.

Goods entered in the hot bath and boiled till the full sericin is removed. If necessary the treatment is repeated in another bath and washed well in a slightly ammonical bath. For tussah silk with its lower sericin content an alkaline boil-off with soap soda can be done.

Notes

1. The degumming done at 90°C has a superior colour and feel when compared to that treated at boiling temperature.
2. Hard water can reduce the effectivity of soap. Hence, it is a must that soft water has to be used.
3. When metal salts are present in the water, soap can form insoluble complex with them and it can deposit on the degummed fibre, thereby reducing the handle, lustre and colour of the fibre.
4. The residual soap in the silk causes problems such as uneven dyeing, yellowness and embrittlement.
5. Soap required for degumming is quite high also it is not environmental friendly.
6. The quantity of soap can be reduced in a soap and alkali combination in degumming with superior results and less susceptible to hard water and metal salts. Undoubtedly, alkali usage is very important due to obvious reasons and the alkalis used are sodium silicate, sodium sesqui carbonate, sodium carbonate, sodium phosphate, sodium hydroxide and a buffer mixture of sodium carbonate and sodium bisulphite.

Recipes for mulberry silk (soiled and difficult to degum)

Process: Cleaning → Steep →Boiling-out 1 and 2 → Rinse warm → Rinse warm → Rinse cold.

Cleaning (For soiled goods)

Guideline recipe for cleaning solution

Quantity	Unit	Chemicals
2	g/l	Ultravon HD (Ciba) or equivalent
3–5	g/l	Silvatol FL (Ciba) or equivalent

30–60 min at 60°C.

4.3.3.3   Degumming

a. Steep, if necessary

Guideline recipe for degumming silk

Quantity	Unit	Chemicals
2	g/l	Silvatol FL (Ciba) or equivalent
2	ml/l	Ammonia 25%

pH at 9–9.5, soak overnight at 40–50°C without rinsing.

b. Boiling out treatment 1 and 2

Recipe for degummimg silk

Quantity	Unit	Chemicals
15	g/l	Neutral soap

pH max 9.5.

1st bath 2–4 h at 98°C.

2nd bath 2–4 h at 98°C.

c. Rinsing

1st bath

1 ml/l Ammonia 25% for 20 min at 50°C.

2nd bath

1 ml/l ammonia 25% for 20 min at 50°C.

3rd bath rinse cold

3rd bath rinse cold

Recipe for tussah silk

Quantity	Unit	Chemical
2–5	%	Neutral soap
4–8	%	Soda ash

pH at 9.5 boil for 1–2 h.

2nd bath

Recipe

1st bath

Quantity	Unit	Chemical
2–5	%	Neutral soap
4–8	%	Soda ash

pH at 9.5 boil for 1–2 h.

Use of soap for degumming has come uneconomical and researches was undertaken by various auxiliary manufacturers to find out cheaper alternatives.

4.3.3.4   Other degumming methods

Other than degumming with soap the alternate process considered are as follows:

- Alkaline degumming
- Synthetic soap degumming
- Enzymatic degumming
- Acidic degumming
- Organic amine degumming.

Alkaline degumming

In alkaline degumming, it hydrolyse proteins by attacking the peptide bonds. The amino acids reported to be mainly affected are cysteine, serine, threonine and arginine. It is highly dependent on pH, and of course temperature and duration of the treatment. The degumming process below pH 8.5 the removal of sericin is slow and ineffective and pH above 10.5 the action of alkali is severe and it can affect fibroin causing high degumming loss. Hence, the ideal pH range is 9.5–10.5. Hence the process has to be carried out under controlled conditions to avoid over degumming. The commonly used alkalis are: sodium hydroxide, sodium carbonate, potassium carbonate, sodium silicate, trisodium phosphate, sodium pyrophosphate, sodium bicarbonate, borax and ammonia. Sodium carbonate is the preferred alkali. As the pH has to be controlled it is advantageous to use a buffer alkali than a single alkali. The following combinations are found suitable: potassium carbonate–sodium bicarbonate, sodium carbonate–sodium bicarbonate, disodium hydrogen phosphate–trisodium phosphate and potassium tetraborate–boric acid. Carbonate –bicarbonate combination is capable of maintaining a pH of 10.2 and hence more acceptable. An addition of a surfactant improves the performance whereas presence of any electrolytes adversely affect the degumming process. The addition of metal-chelating agents such as $Na_3HP_2O_7 .H_2O$ (sodium hydrogen pyrophosphate) prevents damage to the silk during degumming with alkalis (protective chemical).

Degumming with synthetic products

P–400 (Hepatex, Mezzera) developed by Hepatex AG (Switzerland) and Silk Research Laboratory in Lyons, France. It is based on a synthetic soap from organic and inorganic materials. These chemical allows controlled degumming of all qualities of silk on all machines without any damage to the fabric. Miltopan SE (Henkel) is a liquid alkaline mixture of polyacrylic acid, fibre protectant, sequestering and wetting agent, buffer salts, alkali.

These chemicals penetrates sericin very quickly and evenly and hydrolyses it completely. Addition of fibre protective chemicals helps to preserve the fibre without any damages. It can be used in conventional machines like star frames, winches, jigs, jets, etc. The degumming bath can be used several times makes it further economical. Further, one bath degumming and bleaching or degumming and dyeing also can be practiced. Another advantage of these products are large batches of silk can be degummed continuously on open-width washing machines. This method is particularly attractive for printers.

Another method of degumming consists of treating with tartaric acid or citric acid at elevated temperatures. It has been found that complete degumming with no apparent strength loss can be achieved. See table below:

Recipe

Method of degumming	% Weight loss	Tensile strength Deca Newton = 1 Kilopond		Elongation	
		Actual	% Relative	Actual	% Relative
10 g/l Marseilles soap					
1 g/l Soda ash					
1 g/l Seq. agent					
60 min at 98°C	17.5	8.25	94.4	53.9	90
10 g/l Marseilles soap					
1 g/l soda ash					
1 g/l Seq. agent					
60 min at 98°C	25.3	8.34	95.4	44.8	74.8
Water only 20 min 130°C	18.4	8.49	98.7	59.9	89.3
8 g/l tartaric acid , 60 min 98°C	18.6	8.57	98	55.3	92.3
8 g/l tartaric acid 20 min 130°C	25.6	8.41	96.2	47	75.5
8 g/l citric acid 20 min 130°C	21.6	8.47	98.5	45.7	68
Raw yarn		8.74	100	59.9	100

Silk can be degummed as yarn in an in hank or wound package form or as piece goods. The effect must be even, otherwise levelling problems can arise during dyeing. The machinery has to be selected as per practical experience, type of silk, quantity to be degummed. Star frame and tank degumming is one of the oldest and versatile method of degumming even today mainly because it is very gentle on all qualities and give even treatment.

After degumming the material can be dried without tension to avoid any distortion on festoon driers.

Enzymatic desizing

Even though soap boil, soap alkali boil and alkali boil degumming is the most commercially practiced degumming process, there are many disadvantages main among it being the effluent load and difficulty in controlling the process. The alternative method to select was the eco-friendly enzymatic desizing.

Commonly used enzymes for degumming are trypsin, papain and bacterial enzymes, which are capable of degrading proteins, producing polypeptides, peptides and other substances by hydrolysis of the –CO–NH– linkage. They mainly attack lysine or arginine and the amino group of the adjacent amino acid residue in sericin. Even though both silk fibroin and sericin are proteins we can select enzymes such as proteolytic enzymes (e.g., chymotrypsin) which can hydrolyse the mainly amorphous sericin without much difficulty but not the highly crystalline fibroin. Because sericin and fibroin contain different proportions of some important amino acids, the enzymes attack them to different extents. Investigation has shown that alkaline proteases performed better than neutral and acidic proteases in selective degradation of sericin in terms of weight loss up to 24% and has a linear relationship with the amount of alkaline protease used during the enzymatic degumming. Therefore the enzymatic process can be controlled through the enzyme dosage and treatment time. After enzyme treatment, the enzyme is inactivated at 80°C for a short time and this enhances dissolution of the partially hydrolysed sericin fractions from the silk. Even highly twisted yarn fabric such as crepe' can be degummed completely using enzymes. A certain level of mechanical agitation during the degumming process might be needed to enhance enzyme penetration and to make sericin removal complete.

Recipe for enzymatic degumming

Quantity	Unit	Bath additions
1 or 2	% (owf)	Enzyme
0.5	%	Wetting agent
		Sodium bicarbonate to pH 8.5–9.0

MLR: 1:10.

Temperature: 55°C.
Time: 45 min.

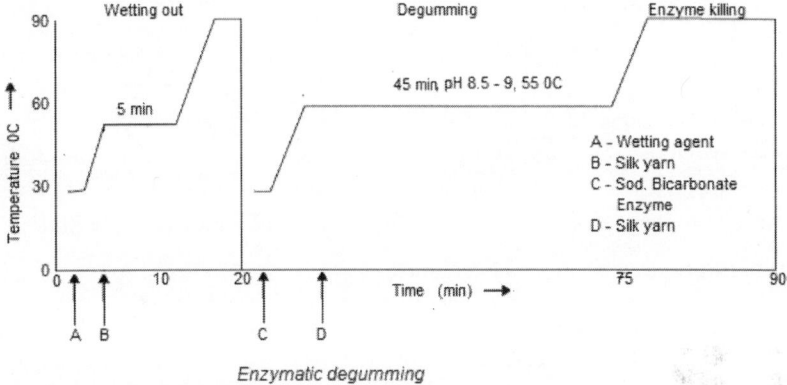

Enzymatic degumming

Figure 4.15. Process for enzymatic degumming

pH: 8.5–9.0.

Notes

Since the enzyme doesn't affect silk adversely, we can safely increase the concentration in degumming bath from 1% to 2%.

Temperature above 55°C renders the enzyme inactive.

Alternate process

Treat the material with a weak alkaline solution of sodium silicate or sodium carbonate.

Degum with the following recipe

Guide recipe for alkaline enzymatic degumming

Quantity	Unit	Chemicals
4	% (owf)	A mixture of $NaHCO_3$:Na_2CO_3:Na_2SO_3 in 2:2:1 ratio
0.3	%	Alkali protease (Trypsin)

Temperature 50–60°C.
Time 10–20 min.

Note

1. This process is supposed to give better results than the previous recipe.
2. Papain enzyme also is used for degumming. A typical process is to first soak the material in hot water. Further, perform degumming with

papin enzyme at 50–70°C in the presence of sodium hydrosulphide, sulphoxylate or sulphite.

Acidic degumming

The acidic degumming is based on the hydrolysis of the proteins of some amino acid residues. Since these are found in greater proportions in sericin than in fibroin selective acidic hydrolysis of the amino acid residues in fibroin can be used for removing sericin from fibroin. Acids used are sulphuric, hydrochloric, tartaric and citric acids. Organic acids are not very effective. Tartaric and citric acids give good results. A typical recipe is as follows:

Guideline recipe for acidic degumming

Quantity	Unit	Chemicals
0.05	mole/l	Acid
3	g/l	Non-ionic detergent

Temperature–Boil (100°C).
Time–60 min.

Degumming with bacterial enzymes

A bacterial enzyme alcalase (NOVO) can completely hydrolyse sericin in 1 h at 60°C and pH 9. This degumming is found more effective than trypsin and papain and even the physical and chemical properties of the degummed material are also better. Another bacterial enzyme named Degummase (Advanced Biochemicals, India) is also a good but effective at 50°C and a pH of 8.75.

Note

1. These enzymes are safer but gives better results and then chemical degumming. However higher concentrations and longer treatment can affect fibroin tendering the material weaker.

A comparison of the silk degummed by traditional method and enzyme method is given below:

Degumming results of different methods of desizing

Property	Soap-soda method	Enzyme method with 1%	Enzyme method with 2%
Weight loss (%)	25	19.5	23.2
Whiteness index	66	60.5	65.8
Breaking load (g)	336.2	342.3	345.2

Elongation (%)	17	17.8	16.9
Tenacity (gpd)	4.98	5.21	5.18

4.3.3.5   Testing of degumming efficiency

The degree of degumming can be checked by

- Assessing the handle
- Checking weight loss
- Staining
- Microscopy
- The ninhydrin reaction.

A simple staining test (Knot et al.) using C.I. Direct Red 80 gives a qualitative assessment of the degree of levelness of gum removal. The test is based on the fact that C.I. Direct 80 stains the sericin but not fibroin. We perform this test by performing a dyeing with the following recipe.

1 g/l Solophenyl Red 3BL 140% (C.I. Direct Red 80)
Liquor ratio 200:1.
Temperature 100°C.
Treatment time 2 min.
Rinse cold.

4.3.4   Bleaching

Dyeing is done directly on degummed fabric and hence bleaching is usually done for full white requirements. This is true in the case of mulberry silk, which is only cream colour after degumming. But as explained earlier in case of Tussah silk the degummed fabric/yarn looks almost light brown and cannot be dyed directly for bright shades. Full white is not possible in case of Tussah silk.

For full whit fabric/Yarn oxidative bleach is given first and then OBA is applied from an oxidative bleach bath.

4.3.4.1   Oxidative bleach

Procedure

Recipe I (using hydrogen peroxide)

Quantity	Unit	Chemicals
10–20	ml/l	Hydrogen peroxide 35%
X	g/l	$Na_2P_2O_7$
1–2	ml/l	Peroxide stabiliser

MLR 1:30.
Enter goods at 50°C.
Raise bath temperature to 80–85°C over 60–90 min.
Leave overnight in the cooling bath.

Recipe II (using sodium peroxide)

Quantity	Unit	Chemicals
2–4	% (owf)	Sodium peroxide
0.5	% (owf)	Magnesium sulphate
3	% (owf)	Sodium silicate
2–4	% (owf)	Sulphuric acid
0.5	% (owf)	Sodium bicarbonate

Treat with the above recipe at an MLR of 1:30 at 60°C for 4–5 h.

4.3.4.2   Reductive bleach
Procedure

Guideline recipe for Reductive bleach

Quantity	Unit	Chemicals
2–4	% (owg)	Stabilised hydrosulphite (e.g., Blankit IN–BASF)
1–2	% (owg)	Formic acid (85%)

Enter the goods cold and raise the temperature of the bath to 60°C over 30 min. Following this, the material is thoroughly washed.

4.3.4.3   Bleaching with sulphur dioxide
Bleaching of silk materials can be done by hanging degummed silk goods in an atmosphere of sulphur dioxide gas or by immersing in a solution of sodium bisulfite in water for 4–6 h. Approximately 5 kg of sulphur is required to bleach 100 kg of silk.

This method of bleaching is not generally practised as it causes air pollution.

For full white a suitable OBA can be added to the reduction bleach bath as per manufacturer's recommendations. Various OBAs are available in the market with different tone and fastness properties and can be selected as per customer requirements.

Alternate recipes for discontinuous degumming scouring and washing

Machines	Yarn dyeing machine, Winches MLR 10:1–30:1			
Silk quality	**Degummed**	**Souple'**	**Schappe and Bourette**	**Tussah**
Procedure				
Wet-out				2–3% wetting agent like Sandopan DTC, 40°C, 30–45 min
Soak		5–10% soap, 5% solvent based amphoteric scouring agent		
Desize				3–5% Bactosol C liquid, 1–3% Common salt, 2–3% Sandopan DTC liquid 70°C, 1 h
Intermediate rinse		Soft water 40°C, 20 min, 12.5% H_2SO_4, 5% Glaubers salt, 80°C, 1 h		
Acid treatment (Souple)		12.5% H_2SO_4, 5% Glaubers salt, 80°C, 1 h		
Boil-off with soda ash	25–30% soap, 0.5 g/l soda ash, 90–95°C, 2 h	25–30% soap, 0.5 g/l soda ash, 90–95°C, 20 min	25–30% soap, 0.5 g/l soda ash, 90–95°c, 2 h	25–30% soap, 0.5 g/l soda ash, 90–95°C, 2 h, if necessary add 10% sodium perborate
Boil with soap	25–30% soap, 0.5 g/l soda ash, 90–95°C, 2 h			
Boil-off with soap	15–20% soap			

| Rinse | Soft water + 1 ml/l ammonia 50°C, 20 min, soft water 40°C, 20 min, soft water cold, 20 min | Soft water + 1 ml/l ammonia 50°C, 20 min, soft water 40°C, 20 min, soft water cold, 20 min | Soft water 40°C, 20 min, soft water cold, 20 min | Soft water 40°C, 20 min, soft water cold, 20 min |

Bio-washing of silk

Enzyme bio-washing can be employed for removing protruding fibres and polishing the spun yarn/fabric appear smooth. Alkaline protease enzymes are generally used for protein fibres like silk and wool. Jet or soft flow can be used for better agitation for the process.

Recipe

Quantity	Unit	Bath additions
1	% (owf)	Enzyme
1	% (owf)	Wetting agent and lubricating agent
		Sodium bicarbonate to pH 8.5–9.0

MLR : 1:10.
Temperature: 55°C.
Time : 40 min.

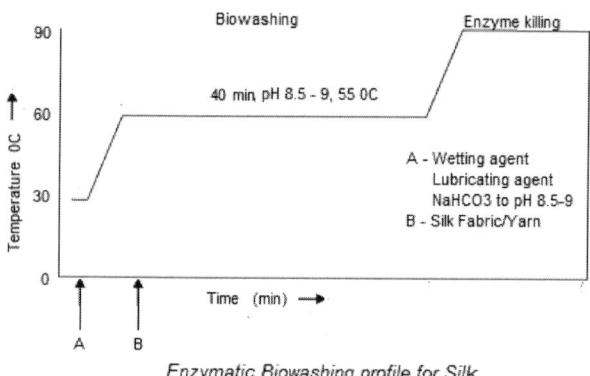

Enzymatic Biowashing profile for Silk

Figure 4.16. Enzymatic Biowashing profile for Silk

Notes

After the bio-washing process the temperature of the bath is raised to 85°C to deactivate the enzyme and dropped.

There will be a marginal drop in the strength of the fabrics once it attacks the protein.

Prolonged action of the enzyme may harm the fabric hence the time treatment has to be strictly followed or prior trials has to be done and fix the time.

It cleanses the protruding fibres from the fabric, thereby reducing its pilling tendency and improving its sheen. Such enzymatic treatment also reduces the stiffness of the fibres, which results in improved soft hand of the fabrics.

PART V

5

Pretreatment of the substrates – synthetic fibres

5.1 Preparation of synthetic fibres and their blends

5.1.1 Pretreatment of Polyester

As in the case of any synthetic fibres, polyester is also more or less a clean fibre and more or less white does not need bleaching before dyeing. However the fibre can contain on the surface softeners, sizes, lubricating agents etc. which normally added during the manufacturing process. The preteatment processes for polyester are mainly to remove these additives and soiling which can interfere with the dyeing process. The process generally involves in washing of the material with the active non-ionic surfactants in an acid, neutralor alkaline medium depending upon the degree of soiling. In alkaline preparations synergetic mixtures of non-ionic/anionic components. A solvent like tetrachloroethylene emulsions may be added assist in the removal of stains.

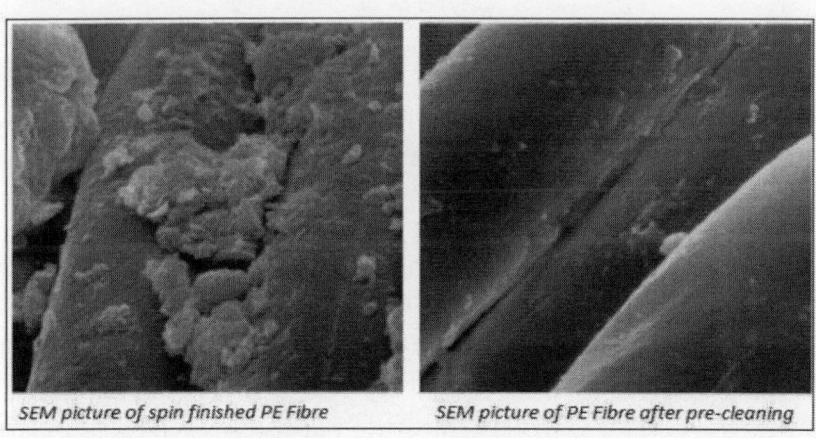

SEM picture of spin finished PE Fibre   *SEM picture of PE Fibre after pre-cleaning*

If the fabric is not very highly soiled or contaminated by oil the material can be washed on even beam dyeing machine. The greige fabric is simply batched on to the beam of a high temperature beam dyeing machine, where it is scoured before dyeing. However, if the fabric is dirty it must first be scoured in a conventional manner.

Openwidth continuous scouring

Most commonly used scouring machine is washing machines. A number of full-width washing machines are used which run tensionless with all driven guide rolls provided with load cells for tension free running can be best considered for scouring of polyester without stretching or distortion.

In continuous open-width scouring machines the scouring boxes should be set at 95–98°C with an alkali/detergent mixture of approximately half the concentration recommended for jig working. To achieve good scouring the speed of the machine should be adjusted to allow a scouring period of up to three minutes. The last two boxes should be set to give first a hot and then a cold rinse. With some machines where the no of washing compartments are less it may be necessary to run the cloth through twice in order to extend the scouring period sufficiently.

The provision of agitating devices in the scouring boxes helps to improve the efficiency of the scouring action.

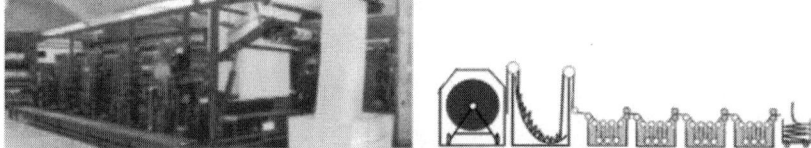

Openwidth washing / scouring machine

The exchange of substances in washing can be divided into three stages:
1. Substance dissolving zone, in which the foreign substances are swollen or dissolved so that they can be carried away by the wash liquor.
2. Zone of substance transfer in the interfacial flow layer near the substrate. In this zone, the transport of substance is mainly caused by diffusion, because the convective flow cannot develop its efficiency.
3. Substance removal zone. The convective flow of liquor results in an intensive exchange of substance.

After scouring the goods mangled or more efficiently hydroextracted on a suction-slot machine, and then stenter dried at 120°C.

Another method is to scour on jigs in open width preferably on a tensionless jigs.

For jig-scouring it is recommended that modern enclosed low-tension machines be employed. Batches of up to 1000 m may be processed, running at the minimum tension which is necessary to maintain a good batch shape. To preserve the natural woven crimp of the warp yarns it is essential to ensure that the running tension between the draw-rollers is insufficient to extend the fabric in length.

Recipe

Quantity	Unit	Additions
5-10	g/l	Soda ash or
1-2	g/l	Caustic soda
2-5	g/l	Non-ionic detergent

The scouring bath is set at 40–50°C with the above recipe. The temperature is raised to the boil and four to eight ends are run at this temperature, depending on the condition of the greige fabric. Heavily soiled fabrics require longer processing, and it may be advisable to drop the soiled scouring liquors after the first four ends and to continue scouring in a fresh bath. Fabrics containing fog marks or other widespread persistent soiling should be treated at the boil for four ends in a bath set with 5 g/l caustic soda (100%) and then re-scoured without rinsing.

J-box scouring is also popular, where after the goods have been impregnated with the treating liquor in a saturator or a roller vat and squeezed to retain a certain amount of moisture, they are placed in the J box. There are two main types of J boxes, one for treating the goods in the rope form, and the other for goods in full width, and both types can be heated. Rope form J-box is seldom used these days because the creases formed in these machines cause trouble in subsequent dyeing or printing. Trouble with crease formation is also liable to occur with OW J boxes, those in which the goods are treated in full width, unless care is taken in plaiting-down to prevent crease formation in the bottom layers of cloth.

J box scouring range

The goods are plaited down in full width, as in flat J boxes, but crease formation is prevented by plaiting the cloth down on horizontal conveyor belts. The liquor sinks into the lower parts of the loops of cloth, but this can be rectified by running the conveyor belt system at several levels and turning the cloth upside down whenever it passes from one belt to the next. With this system, too, the danger of crease formation cannot be completely eliminated, particularly if the cloth is of a delicate quality.

Heat setting

In order to achieve good shape retention and to prevent the marking of running creases during wet processing, polyester fibres must be stabilized; in other words, tensions within the individual fibres and in the fabric are levelled out by relaxation. Setting is carried out at high temperature, usually with hot water, steam, or dry heat. The selection of the setting method depends on the textile material itself and the desired setting effect, and very often, of course, on the equipment available. Relaxation of tensions within the textile material results in shrinkage. Excessive shrinkage is liable to cause trouble in subsequent finishing. After setting, the material should not undergo any further shrinkage during wet processing; this is particularly important in cases where the material is dyed or finished in the batched state.

Fabrics which require heat-setting should be treated for 30 seconds at 180°C on a pin stenter. The air flow should be balanced to support the cloth and prevent sagging. The fabric should be tensioned lightly in the direction of the standard filament yarn and allowed to relax in the other direction. For example, a standard type 100 filament warp filament weft fabric is stentered without overfeed and allowed to relax in the weft direction.

Heatsetting is exclusively done on stenter these days, since it permits good control of the width of the goods and can be operated with a given overfeed. The chains should be equipped with pins, because clamps/clips, even when they are heated, are liable to cause variations in temperature across the width of the goods, particularly near the selvedges. Stenter for knitgoods and loosely woven fabrics like curtain cloths should be provided with a throttle valve to control the air circulation, so that delicate fabrics do not flutter or become distorted. The goods must virtually be carried through the stenter by the air current. The use of a perforated belt to transport the goods is certainly beneficial for the textile material like knitted, but this is liable to reduce the efficiency of the stenter. Stenters to be used for knitgoods should be provided with an aligning instrument to ensure that the fabric runs with straight wales and courses into the machine. Before a stenter is taken into operation for setting polyester fabrics, it is necessary to check the temperature in the stenter across thewhole width of the goods. Any variations in temperature in setting are bound to cause variations in the dyeing behaviour of the material. The recommended method of checking the temperature is to secure thermoelements or Thermopapers* to piece goods passing through the stenter.

Stenters or similar setting systems with direct heating, i.e. machines in which the waste gases from firing are used directly as heating medium, are occasionally used. If these gases contain nitrogen or sulphur oxides, they are liable to destroy disperse dyes. In continuous dyeing, this may cause trouble if unfixed dye on the surface of the fibre is exposed to the attack of these waste gases. Finished dyeings normally undergo no shade change in after setting. Optimum setting effects can only be achieved by cooling the hot goods very suddenly as they leave the setting zone, which is provided in most of the setting stenters.

Notes
1. Oil stains frequently contain deposits of finely divided iron, lead or other metals which may remain entrapped between the fibres after scouring. Iron stains are removed by an acid scour at 60°C in a bath set with 2 g/l non-ionic detergent and 2–3 g/l acid (a mixture of 5 g/l of oxalic acid and 5 g/l of hydrochloric acid). In special cases hydrflourinc or nitric acid (for lead stains).
2. Loom stains may require spotting with a concentrated solvent/ detergent mixture before scouring.
3. Prolonged storage of "spotted" material may result in an ageing effect which produces streaks or blotches in the spotted areas after dyeing. It is therefore recommended not much before the material is taken for scouring.
4. Where oil staining is widespread and heavy, the fabric may be padded with a solvent/detergent mixture, batched, and allowed to rotate on a stand for some hours before scouring.

Pretreatment of worsted spun polyester yarn fabrics
The finishing sequence recommended for fabric containing worsted spun yarn polyester fibre is as follows:
Remove heavy oil stains → Pre-set → Scour → Piece dye if required → Stenter → Brush and crop →Heat-set → Steam or damp →Decatise → Press
Or
Remove heavy oil stains → Pre-set (for fabric to be piece dyed) → Scour Piece dye if required→Stenter → Brush and crop → Heat-set → Singe Steam or damp → Decatise → Press

In either routine heat-setting may be carried out before dyeing, since it confers a very good set on the fabric and therefore reduces considerably the possibilities of running marks when dyeing in rope form. The risk of dye sublimation is also eliminated.

The above routines are independent of the yarn type and of the polyester content of the blend within the recommended blend compositions, including 100% polyester.

Brushing and cropping

Even when the optimum fabric construction has been used pilling will still occur unless all the surface fibre is removed. Wherever necessary, brushing and cropping is recommended. The fabric should be brushed to raise the surface fibre, which is then removed by cropping, where practicable. Fabrics heavier than approximately 169 g/m² can usually be cropped satisfactorily. Brushing and cropping reduce the length of the surface fibres and hence the size of the fused polymer beads. In this way a harsh fabric handle, caused by the beads, is avoided.

Singeing

Singeing is essential for all fabrics containing polyester short staple fibre if pilling is to be eliminated. It is not possible to make precise recommendations for the singeing process, since conditions and machinery vary widely throughout the world. As a guide, gas singeing is preferable to plate singeing due to various reasons. To ensure the complete removal of all projecting surface fibres it is essential to carry out a thorough singeing treatment. Two or three passages through a moderate flame, rather than a single passage through a severe flame, are preferred in order to produce a clear surface without risk of damage to the fabric. A steam quench should be used where repeated singeing processes are carried out. A light brushing treatment with a soft nylon brush prior to singeing facilitates the complete removal of surface fibre. Regular cleaning of the burners is required to ensure an even flame, and in order that burner settings shall be reproducible constant-pressure governors and sensitive mixture controls should be fitted to the gas and air supply. The beads formed in singeing have a higher dye uptake than the normal polyester fibre in the body of the cloth, and therefore singeing should not be practised before dyeing, otherwise a speckled appearance will result. Singeing, in common with other heat treatments, can make the removal of dirt from polyester fabrics more difficult and it should not be carried out before the fabric has been scoured clean. These two considerations determine the position of singeing in the finishing process and preclude the practice common with cotton goods of singeing grey cloth. Fabrics made from 100% polyester fibre require even greater care in singeing than do blends. The moisture content of these fabrics is very low, and the singeing conditions must be adjusted carefully to avoid any danger of the body of the fabric being raised to the fusion temperature.

The pretreatment of Polyester Filament fabrics

The characteristics of filament fibres are to different than normal polyester fibres and hence one has to follow different processing routes taking into account of all these properties of the filament fibres.

There can be three different processing routes one can select.

Remove heavy oil stains by spotting or other methods → Scour → Heatset Piece dye → Stenter finish.

Remove heavy oil stains by spotting or other methods → Heatset → Scour → Piece dye → Stenter finish.

Remove heavy oil stains by spotting or other methods → Scour → Piece dye → Heatset, Finish

After-waxing agents which may have been applied to sized warp yarns, loom stains and other forms of soiling or contamination are difficult to remove from cloth which has been heat-set. For this and other reasons first process is the most generally useful; the cloth is presented to heat-setting in a clean condition, shrinkage during dyeing is eliminated, and the cloth is not subjected to high temperatures after dyeing. Second process eliminates a drying process and is suitable for fabrics which are perfectly clean in the loomstate. It is, however, rarely used for 'polyester fabrics, with the possible exception of curtain nets. When heat setting is the final process, as in 3rd process, some stiffening is likely to be produced. The degree of stiffening depends on the construction of the fabric and on the tensions which are developed in setting. In other respects the routine is attractive, and it is limited only by the fastness of the available dyestuffs to sublimation under heat-setting conditions.

Scouring

The fabric made of filament yarn may be strong and having good lustre it has a disadvantage of a papery feel which is not very well accepted by the customer. To overcome this drawback and make it more soft the filament yarn is usually 'crimped' the process of which is called texturizing. Texturising is possible for synthetic fibre because of its thermoplastic nature. Texturising improves the elasticity, bulkiness, resiliency, crease resistance etc. of the fabric.

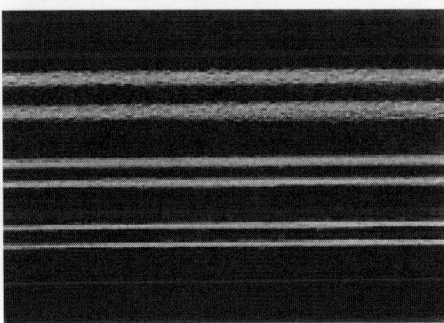

Figure 5.1. Texturised yarn

There are different methods of texturizing. The textured woven and knitted goods possess latent heat of torque which is already set during texturising process. The warp yam is applied with a large amount of sizing and oiling agent and thus the grey fabric loses its crimp by mechanical tension, becoming in a flat state. Therefore, it is necessary to restore the crimp of the yarn by relaxing treatment in water or steam, but the relaxation is better in the former. Thus textured fabrics are relaxed in water at 80–95°C in a tensionless condition for 20–30 min and for high twist yarn 30–45 min. The temperature increase has to be done in a gradual manner and the fabric even though it is tensionless, it should not be allowed too loose to avoid any wrinkling.

'Terylene' filament woven fabrics are rarely scoured in rope form. Fabrics of relatively open structure such as voiles, marquisettes and lenos may be scoured in a shallow winch provided that the temperature does not exceed 60°C. But nowadays it is done in various tensionless machines in open width form. Scoring can be done with the following recipe to remove all spinf finishes, lubricating agent. Lubricating agents are normally self emulsifying in the presence of water, but it is better to use detergent and emulsifying agents better cleansing

Recipe

Quantity	Unit	Additions
2–3	g/l	Soda ash
1–2	g/l	Nonionic detergent

For jig-scouring or any other tensionless machine like open wodth washer with enough compartments to give enough time for the scouring operations can be used. When running on tensionless jigs batches of up to 1000 metres or more may be processed, but care should be taken to see that it is running at the minimum tension which is necessary to maintain a good batch shape. To preserve the natural woven.

In case of texturised fabric following precuautions may be taken:

1. It is better to adjust the tension (even though tensionless in principle but will need some tension for the running of the machine) in such a way as there is no extension in the fabric length, lest the crimpiness will be lost.

2. If the spin finishes used for textured polyester yarns and knitted fabric is oiling agent of self-emulsified type it can be readily removed by rinsing at 50–60°C without any addition of detergent.

3. Spin finishes that do not contain emulsifying agent are removed by washing with 1–2 g/l a combination of non-ionic and anionic detergent at 60–70°C The temperature of the bath should not be raised more than 2°C to avoid production creases.

4. Relaxation and scouring are done together for circular knitted fabrics or garments in special machines like paddle scouring, where they are loosely packed into polyester mesh bags and scoured/relaxed using 1 g/l non-ionic or anionic detergent, 0.5 g/l sodium hexametaphosphate and 1 g/l solvent scouring agent at 60°C for 30 min. The liquor is cooled, fabric in rinsed, hydroextracted and dried without tension at 60–80°C.

5. In special case, solvent scouring can be done in rotary drum machine, restricting the load to 50% of the machine maximum capacity and the centrifuging time to 1 min to avoid the formation of pressure creases.

5.1.2   Preparation of polyamide

Polyamide is a synthetic protein fibre. The most common sizing agents used for nylon are based upon PVA (poly vinyl alcohol), gelatine, casein or variety of vegetable and mineral oils. When water soluble PVA is used as sizing agent in weaving, no desizing of synthetic fibre fabrics is necessary. A normal washing with a non-ionic detergent (by boiling) is sufficient to remove the PVA size.

The preparation route can be as:

Heat setting (Thermo fixation) → Washing → Bleaching.

5.1.2.1   Thermo fixation

Heat setting is done on stenters at the following temperatures.

PA 6	Perlon-type	8–20 s at 190–192°C
PA 6.6	Nylon-type	8–20 s at 205–215°C

5.1.2.2   Washing of polyamide

Guideline recipe

Quantity	Unit	Chemicals
1	ml/l	Non-ionic and low-foaming stain remover and washing agent
0.5–1.0	g/l	Sequestering agent
0.5–1.0	g/l	Soda ash

Temperature 80–90°C.
Time   20–30 min.
Rinse   hot and cold.

Alternate recipe

Machine: Open soaper, J-box, overflow.

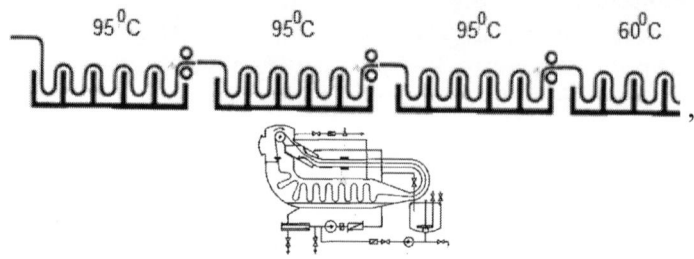

Figure 5.2. Washing OW or Rope form (Circulating Machine)

Recipe

Quantity	Unit	Chemicals
3–5	g/l	Non-ionic detergent

Temperature: 95–100°C.
Time: 20–30 min.

Note: PVA of higher molecular weight (1,00,000 above) may not be removed by washing alone.

5.1.2.3   Bleaching of polyamide

Conventional bleaching processes.

Chlorite bleach
Chlorite bleaching is highly suitable for polyamide materials. A stabiliser for chlorite bleaching for an optimum utilisation of the activated chlorite for bleaching has to be used. If the auxiliary doesn't contain anti-corrosive agent to prevent secondary reactions and keeps the pH value constant, a separate anti-corrosive agent has to be used.

Hypochlorite bleach is not suitable for polyamide.

Guide recipe

Quantity	Unit	Chemicals
1	ml/l	Non-ionic and low-foaming stain remover and washing agent
0.5–1.0	g/l	Stabiliser for chlorite
6–10	g/l	Sodium chlorite 30%
		pH 3.5–4 with formic acid

Temperature 90°C.
Time   30 min.
Rinse   hot and cold.

Reductive bleach

Reductive bleach is suitable with good effects. A reducing agent based on hydrosulphite. It is very often used for polyamide, especially in case of bleaching and optical brightening in one bath.

Guideline recipe for reductive bleach

Quantity	Unit	Chemicals
1	ml/l	Non-ionic and low-foaming stain remover and washing agent
3	g/l	Stabilised reducing agent

Temperature   90°C.
Time   30 min.
Rinse   hot and cold.

Peroxide bleach

Peroxide bleach is suitable for polyamide, but without considerable bleaching effect. In case of 100% PA the peroxide bleach is not recommended. But when blended with cotton the peroxide can be used with reduced amount of caustic soda. To avoid oxidative degradation of polyamide during peroxide bleaching, especially in case of PA 6 (the fibre protection agent which protects polyamide fibres) helps in preventing loss in tensile strength.

Guideline recipe for polyamide/cotton

Gudeline recipe for polyamide/cotton blends

Quantity	Unit	Chemicals
0.5–1	ml/l	Non-ionic and low-foaming stain remover and washing agent
3	g/l	Stabiliser for peroxide
0.3	g/l	Fibre protective agent for PA
2–3	g/l	Caustic soda 50%
3–4	g/l	Hydrogen peroxide 35%

Temperature   98°C.
Time   30 min.
Rinse   hot and cold.

Neutral bleach

A peroxide bleach without the addition of alkali gives good bleaching effect on polyamide. Especially on PA fibres which has been yellowed due to heat setting neutral bleach can give good whiteness. As the bleaching has to be done at neutral pH, we may need an activator and buffer in the bleaching bath to catalyse the bleaching action of peroxide and control the pH at the bleaching temperature of 60–75°C. With these auxiliary we can bleach at pH 6.5–7.5.

Guide recipe

Quantity	Unit	Chemicals
4	ml/l	Peroxide activator for neutral pH
4	ml/l	Buffer
4	ml/l	Hydrogen peroxide 35%

Temperature   70°C.
Time   45 min.
Rinse   hot and cold/
Comparison of different bleaches

Whiteness achieved by different bleaches

Fabric process stage	Whiteness in Berger
Greige	75
Heat set (200°C, 30 s)	13
Neutral bleached	49
Alkaline peroxide bleach	29
Reductive bleach	27

5.1.3   Preparation of polyester goods

5.1.3.1   Desizing

The usual sizing agents of polyester can only be removed by an alkaline washing with a non-ionic detergent.

Recipe

Quantity	Unit	Chemicals
2–3	g/l	Non-ionic detergent
1–2	g/l	Caustic soda or soda ash

The grey fabric washed at 80–95°C, for 20–30 min in an open width soaper or jet or overflow machine.

Alternate recipe

Quantity	Unit	Chemicals
1–2	g/l	Sodium tripolyphosphate
1–2	g/l	Caustic soda or soda ash
2–3	g/l	Non-ionic detergent

Treatment as above.

Notes

It is better to use soft water for this treatment as metal ions may interfere with the removal size or add sequestering agent.

If a fabric with polyester continuous filament warps is sized with pure polyacrylate Sizes, it can be desized readily and completely by a mild wash with 0.5–1 g/l of detergent in weakly alkaline medium, because these sizes are readily soluble in water. An open-width washer can be used for this.

In case of spun polyester fabrics, the desizing is not as easy as described above since the size used may not be only soluble synthetic sizing agent but it may be mixed with other common sizing agent like starches. If a pure size is used as an enzymatic desizing explained earlier may have to be used. But if starch derivatives, CMC (Carboxy Methyl Cellulose), etc. are used as additions along with synthetic sizes,. aA washing process allows sufficient time for swelling of these additives may be adequate to remove most of the size. In a typical process a woven fabric in full width may be first impregnated with concentrated detergent solution, and then left standing for some time to let the size swell. The impurities and foreign matter are then removed by washing in the full-width washer at elevated temperature. Using a detergent with dispersing agent also ensures complete removal of fats, oils and waxes that contain no emulsifying agent.

Recipe

Quantity	Unit	Chemicals
5–15	g/l	Detergent
1–2	g/l	Sequestering agent
0–5	g/l	Soda ash

Pad at 20–60°C with a liquor pick-up of 60–100% and allowed to soak and wash in a 4–5 compartment soaper.

1st tank: water at 60–95°C.
2nd tank: water at 60–95°C.
3rd tank: water at 40–60°C.
4th and 5th tank: rinse at approximately 40°C.

5.1.4   Pre-treatments of polyester knitted fabrics

5.1.4.1   Processing sequences for circular knitted fabrics

Some common processing sequences for weft knitted fabrics are given below. Some of these operations can be combined or dispensed with, depending on the type of textile goods, the equipment available, and the desired quality of the end product.

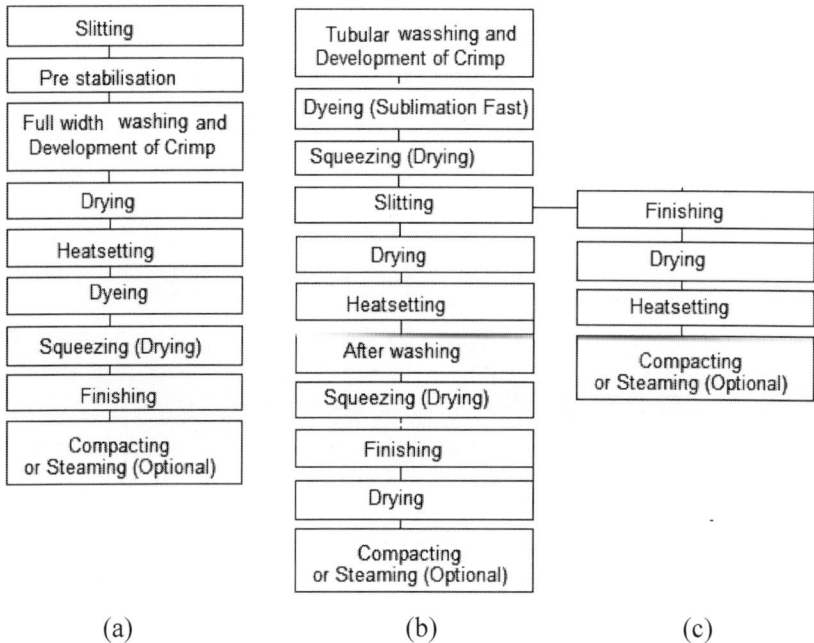

(a)   (b)   (c)

Figure 5.3. Process sequences for circular knit fabrics

5.1.4.2   Processing sequences for warp knitted fabrics

Tubular goods are normally turned inside out prior to wet processing. Circular knit fabrics should be slit as soon as possible after knitting (within 24 h at most). If they cannot be further processed immediately, they must be rolled flat, completely free from creases. If the goods are stored in tubular form for some time, the folds will become fixed.

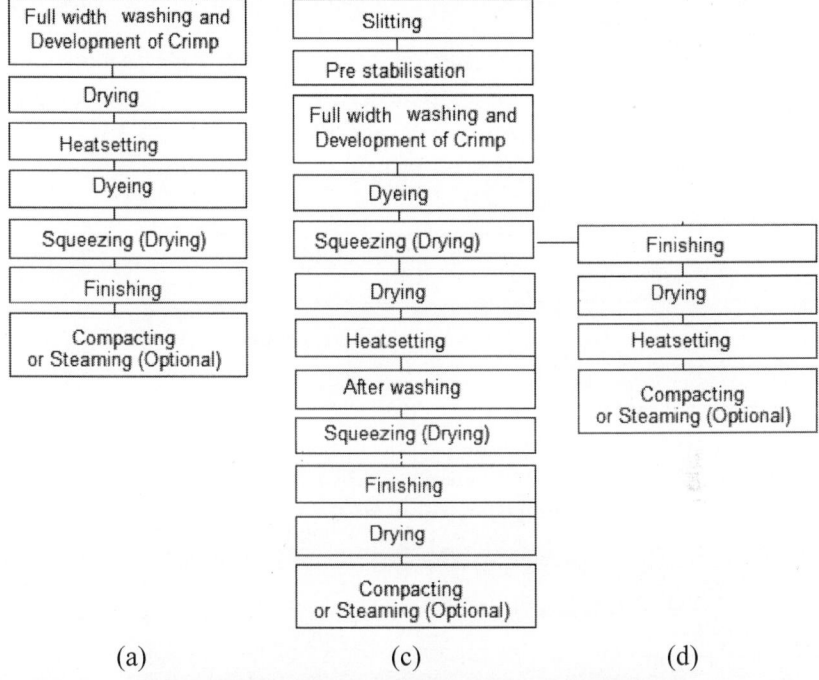

(a)   (c)   (d)

Figure 5.4. Processing sequence for warp knitted fabrics

5.1.4.3   Pre-stabilisation

Pre-stabilising is necessary if the goods are liable to become creased during pre-cleansing or if the selvedges have a tendency to curl. This is done by running the goods in a tensionless state over a steaming table or through a steaming chamber (saturated steam, 100°C); the fabric relaxes and the fibres begin to crimp. If necessary, the goods are then set lightly on a stenter at approximately 130°C. Loom-state goods should not be set at a higher temperature, because the spin finish would then become fixed in the fabric and the crimping contraction would be blocked, particularly in the case of knit goods. Prior to dyeing, after the spin finish, machine oil and other impurities must be removed. At the same time, washing must develop the crimp that the yarn has lost in knitting. The simultaneous action of mechanical motion, water and heat causes a relaxation of the inner tensions in the loom-state goods; the fibres regain their crimp, and the fabric obtains bulk and elasticity. This naturally involves some shrinkage of the fabric in length and width. The handling and the weight per square metre of the finished cloth are properties that are the result of this first operation; they cannot be influenced subsequently to any significant extent. The pre-wash and the development of the crimp are carried out in full-width washers that subject the goods to low-tension, or in winch becks.

Even though spin finishes containing emulsifying agent and sizes can be removed by rinsing at 40–50°C in most cases it is always better to give a wash at higher temperature with a detergent or emulsifying agent which will remove the spin finish which doesn't contain emulsifying agent.

Recipe

Quantity	Unit	Bath additions
0.5–1.0	g/l	Detergent
		Acetic acid to pH 6

Wash at a temperature of 60°C, for 10–15 min. The material should be neutralised to ensure that no alkali present in the fabric.

Crimp development and washing process can be combined in a 4 compartment soaper as follows:

Recipe
1st compartment.
Water at 40°C.
2nd compartment.

Recipe for 2 nd compartment

Quantity	Unit	Bath additions
0.5–1.0	g/l	Detergent
0.3–0.5	g/l	Sequestering agent
0–3	g/l	Soda ash

At 70–90°C.

3rd compartment.

Recipe for 3rd compartment

Quantity	Unit	Bath additions
0.3–0.5	g/l	Detergent

At 70–90°C.
4th compartment
Rinsing and neutralisation with acetic acid at 40°C.

Notes

1. In an open-width washing it is better to have a swelling or relaxation section without any tension lengthwise (e.g. a scary) to allow swelling and shrinkage which can be of 20% in some cases.

2. In case a winch, jet or soft flow is used the distance between the winches a liquor has to be minimised to reduce the longitudinal tension on the fabric.

3. The liquor must be cooled slowly from the washing temperature to below 50°C before draining to avoid fixation of running creases.

4. Yarn-dyed textiles of pure polyester fibre are often washed in the presence of a reducing agent, e.g., 0.5–1 g/l of hydros to remove dye residues retained in the film of winding oil.

5. Anthraquinonoid dyes are not destroyed, but only removed, by the treatment; they are then liable to stain pale parts of the fabric. For this reason, preliminary trials should always be carried out.

6. Any metal filings (iron and iron oxides) and iron stains should be removed with oxalic acid by the recipe given.:

Recipe

Quantity	Unit	Bath additions
1	g/l	Oxalic acid
0.5–1.0	g/l	Detergent

At 50°C for 30 min.

7. Washing and development of the crimp can also be carried out in an organic solvent (perchlorethylene).

Next, the material should be squeezed and dried.

5.1.4.4   Heat setting

All materials that are liable to undergo fixation of the running creases must be heat-set prior to dyeing in winch becks, jet or overflow dyeing machines. Heat setting prior to dyeing is a good method of ensuring a definite weight of finished material per square metre. Whenever a new type of textile material is to be finished, preliminary trials must be carried out to determine the most economical method. If sufficient time is not available, the best solution is to heat-set most of the first batch prior to dyeing, and then dye it together with a small amount of fabric that has not been pre-set. A comparison of the results obtained then shows whether it is possible to dispense with intermediate drying and heat setting and economise the process. The goods are heat-set on a stenter or a perforated-drum system with expander chain for 20–30 s with hot air at 180–190°C, according to the type of fibre and quality of goods.

Notes

1. Before heat setting, the goods have to be dried uniformly as variations in the moisture content cause variations in heat setting, resulting in uneven dyeing. If the stenter is having enough chambers, sometimes fabric is heat

-set by feeding a wet cloth (avoiding one drying) squeezed through water ensuring equal moisture along the length and width of the fabric.

2. Theoretically, higher the setting temperature, the better the dimensional stability of the treated goods, but the handle becomes flatter, harsher and yellowish if the setting temperature is too high. It is better to follow recommendations for heat setting given by the suppliers of the fibre or prior trials has to be done before fixing the process.

3. The goods should not be stretched more than 5% wider than the desired end width.

4. The overfeed is adjusted so that the cloth leaves a smooth, stretched state and has the desired weight per square metre. An overfeed of up to 20% is commonly used.

5. The air circulation inside the stenter should not cause any fluttering of the goods.

6. The sagging of very wide cloth in the stenter can be prevented by installing a conveyor belt of metal netting or by adjusting the air nozzles accordingly so that the centre to selvedge density variance can be avoided.

5.2   Preparation of synthetic blends

5.2.1   Pre-treatment of polyester/cotton (cellulosic) blends

Polyester/cotton is one of the most commonly used blended fabric used in textiles. Mainly used in almost all garments like rain coating, outerwear, shirting, blouses, dress goods and casual wear. Although the majority of polyester/cellulosic fibre blends go into woven goods, a significant share is also used for knit goods. The cellulosic fibre is usually cotton, spun rayon or modal fibres, sometimes even linen. For knit goods, polynosic fibres are usually preferred owing to their high lustre. The commonly used blends (by weight) are 67:33 or 65:35, 70:30 and 50:50. Polyester/spun rayon blends often have the blend ratio 70:30, for dress goods also 50:50. Polyester/linen blends are usually 70:30 or 50:50. The warp and weft yarns of some blended fabrics may be composed of different blends.

Fabrics of polyester/cellulosic fibre blends must be subjected to a number of pre-treating finishes to ensure good results in dyeing and an improved appearance of the goods. As polyester doesn't need much preparation and any preparation process of cotton will clean them, the preparation process has to be set according to cellulosic fibre contained in the blend. The procedure and conditions must therefore be adapted to suit the type of cellulosic fibre concerned. At this point, we must remember that polyester is destroyed hydrolytically by strong alkali. It is therefore not always possible to use alkali concentrations that are desirable for the pre-treatment of cotton. This applies particularly to prolonged processes.

5.2.1.1   Pre-treatment possibilities

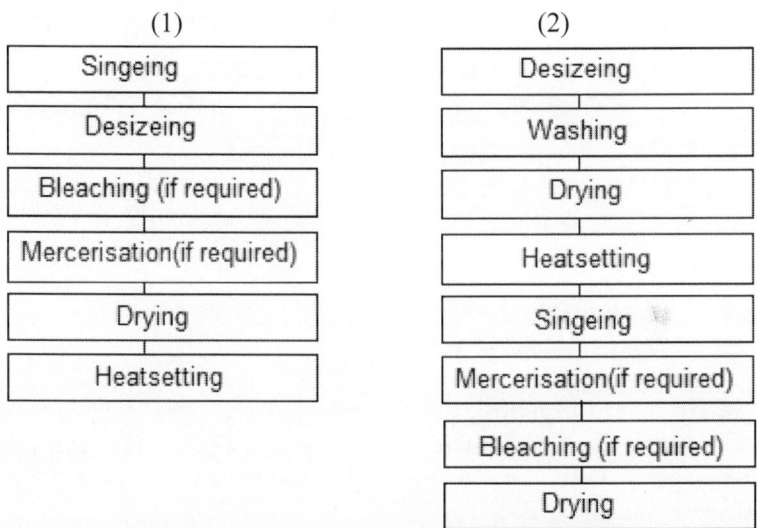

Figure 5.5. Pretreatment routes for PE/C blends

If the polyester fibre contained in the blend is to be dyed by the exhaust process, the fabric is not singed until it has been dyed. If the goods are dyed, however, by the thermosol process, it is advisable to singe during the pre-treatment. The balls of melt which form at the ends of the polyester fibres during singeing, dye to a darker shade in the exhaust process than the fibres themselves. This causes the fabric to appear sooty. The effect does not occur in thermosol dyeing. The pre-treating route given under (2)(see figure) is somewhat more costly but, as a rule, it produces goods with which best results can be obtained in dyeing, particularly by continuous methods, if the fabric is desized before singeing, the fibre ends project out of the fabric for better removal by singeing.

5.2.1.2   Desizing

If size contains only water soluble sizes a good weeting out and washing may be sufficient giving proper time for soaking. But if starch is contained in the sizing any one of the cotton desizing process may be followed to degrade the size and make it water soluble. A washing after desizing may be followed in the dyeing machines or in a continuous washing range.

Enzymatic desizing

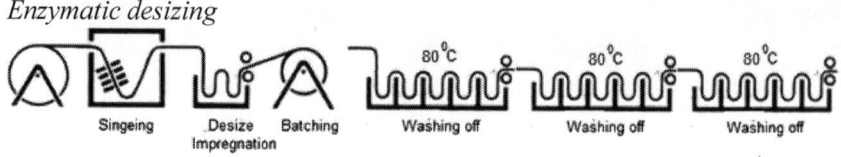

Singeing   Desize   Batching   Washing off   Washing off   Washing off
        Impregnation

Figure 5.6. Enzymatic desizing process

Table 5.3. Guideline recipe for enzymatic desizing by pad-batch process.

Quantity	Unit	Bath additions
5–10	g/l	Enzymatic desizing agent
5	g/l	Common salt
5–10	g/l	Detergent

Padding temperature approx. 60°C, batching time at least 3 h.

Note
The temperature of treatment, time for reaction, etc., will basically depend on the enzyme used, since they are temperature, and pH specific. The recommendations from the enzyme manufacturers may be strictly followed for effective results.

The main advantage of enzymatic desizing is that it is gentle and mild treatment, hence no adverse effect on cotton or polyester.

Bromite desizing

Guideline recipe

Quantity	Unit	Bath additions
1–3	g/l	Active bromine
20–30	g/l	Caustic soda
5–10	g/l	Detergent

Pad and batch for 60–90 min at a temperature of 90–95°C.

Notes
The advantage of this method lies in the rapid reaction but the disadvantage is that the fibre is liable to be damaged if the required conditions, i.e., pH value, time and temperature, are not accurately controlled. The lower the pH, the higher the speed of reaction and the more severe the damage to the fibre.

Desizing with persulphate and peroxide

Guide recipe

Peroxide de-sizing recipe	Persulphate desizing recipe	Unit	Bath additions
10	3–6	g/l	Sodium or ammonium persulphate
40		ml/l	Hydrogen peroxide 35%
10		ml/l	Sodium silicate 38°Bé
5	8–10	g/l	Caustic soda
5		g/l	Peroxide stabiliser
5	5–10	g/l	Detergent/wetting agent

Process for persulphate desizing: Batching time 60 min.
                                               Batching temperature 90°C.
Procedure for peroxide desizing: Padding temperature: 20–40°C.
                                               Batching time: 6–20 h.

Recipe for jigger washing

Quantity	Unit	Bath additions
1–2	g/l	Detergent
1–2	g/l	Soda ash

30 min (4 ends) at 70–80°C, rinse and neutralise with acetic acid.

The washing can be done in rope form in winch (for some special reason such as to allow the goods to relax or undergo a certain amount of residual shrinkage) it has to be done below 40–50°C. If the wash temperature is higher, running creases are liable to become permanent. Wherever possible high temperatures, strong squeeze-off and the use of spray tubes facilitate removal of the size.

5.2.1.3   Heat setting

As required in case of polyester fibres, polyester/cotton blends also needs stabilisation to improve its qualities of dyeing, shrinkage, creasing, pilling, handle, etc. If the material has to be pre-treated or dyeing in the rope form setting has to be done before hand so that the creases are not set during these processes. In the open- width form of beam or jig dyeing also, heat setting has to be done before dyeing to prevent shrinkage during high temperature dyeing in beam or lengthwise tension in jig and cause dyeing problems. The material is heat set, e.g., in the stenter with hot air. The conditions are usually as follows:

      20–40 s at 190–210°C.

If the blend fabric is to be dyed by the thermosol method later on, pre-setting is normally unnecessary unless the fabric displays poor running properties. It has also been found advisable to pre-set in a stenter if the actual thermosolising process is carried out in units without width control (most of thermosoling may not have a width control).

Notes
1. Heat setting will usually bone dry the fabric rendering rewettability a problem. Hence care has to be taken in the scouring or pre-treatment should be done with proper absorbency or probably with a rewetting agent.

2. It is naturally possible to improve liquor uptake by adding a wetting agent during one of the subsequent continuous processes but this is limited because of the foaming properties of wetting agents.

5.2.1.4   Scouring
Normally alkaline scouring is followed. Caustic soda, soda ash, trisodium phosphate are the main alkali used in scouring.

Batch wise scouring
Scouring on jigs

Recipe

Quantity	Unit	Additions
3	%	Caustic soda (or 5–10 g/l soda ash)
2–5	g/l	Detergent

Run at 70°C for 90 min or at 100°C for 15 min.

Scouring on pad-steam

Recipe for scouring on pad-steam

Quantity	Unit	Additions
5–10	%	Soda ash
2–5	g/l	Detergent

Pad with 100% pick-up and steam for 30–60 min at 100°C.

Continuous scouring
Continuous scouring and washing of cellulosic and synthetic fibres and their blends

Table 5.4. Guide recipe

Form	Woven	
Process/machine	P/B, OW washer	OW washer
Type of treatment	With pre-swelling	Without pre-swelling

Guide recipes

Pad-batch (possibly with singeing

Saturator Scouring/stain removing agent Soda ash	3–5 g/l 1–2 g/l	
Impregnate Pick-up Temperature	50–80% 20–30°C	
Batch Time Temperature	1–3 h Room temperature	

OW washer (e.g., 6 compartments)

Compartment 1 Water overflow Temperature °C	50–60°C	50–60°C
Compartment 2 Scouring/stain removing agent Soda ash[1] Temperature	1–2 g/l 1–2 g/l 60–80°C	2–4 g/l 2–4 g/l 60–80°C
Compartment 3	Same as above	Same as above
Compartment 4 Water, Temperature °C	80	80
Compartment 5 Acetic acid 85% Temperature	1–2 ml/l 30–40	1–2 ml/l 30–40
Temperature 6 Water Temp °C	30–40	30–40

[1]Addition of liquor 10–20 times. Strengthening liquor addition depends on speed, degree of soiling, detergent strength.

5.2.2   Preparation polyester/acetate blends

Since the acetate fibre is less resistant to scouring chemicals and heat, the pre-treatment process has to be designed bearing in mind the presence of acetate in the blend. Sizes used being water soluble synthetic polymers, it can be removed by a usual washing with a detergent, during which the spinning oil, stains also can be removed by making the scouring bath little alkaline by the addition of ammonia.

Recipe

Quantity	Unit	Additions
0.5–1.0	g/l	Detergent
2–3	m/l	Ammonia 25%

Procedure: Run for 15–30 min at 65–75°C, rinse warm, rinse cold and neutralise.

Heat setting
Heat setting has to be done before dyeing on a hot air stenter at 180–190°C, for 20 s which will ensure the setting of polyester.

Bleaching or whitening can be done as per methods given under polyester/wool pre-treatment.

5.2.3   Pre-treatment of polyester/triacetate blends

As in the case polyester/acetate blend the material is washed to remove spinning oil, size and stains. This washing is sufficient for dyeing.

Recipe

Quantity	Unit	Additions
0.5–1.0	g/l	Detergent
0.5–1.0	g/l	Soda ash

Wash at 70–80°C for 20–30 min, wash and dry. After washing operation the material may be heat-set for 20–30 s at 200–210°C in a stenter.

Heat setting tenders both polyester and triacetate level dyeing so that even darker shades can be dyed without much problem.

5.2.4   Pre-treatment of polyester/acrylic blends

The polyester/acrylic blend may be compared with polyester/wool blend since it has good shape retention, woolly nature, easy care properties etc. It is used for manufacturing ladies' and men's outerwear fabrics, light coat material for summer fashions, children's wear and casual wear, etc.

Both staple and texturised polyester acrylic blend are used in outerwear garments and furnishings. In intimate mixtures polyester and acrylic percentages can change between 40%and 60%. The main advantage of polyester/acrylic blends are it has the stability of polyester along with wool like fleeciness of acrylic.

Pre-treatment or dyeing most of the processes are similar to polyester/ wool blends. However, the anionic character of the acrylic fibre, that have to be taken into account in planning any process.

Polyester/acrylic blends, let it be yarn or fabric has to be given gentle washing before dyeing to remove spinning oil and other contaminants. Yarns can be washed in hank form or cheeses in the dyeing machines before dyeing. In case of fabrics, it can be done in open-width soaper, in a jet washing machine, or even on a winch-beck. Washing can be done weakly alkaline or acid medium.

Guideline recipes

Quantity	Unit	Additions
0.5–1.0	g/l	Non-ionic or anionic detergent
0.5	Ml/l	Acetic acid to adjust pH 4–5

Wash at 60°C for 20–30 min. If the material is more soiled the detergent quantity can be increased or can be soaked in a 10–20% solution for some time and washed.

Recipe

Quantity	Unit	Additions
0.5–1.0	g/l	Non-ionic and anionic detergent
0.25–0.5	g/l	Ammonia 25%
0.2–0.4	g/l	Soda ash
		Check the pH to be 8–9

Wash at 40–50°C for 20–30 min. Quantity of detergent depends on the soiling of the fabrics. After washing it should be neutralised, probably in the last compartment of the soaper or separate bath in jet or winch.

Notes

1. In washing in rope washing machine, it should be done with low upper bowl pressure.

2. In winch becks, jet or overflow machines, the goods must be washed in tubular form, otherwise running creases will become set. Easily creased goods are best washed in open-width if they cannot be set before washing.

5.2.4.1   Heat setting

Polyester/acrylic blended spun fabrics are set before dyeing to impart the shape and dimensional stability needed for dyeing and in later wear or use; pilling is also diminished. The material is set in a stenter at 190–200°C for 30 s.

When texturised fibres are present, the setting conditions depend on the kind of polyester and are of the order of 150–165°C (for 30 s).

Notes

1. Heat setting time must not exceed 30 s at 180°C or 5 min at 135°C, otherwise there are chances of irreversible yellowing, hardening of the acrylic fibre and problems of the dyeing may increase.

2. The overfeed generally depends on the fabric, but as a general rule, woven may be given 5–7% off and knitted fabrics 6–10%.

3. If the heat setting is done before dyeing width is adjusted, higher than the final width considering the shrinkage in dyeing and other processes after heat setting, but if done after dyeing the heat setting can be done to the final required width.

Crabbing can be done as in case of wool fabric even though it is not that effective in acrylic fibres, but definitely helps in better dyeing.

5.2.4.2   Full white

Polyester/acrylic fibre blend materials can be bleached under the same conditions as given under pure polyester fibres, i.e., with

Recipe for full white

Quantity	Unit	Additions
1–2	g/l	Sodium chlorite 80%
0.8–1.5	g/l	Stabiliser for sodium chlorite
0.2–0.5	g/l	Wetting and detergent
1–2	ml/l	Formic acid 85% (pH 3–3.5)
		Adjust the pH to 8–9 with formic acid 85%

Boil for 30–45 min at a liquor ratio 10:1–40:1.

After this treatment, the liquor is not dropped until it has first been cooled slowly (1°C/min) down to 50–60°C the fabric is then rinsed.

If suitable OBA which brightens both polyester and acrylic (e.g., 0.5–2% Ultraphor NA – BASF) is added to the bleach bath, the goods not only undergo a bleach but are also optically brightened. Ultraphor NA brightens both the polyester and acrylic fibres contained in the blend.

Alternately polyester can be separately whitened by suitable method without affecting acrylic component (e.g., Palanil white liquid BASF) and acrylic can be separately whitened with suitable OBA (e.g., Ultraphor PAN) which reserves polyester.

5.2.5   Pre-treatment of wool/acetate blends

When diacetate and triacetate are blended with wool the processing has to be followed mostly as per wool. Scouring and milling are done with lower alkali content and temperature of treatment should go higher than 50°C, pressure on the rolls should not be high. It is advisable to crab the fabrics before dyeing treatment before dyeing.

5.2.6   Pre-treatment polyester/polyamide blends

The usual blend used is 50/50. This blends not so common due to the difficulties in dyeing.

5.2.6.1   Pre-treatment

The pre-treatment of polyester/polyamide is same as the treatments explained under polyester. Polyamide can withstand any polyester treatments.

5.2.6.2   Washing
Recipe

Quantity	Unit	Additions
1	g/l	Detergent
0.5–1.0	g/l	Sequestering agent
0.5–1.0	g/l	Soda ash

Temperature   80–90°C.
Time   20–30 min.
Rinse hot and cold.

5.2.6.3   Heat setting

Woven and knitted polyester/polyamide blend fabrics are set before dyeing or bleaching to avoid undue shrinkage and subsequent dyeing problems. The heat setting temperature should depend on the softening temperature of the polyamide used in the blend.

   a) Polyester/nylon 6 fibre blends
       20 s at 192–195°C; (e.g., Diolen/Perlon blend fabrics).

b) Polyester/nylon 6.6 fibre blends
    20 s at approx. 210°C.
Alternatively, polyamides can be hydroset at 130°C for 20 min. But it should be borne in mind the hydrosetting process is not as effective to the polyester as in case of hot air setting.

For full white polyester/polyamide can be bleached with sodium chlorite, as given in the polyester section or polyester/acrylic section. Whitening can be done OWA types applicable for both polyester and polyamide, applied to both fibres under HT conditions or in the presence of carriers.

Recipe

Quantity	Unit	Additions
0.5–1.0	g/l	Non-ionic detergent
1.5	g/l	Stabiliser for sodium chlorite 30%
		Adjust pH 3.5–4 with formic acid

Temperature   90°C.
Time   30 min.
Rinse   Hot and cold.
Polyester/polyamide can be given reductive bleach also with good results.

Recipe

Quantity	Unit	Additions
3	g/l	Reducing agent (Hydrosulphite)
0.5–1.0	g/l	Non-ionic detergent/wetting agent

Temperature   90°C.
Time   30 min.
Rinse   Hot and cold.

Notes
1. The reductive bleach is very often used for polyamide, especially in case of bleaching and optical brightening in one bath.
2. Hypochlorite bleach is not suitable.
3. Peroxide bleach is not recommended for polyamides, but in blend form peroxide bleach can be done in the presence of a fibre protective agent (e.g., CHT fibre protection PA liquid), if necessary.

5.2.7   Pre-treatment of polyester/wool blends

Polyester/wool blends have established its importance in the field of outer wear due its many qualities compared to other blends like shape retention, light weight, easy care properties, etc. Often other than wool, mohair, cashmere and other hair fibres are also fused in blending with polyester.. The polyester fibres are mostly used in the form of staple fibres and very seldom as continuous filament, i.e., polyester staple fibres of the normal type, minimum-pilling type and also textured yarns. On basis of technological tests and wearing tests, it has been found that blend ratios of 55% polyester and 45% wool are particularly suitable. Other blends which is also available in the market are 20:80, (polyester/wool is woven from 55:45 warp and a pure weft yam) 70:30, etc.

The polyester wool materials has to be washed before bleaching or dyeing. For brilliant shades the material may be bleached. Even after bleaching, the brilliancy of shade may be affected by the yellowing of wool unless the dyes are selected for bright shades. It is better to give an anti-felting treatment for wool rich blends to avoid shrinkage (see wool section)

Removal of Oil Stains

Loom stains and other oil stains can be removed from polyester/wool fabrics provided they have not been fixed as a result of pre-setting the fabric before the removal. If fabrics are subjected to a heat treatment before the removal of heavy stains is attempted, difficulty may be experienced. Stains may be removed from loomstate fabric by the use of proprietary stain cleaning products, followed immediately by a vigorous local hand scour before the solvent evaporates. Cold organic solvents may also be used, suit- able ones being perchloroethylene, trichloroethylenrtrichloroethane or white spirit directly or along with emulsifier. Solvent which may remove dyestuff, such as chloroform or methylene chloride, must not be employed. Careshould be taken to avoid any ring marks, especiallyon fabrics which are to be piece dyed to pastel shades,and the fabric should be scoured immediately afterapplication of the solvent.

Alternatively, the soaking of oil stains in a concentrated solution of a non-ionic detergent overnight followed by avigorous local hand scour should remove oil markssuccessfully. If this technique is used on fabrics to bepiece dyed, the immersion time should not be longerthan 24 hours, otherwise the treated areas may havea different dye uptake from that of the remainder ofthe fabric.

Pre-setting

Polyester/wool fabrics may be pre-set before scouring or dyeing in order to reduce cockling, rope marks and shrinkage. Pre-setting should be carried out if similar fabrics made from all-wool would normally be pre-set. Examples are lightweight plain-weave fabrics, check or tartan fabrics which may contain yarns of different origin, gaberdines, and fabrics containing crossbred wools.

Wet blowing, dry blowing, crabbing and torpedo setting have all been used satisfactorily for setting polyester/wool fabrics, and these processes differ principally in the magnitude of the effect they produce. Dry blowing and wet blowing is not very popular today. is recommended, since disperse dyestuffs. Whatever process is selected it is important to ensure that the fabric is evenly treated. It is recommended, therefore, when dry or wet blowing or steam crabbing, that several layers of wrapper should be run on to the steam roller before running on the polyester/wool fabric. Dry blowing is most satisfactorily carried out for a total time of 10 minutes with the steam-pressure gauge registering 1.04-2.8 kg/cm2 immediately prior to entering the rotating steam roller. In general, fabrics should be ended and blown for 5minutes in both directions.

Two-bowl crabbing is satisfactory for many polyester/wool fabrics, but for those requiring a moresevere set the steaming roller should be used. It isrecommended that the water temperature in the firstbowl be 80-90°C whereas in the secondbowl the water should be at the boil. A period of3-5 minutes in each bowl, followed by a similarsteaming period if considered to be necessary, willgive satisfactory results. This is further explained below.

5.2.7.1   Desizing

Generally, polyester wool contain only synthetic water soluble sizes, which can be washed-off easily. In case any protein based or starch sizes are used, trials may be taken for the methods for removing them.

5.2.7.2   Washing

The material is washed before dyeing or bleaching, usually in the machines where the bleaching or dyeing process is going to take place or separately in an open width washing machine.

Recipe

Quantity	Unit	Additions
0.5–1.0	%	Non-ionic detergent
		Adjust pH 8–9 with about 0.25–0.5 ammonia 25% or 0.2–0.4 g/l soda ash

Procedure: Wash at 40–50°C for 20–30 min and rinse.

Soap soda scour

It is advisable to follow soap soda scour heavy stained and lower quality wool containing fabrics. The process is done in two treatments:

First bath
4–6° Tw soda ash (1.9–2.8% sodium carbonate solution)
2% Non-ionic detergentor

Liquor ratio 4 : 1
Time 20 minutes
Temperature 40°C
The liquor is run off and the fabric washed in warm water at 35–40°C for 10 minutes.

Second bath
2% non-ionic detergent on weight of material
Liquor ratio 4 : 1
Time 40 minutes
Temperature 40°C

The fabric is finally washed off for 45 minutes at 35–40°C and then cooled down gradually. It is good practice to acidify coloured fabrics with either acetic or formic acid during the final stages of washing off. Where light-coloured or undyed wool is present it is recommended that the fabric be finished to a pH of about 3 using a non-volatile acid such as tartaric.

5.2.7.3   Bleaching

Even though polyester fibre do not need any bleaching before dyeing it is better to bleach the wool fibre to avoid yellowing during dyeing or later on. Wool can be bleached by oxidative or reductive methods or a combination of both.

Oxidative bleach

Recipe

Quantity	Unit	Additions
10–20	ml/l	Hydrogen peroxide 35%
2–4	g/l	Stabiliser
0.25	g/l	Wetting agent
		pH 5.5–5.7

Procedure: Treat for 45–60 min at 80°C or 150 min at 65°C; then give warm and cold rinses.

Notes

1. This bleaching is much better than alkaline bleaching with the addition of sodium pyrophosphate. The bleaching is much faster, bleaching is less severe and the handle of the bleach fabric is much softer.

2. Whiteness achieved by this acidic bleaching is comparable to alkaline bleaching and less tendency to felt.

Reductive bleach

Reductive bleach is carried out by sodium hydrosulphite. Stabilised hydro's like Blankit N (BASF), Rongolite C, etc., gives much better than common sodium hydrosulphite and can give even better whiteness and light fastness than material where optical brighteners are used.

Recipe

Quantity	Unit	Additions
2–3	g/l	Stabilised hydros (e.g., Blankit N)
0.25	g/l	Combined wetting/detergent agent

Procedure: 90–120 min at 50°C, then rinse and acidify with acetic acid or add about 0.5 ml/l hydrogen peroxide (35%) to the last rinse bath.

A reductive bleach along with an optical brightener or a reducing agent with OBA (e.g., Blankit D, Blankit DA) gives a full bleach effect.

Recipe

Quantity	Unit	Additions
1–5	g/l	Blankit DA
0.3	g/l	Wetting agent
		Adjust pH to 4 with acetic acid

Procedure: Run for 10–20 min at 100–105°C. Then give a good rinse with warm water.

If polyester in the blend also has to be whitened to give a bright full bleach effect an optical brightener for polyester which can fix at 105°C along with a carrier can be used in the same bath.

Quantity	Unit	Additions
3–5	g/l	Blankit DA
0.1–0.2	g/l	Suitable optical brightener (e.g. Palanil white R liq.)
1.5	g/l	Carrier (e.g., Palanil Carrier B)
0.5	g/l	Wetting agent
		Adjust pH to 4 with acetic acid

Procedure: Heat within 30 min to 105°C and treat for 15 min at this Temperature, then give a good warm rinse.

Notes:

1. Use of optical brightener should be adjusted as per the percentage of that fibre in the blend. The recipe given in guideline is a recipe suitable for a 50/50 blend.

2. The optical brightener quantity in case of wool also depend on the kind of wool used in the blend.

3. Higher concentration of the OBA (5 g/l) in the above recipes are for higher MLR of the range 1:20. As it is known, higher quantity of OBA, than required tends to yellow the final shade.

4. Wetting/dispersing agent should prevent precipitation of OBA (e.g., Leophen U).

5.2.7.4   Crabbing

Crabbing is an operation described under wool. But in case of polyester/wool to obtain an optimum setting effect, wool may be given a hot water setting (crabbing) and polyester has to be set hot air treatment (heat setting) but bearing in mind that the material undergoing heat setting contains wool which can be affected by high heat treatment.

Crabbing can be done as in case of wool without any restriction as this treatment has hardly any effect on polyester. Normally, goods that are to be piece dyed are crabbed at the boil after they have been washed. The water and the goods should have a neutral or weakly acid reaction.

Process: Enter the fabric in open-width with low-tension and lowered pressure roller into the water at 50–60°C. Then heat the liquor to the boil. Run for a few minutes and raise the pressure roller. Treat the goods for a further 15–20 minutes. If the fabric is of a heavy quality, it is more advisable to crab by giving two ends with a running time of 10 min each in the boiling bath. Otherwise, uniform setting over the entire length of fabric is not ensured. This also helps to prevent tailing in subsequent dyeing. Then cool the heated fabric suddenly by giving a cold water passage or allow it to cool slowly for 4–6 h on the cooling roller (if necessary overnight). The process used depends on the type of fabric and the effect desired.

Notes:

1. If coloured woven pieces are to be crabbed, it is necessary to consider the wet fastness of the dyed shades. It is advisable to crab at about 60°C and to acidify with some acetic acid or formic acid.

2. Crabbing does not have the lasting effect on the wool as heat setting on the polyester fibre contained in the blend. It is therefore advisable to repeat the crabbing operation during the course of subsequent wet

finishing treatments in order to smooth out creases or to improve the handle of the blend fabric from the side of the wool contained in the blend. The procedure of the second crabbing or smoothing treatment is similar to that of crabbing coloured woven s mineral oil goods, i.e., at a temperature of about 60°C.

3. After any process, including drying has to be done at not more than 100°C, with minimum tension. During drying the width may be stretched minimum (say, 1–1.5 cm more than the wet width with an over feed of 4–6% to remove any creases on the fabric.

5.2.7.5   *Heat setting*

Polyester/wool blend fabrics that are to be piece dyed are normally heat set before dyeing. The object of heat setting is to improve the handle, resiliency, crease resistance and the dimensional stability and to reduce shrinkage and the tendency of polyester fibres to pilling. A similar effect is achieved on the wool fibre contained in the blend by crabbing (see above).

Procedure: A hot air treatment of 30 s at 180–195°C in the stenter is usually adequate.

Notes

1. To achieve god dyeing without streakiness and listing one should ensure that the setting effect obtained across the width and length of the fabric is uniform. For this, the fabric should have uniform moisture throughout the fabric, they should not sag or flutter inside the stenter and there should not be undue stretch width wise.

2. It is also necessary to allow the fabric to shrink freely throughout the entire hot air treatment by suitable adjustment of width and overfeed.

3. Depending on the type of fabric, the warp and weft wise shrinkage is usually of the order of 3–8%. Very densely woven fabrics display less shrinkage than loosely woven.

4. The handle deteriorates when the temperature is raised or the treating time prolonged whereas the resistance to creasing, pilling and to form mill rigs suffers when the temperature is lowered or the treating time reduced. A heat setting treatment carried out under tension, i.e., with less shrinkage, results in poor crease recovery and increased tendency to pilling; in addition, such a fabric is liable to display higher residual shrinkage later on.

5. If dyed fabric requires heat setting may be done for 30 s at 170–180°C if faultless results are to be obtained. In such cases it is always necessary to select disperse dyes which withstand the conditions of heat setting.

6. Since dyed goods would have undergone a shrinkage in dyeing process they usually display less shrinkage. Hence care should be taken in adjusting width during heat setting such cases.

7. The optimum setting effect is achieved if the hot fabric is cooled by a shock treatment in the cooling zone immediately after leaving the last setting zone.

8. Setting always imparts to the fabric a certain degree of harshness which disappears again in subsequent dyeing so that an intermediate wash is not necessary for goods that are to be piece dyed. Densely woven fabrics are an exception. Such fabrics become particularly stiff on heat setting and have to be washed again to facilitate handling before they are dyed.

9. In case of dyed goods, the stiffness must be removed by an after wash to improve the handle, crease recovery and resiliency. In this case, the fabric is treated in an open soaper, rope or jet washing machine or on the winch with

> 0.5% detergent/wetting agent
> x% softener (optional)

at pH 5–6 with 0.5–1 ml/l acetic acid 30% for 20–30 min at 45–60°C. Rinse and dry.

5.2.8   Pre-treatment of acrylic/cellulosic blends

The acrylic cellulosic blend have got the advantage of the good shape retention, abrasion resistance and tear strength of acrylics and the moisture retention of cotton. The blends of acrylic/cotton and acrylic/viscose are mainly used for inexpensive outerwear, plush articles. Articles of which the pile consists of acrylic fibres and the backing material of cellulosic fibres are very suitable for furnishings. Plush articles with a long pile are used for imitation fur, floor coverings with a fur-like property or as plush material for toys. Blended yarns of acrylic/cellulosic fibres are prepared on the woollen system.

5.2.8.1   Pre-treatment

The material is first scoured washed to remove the spinning oil handling stains etc.

Recipe for pretreatment

Quantity	Unit	Chemicals
1	g/l	Non-ionic detergent
0.5–3.0	g/l	Disodium phosphate (or 1 g/l trisodium phosphate)

Run for 20–30 min at 60–70°C and rinse. It is advisable to add some acetic acid to the last rinse bath to remove all alkali from the goods.

5.2.8.2   Bleaching

If necessary, the acrylic/cotton blends can be conveniently bleached using sodium chlorite since it can bleach both cotton and acrylic. The bleaching may be done in a long liquor with the following recipe.

Guideline recipe for chlorite bleach

Quantity	Unit	Chemicals
1–2	g/l	Sodium chlorite 80%
1–2	g/l	Sodium chlorite stabiliser
1–2	ml/l	Formic acid (pH 3.0–3.5)

The goods are entered into the cold bleaching bath which is heated within 20–30 min to the boil and the fabric treated at this temperature for approx. 30 min.

Full white material may be optically brightened by adding suitable whitener for acrylic in the above bleaching bath. Further whiteness can be achieved by treating with OBA for cellulosic fibre in a separate bath after sodium chlorite bleach.

Notes

1. Optical whitening agents acrylic fibre which tints the cellulosic fibre should be avoided as it will affect the light fastness badly.

2. When blends of acrylic fibre with regenerated cellulose fibres are to be bleached, a bath temperature of 80°C should not be exceeded as otherwise oxycellulose is liable to form and impair the cellulosic fibre contained in the blend. In this case, the temperature is raised within approx. 20 min to 80°C and the temperature maintained for approx. 30 min.

5.2.9   Pre-treatment of acrylic/wool blends

Acrylic being wool like it is very suitable for blending with wool and such articles are hardly distinguishable in appearance and texture from pure wool articles. Blends of acrylic fibres and wool are not only similar to wool in many of their properties but also offer advantages. This fibrous blend has better dimensional stability which is very valuable in use. In tensile and tear strengths, too, the fibrous blend is superior to that of pure wool. Another advantage is that practically all colouring effects are possible. In addition to solid shades, it is possible to produce two-colour shades in all imaginable combinations.

When unrelaxed high bulk acrylic fibres are used in blended yarns, it should be completely relaxed before being dyed, either by steaming in an autoclave at 107°C for 10 min after preliminary evacuation of the steamer, or by immersion in boiling water for 5 min.

5.2.9.1   Pre-treatment

The material has to be first washed to remove spinning oils, lubricants and handling stains with following recipe:

Recipe

Quantity	Unit	Chemicals
0.2–0.5	g/l	Non-ionic detergent
0.25	g/l	Ammonia 25%

Treat for approx. 30 min at 40–50°C.
The goods are then rinsed until alkalinity is removed.

Notes

1. The amount of detergent used depends on the degree of soiling and the liquor ratio.

2. Heavily soiled materials may be washed with detergent containing anionic and non-ionic mix with some additional solvents.

5.2.9.2   Bleaching

Acrylic fibre is supplied enough white and may not need bleaching. The wool part can be bleached using peroxide preferably in long liquor.

Recipe

Quantity	Unit	Chemicals
10–15	ml/l	Hydrogen peroxide 35%
1.5–2.0	g/l	Sodium pyrophosphate

Treat the material at 45–50°C, till the required achieved. To achieve the adequate whiteness the material may be treated for 4–6 hours or overnight.
Notes

1. The long bleaching treatment can be shortened by bleaching at higher temperature in the presence of a peroxide stabiliser cum wall protective agent (e.g., Lufibrol W).

Recipe

Quantity	Unit	Chemicals
10–15	ml/l	Hydrogen peroxide 35%
2–3	g/l	Lufibrol W
0.05–1.0	ml/l	Non-ionic detergent

Start the treatment at 35–50°C and the temperature raised within 10–15 min to 80°C. The bath is cooled already after 30 min and the goods rinsed.

The whiteness of wool can be further improved by a reductive bleach after peroxide bleach.

Recipe

Quantity	Unit	Chemicals
1–3	ml/l	Stabilised hydros
0.25–0.5	g/l	Anion-active detergent

At 50°C for 2–2½ h.

If full white effect has to be achieved the OBA may be added along with the reductive bleach. Wool also can be optically brightened separately.

1. If acrylic has to be whitened it may be done with suitable whitener separately.

2. Generally, fatty acid type of lubricants applied in woollen processing should be avoided to prevent discolouration of the acrylic fibres.

3. For worsted yams, scouring can be conducted at a maximum temperature of 60°C for 30 min with 1 g/l non-ionic detergent and 0.5 ml/l acetic acid (80%).

4. Woollen spun yams lubricated with oleine are scoured at 30°C with synthetic detergent and sodium carbonate. The addition of solvent is sometimes advantageous for the removal of oils.

5. Worsted fabrics are crabbed and then scoured in either a Dolly at temperature not exceeding 40°C or in winch.

5.2.10   Pre-treatment of acrylic/polyamide

Polyamide fibres are blended with wool and acrylic for making wind jackets, ski and bathing pants, etc. Acrylic fibre have advantages of having the woolly textile nature, better tensile strength and abrasion resistance than wool and at the same time less costly.

5.2.10.1   Preparation

Spin finishes, lubricants, contamination handling stains, etc., may be removed by washing with following recipe

Recipe

Quantity	Unit	Chemicals
0.2–0.5	ml/l	Non-ionic detergent
0.25	ml/l	Ammonia 25%

Treat for 30 min at 50–60°C. The amounts of detergent required depend on both the degree of contamination, soiling and also the liquor ratio.

If heavily soiled the material may be treated with following recipe:

Recipe

Quantity	Unit	Chemicals
2	g/l	Non-ionic anionic mix detergent
1	ml/l	Ammonia 25% OR
0.05–1.0	g/l	Soda ash

Treat at 60–70°C 30–45 min and rinsed until they are free from alkali.

5.2.10.2   Bleaching

Bleaching of wool/polyamide material can be done by a treatment with sodium chlorite. The goods are entered into the cold dye optical brightening bath which containing following chemicals:

Recipe

Quantity	Unit	Chemicals
1–2	g/l	Sodium chlorite 80%
1–1.5	ml/l	Chlorite stabiliser
0.5–1.0	g/l	Formic acid 85% to make pH 3–3.5
x	g/l	Suitable OBA for wool

At along liquor, heating the liquor in 45 min and holding at boil for about 1 h. If it has to be optically brightened the OBA for wool and polyamide can be added in the bleach bath as per the blend ratio.

Polyamide can be separately optically brightened by treating the material for about 1 h at a liquor ratio of 30:1–50:1 and a temperature of 70–80°C in a bath containing an optical brightener for polyamide (e.g., 3–5 g/l Blankit IN – BASF).

Optical brightening agent for wool which tints the polyamide may not be used as it will hamper the light fastness of the resultant white.

5.2.11   Pre-treatment of acrylic/acetate blends

This blend is not very popular, even though the blends of acrylic fibre and acetate exhibit the woolly nature of the acrylic fibre and also the silky lustre of acetate. The reason for this may lie in the difference in the dyeing behaviour of acrylic fibres and that of acetate which always confronts the dyer with serious problems.

For removing spin finishes, lubricants, handling stains, etc., the material may be washed with following recipe.

Recipe

Quantity	Unit	Chemicals
0.2–0.5	g/l	Non-ionic detergent
0.25	ml/l	Ammonia 25%

For 30 min at 40–50°C. The quantity of detergent may be changed as per the degree of soiling of the fabric. After the cleansing treatment, the material is rinsed till no alkali is left on them.

For bleach ground or full white the material can be bleached oxidation or reduction method or both. For full white fabric OBA may be added to the bleach bath.

Recipe for oxidation bleach

Quantity	Unit	Chemicals
1–2	g/l	Sodium chlorite 80%
1–1.5	g/l	Chlorite stabiliser
1	ml/l	Formic acid 85% for pH 3.5–4

The material is treated for 30–60 min at about 75°C.

For full white 0.5–1.5 g/l of suitable OBA which can tint both acrylic and acetate (e.g., Utraphor N or PAN-BASF) can be added in the bleaching bath.

Recipe for reduction bleach

Quantity	Unit	Chemicals
2–3	g/l	Stabilised hydros
0.5–1	g/l	Cation active detergent cum dispersing agent

The material is treated for about 60 min at 70°C.

Notes

1. To keep the acrylic whiteness safe the treatment should not be done at higher temperature than mentioned above even though acrylic fibre may need the temperature of about 90–95°C.

2. Care should be taken to avoid higher temperature, especially in oxidation bath as the acrylic may be damaged at boiling temperature in the presence of sodium chlorite and formic acid.

3. The above process gives very good whiteness on acetate.

5.2.12   Pre-treatment of acrylic/triacetate blends

The acrylic/triacetate blend has the advantages of the silky handle of triacetate and the fullness and woolly appearance, shape retention, good ironing properties, full handle and easy care properties of the acrylic fibre, which mainly used for fashionable garments. The usual blend ratio used in triacetate/acrylic blends lies between 33:67 and 75:25.

The lubricating agents, spin finishes, soiling of both the synthetic fibres can be washed-off in either acidic or alkaline medium along with a detergent.

Recipe for acid treatment

Quantity	Unit	Chemicals
0.5	g/l	Non-ionic detergent
0.5	ml/l	Acetic acid 30% pH 4–5

The materials are treated for about 30 min at approx. 60°C.

Recipe for alkaline treatment

Quantity	Unit	Chemicals
0.25–0.5	g/l	Non-ionic detergent
0.5	ml/l	Ammonia 25%

The materials are treated for about 30 min at approx. 60°C.

5.2.12.1   S-finish

The hydrophobic nature and anti-static property of the triacetate has some disadvantages for the blend and the garments made of it. To convert the triacetate fibre partly hydrophilic manufacturers suggest a special treatment called S-finish. This treatment involves in the superficial saponification of the surface layers of the fibre to regenerated cellulose, by an alkaline treatment. The regenerated cellulose layer absorbs moisture mand makes the fibre permanently anti-static. This in turn reduces the soiling properties of the triacetate fibre. The S-finish also improves the thermal stability and thus the ironing properties and the tear strength of the goods. The treatment also imparts a soft handle. S-finish is usually done on 100% triacetate but are done on triacetate blends provided the other component in the blend is not negatively affected by the treatment.

Treatment: The pre-washed material is treated with g/l caustic soda for 2 h at 50°C maximum at an MLR of 1:40. The temperature above 50°C, as at this temperature the acrylic may yellow and weamen them.

5.2.12.2 Bleaching

Both component being synthetic and especially acrylic comes with good whiteness may not need a bleaching as such. However, if whiteness has to be improved further it can be given an oxidative bleach with sodium chlorite.

Recipe

Quantity	Unit	Chemicals
1–2	g/l	Sodium chlorite 80%
1–1.5	g/l	Chlorite stabiliser
1–2	ml/l	Formic acid 85% for pH 3–3,5

The material is treated at boil for 30–45 min, at a long liquor ratio (e.g., 30:1). The temperature gradient from 85–100°C should not exceed 1°C/min. The same way after the treatment, the bath is cooled at 1°C/min, to 70–80°C and dropped. Rinse warm, rinse cold.

For full white fabric OBA can be added in the bleaching bath itself (e.g., 0.5–2% Ultraphor NA–BASF is a common OBA for both acrylic and triacetate). If only acrylic has to be whitened one may add the OBA for acrylic in the bath (e.g., 0.5–1.5% Ultraphor PAN–BASF). The same way triacetate can be separately whitened by a polyester whitener (e.g., Palnil White R Liq). Generally. White R cannot be applied along with acrylic OBA in the same bath since the cationic OBA for acrylic may not be miscible with anionic dispersing agent present in the polyester whitener liquid and can cause precipitation. This whitener is applied at higher temperature which may yellow the acrylic and hence not followed. Hence, if at all triacetate has to be optically brightened in this blend it has to be done at 100°C with the help of a carrier.

Recipe

Quantity	Unit	Chemicals
0.05–0.3	%	Polyester OBA (e.g., Palanil Brill. White R Liq- BASF)
x	g/l	Carrier (e.g., 2–3 g/l Levegal PT–Bayer)
0.5–1	ml/l	Acetic acid 30% to pH 5–6
0.5–1	g/l	Dispersing agent

Treat the fabric at boil for 1–2 h. Rinse hot, rinse cold (reduction clear is not a must).

5.2.13   Recipes for pre-treatments for various blends

5.2.13.1   Discontinuous scour boiling

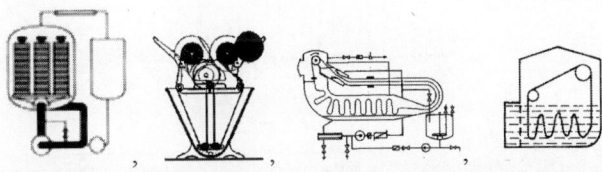

Figure 5.7. Various machineries for discontinuous scouring

Recipe for discontinuous scouring

Form	Loose fibre, yarn, woven and knit goods			
Substrate	Co, PES/Co, PA/Co, Modal/Co, CV			
Machine	Package machine, Kier boiling	Jig	Jet, overflow	Winch
MLR	3:1–5:1	2:1–5:1	**6:1–10:1**	**12:1–20:1**
Guide recipes				
Caustic soda solid g/l	5–20	12–26	6	6
Soda ash g/l			3–5	3–5
Suitable scouring agent ml/l	1–2	2–4		0.5–2
Low-foaming scouring agent ml/l		2–3		
Seq. agent ml/l	2–4			
Working method				
Liquor temperature NT °C	95–100	95–100	95–100	95–100

Liquor temperature HT °C	120			
Treatment time h		4–8	1–3	
min			45–90	45–90

Note

Washing: as hot as possible (over 90°C for the first rinsing bath), then it is advisable to acidify/neutralise with acetic or formic acid for 15–20 min at 30–40°C.

For regenerated cellulose fibres it is advisable to treat with smaller amounts of alkali (ca. 50% of the amount given) and temperatures up to max. 85°C.

5.2.13.2   Continuous scouring of Co, Co/PES, PA/Co, Modal/Co

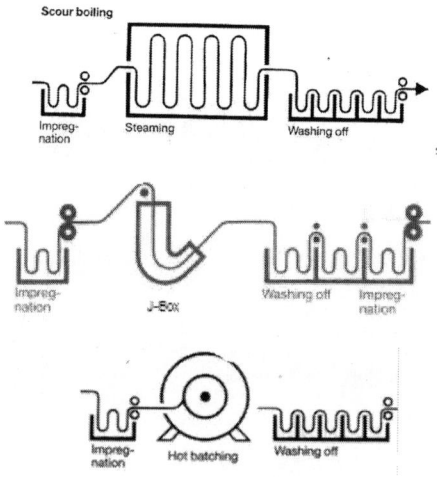

Figure 5.8. Continuous scouring Machineries

Recipe

Form	Woven	
Substrate	Co, PES/Co, PA/Co, Modal/Co, CV	
Machine	Pad-steam NT/HT, Combi steamer, Conveyor	U box, L box, J box, Pad-roll
Type of treatment	Short dwell time	Med and long dwell time

Guide recipes

Caustic soda solid g/l	30–80	30–80	20–60	20–60
Suitable scouring agent ml/l	5–15		5–15	
Low-foaming scouring agent ml/l		3–6		3–6
Seq. agent ml/l	0–10	0–10	0–10	0–10
Working method				
Saturator impregnation				
Liquor temperature °C	20–40	20–40	20–40	20–40
Pick-up %	100	100	100	100
Steaming and dwelling				
Temperature NT °C	100–106	100–106	90–98	90–98
HT °C	110–140	110–140		
Time NT min	5–20	5–20	20–120	20–120
HT min	1–2	1–2		
Washing-off	As hot as possible (at least 90°C) neutralise if necessary			

Notes

1) The amount of caustic soda must be adapted to the substrate and steaming/dwell times.

2) With wet-on-wet impregnation the ideal add-on is 25–30%, metering of feed liquor is essential.

5.2.13.3   Discontinuous scouring of synthetic fibres (PA, PAC, PES, CT, CA)

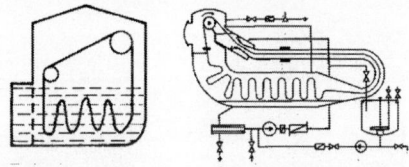

Figure 5.9. Discontinuous scouring machineries

Recipe

Chemicals	PA	PAC		PES			CT	CA
		I	II	I	II	III		
Solvent based scouring agent ml/l	0.3–1	0.3–1.0		0.3–1.0	0.3–1.0		0.3–1.0	0.3–1.0
Suitable detergent ml/l			0.3–1.0			0.3–1.0	0.5–1.0	
Scouring agent for synthetic sizes				0.3–1.0				
Sodium tripolyphosphate g/l								
Tetrasodium pyrophosphate g/l		0.5–1.0			1–2			
Soda ash g/l	2–3							
Formic acid 85% ml/l			1–2			1–2		
Time, min	30	30	30	30–45	30–45	30–45	30	30
Temperature °C	60–70	50–60		50–60	50–60	60	50–60	50–60

Liquor ratio for all cases 10:1–30:1.

PAC Recipe II and PES Recipe II are for cases where cationic agents are present.

5.2.13.4 *Continuous scouring of synthetic fibres (PA, PAC, PES, CT, CA)*

For a continuous scouring of these materials a washing machine can be used. In a 5–6 compartment soaper, first two compartments the following recipe can be taken and further compartments for washing, neutralising and rinsing.

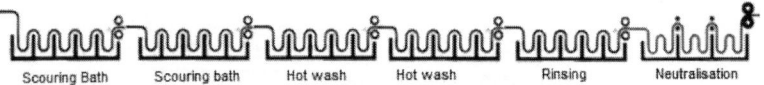

Scouring Bath   Scouring bath   Hot wash   Hot wash   Rinsing   Neutralisation

Figure 5.10. Continuous scouring range

Recipe

	PA	PAC		PES		CT	CA
		I	II	I	II		
Solvent based scouring agent and stain remover ml/l	0.3–1.0	0.3–1.0		2–4		0.3–1.0	0.3–1.0
Amphoteric detergent ml/l			1–5		1–5	0.5–1.0	
Scoring agent for synthetic sizes				0.3–1.0			
Sodium tripoly-phosphate g/l	2–3	0.5–2				1–2	
OR Soda ash g/l	2–3						
Formic acid 85% ml/l			1–2		1–2		
Time, min	30	30	30	30–45	30–45	30	30
Temperature °C	70–80	40–50	50	40–60	50	60	50

5.2.13.5   Continuous scouring and washing of synthetic fibres

Recipes

Form	Tow, yarn, woven and knit goods	
Substrate	PA, PAC, PES, CA, CT	
Machine	Backwasher, suction drum, and other OW washing machines	
Type of treatment	Alkaline	Acid
Guide recipes		
Low-foaming scouring agent ml/l	0.5–2.0	0.5–2.0
Low-foaming scouring agent acid stable ml/l		2.0–5.0
Anionic Seq. agent	0.5–1.0	0.5–1.0
Sodium tripolyphosphate	0.5–3.0	
Soda ash		2–3

Formic acid	1–2
Treatment temperature °C	50–80°C, (Max 60°C for PAC, CA)
Treatment time min	Depends on size of machine and time of passage, longer treatment times improve washing effect
Rinse	Warm and cold

Notes

1. Acid scouring is recommended with cationic preparations.

2. After alkaline scouring it is advisable to treat briefly with acetic or formic acid to neutralise any alkali residues which may interfere in dyeing.

5.2.13.6   *Discontinuous scouring and washing*

(a) Cellulosics and their blends–Loose fibres, yarn, woven and knit goods

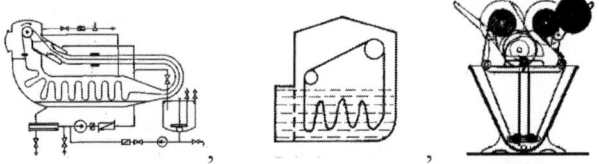

Figure 5.11. Discontinuous scouring machineries

Table 5.5.

Machines	Jet, overflow and other circulation mcs.	Winch	Jig
Liquor ratio	6:1–15:1	8:1–20:1	2:1–5:1
Non-ionic/anionic stain removing and scouring agent with caustic soda resistance up to 98°C ml/l		1–2	2–4
Non-ionic, silicon free, low-foaming, wetting and detergent ml/l	0.75–1.5		
Soda ash g/l	0.5–2.0	0.5–1	2–4
Method			
If necessary rinse with water 40–50°C, 10 min			
In fresh liquor			
Starting temperature °C	40	30–40	30–40

Treatment temperature	60–90		60–90	60–90
Treatment time min	20–30		20–30	45–60

Rinse hot, cold, neutralise with acetic or formic acid

(b) Synthetic fibres (PA, PAC, PES, CT, CA)

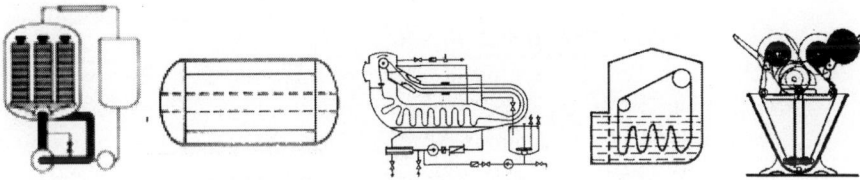

Figure 5.12. Discontinuous scouring machines

Recipes

Type of treatment	Alkaline	Alkaline low-foam	Acid	Acid low-foam	Neutral	Neutral low-foam
Non-ionic/anionic stain removing and scouring agent with caustic soda resistance up to 98°C ml/l	0.5–0.2	1–2	0.5–0.2		1–3	
Non-ionic, silicon free, low-foaming, wetting and detergent ml/l		0.75–1.5		0.75–1.5		0.75–1.5
Soda ash g/l	2–3	2–3				
Formic acid 85% ml/l						

Working method

Treatment temperature: 50–70°C (Max. 60°C for PAC and CA).

Treatment time; 30–45 min.

Rinsing: Wash-off warm and cold, with alkaline treatment neutralise if necessary with acetic or formic acid.

Machines: All machines: Package machines, cheese and beam dyeing machines, jet/overflow, winch, jig, etc.

Liquor ratio: 2:1–20:1.

5.2.14   Optical whitening

Bleaching what we have explained above does not give a white fabric in a customer's point of view. The best white which is obtained by probably a combined oxidative and reductive bleach will be only a yellowish white as far as a garment customer is concerned. The first attempt was to use a bluish or violetish tint to cover up the yellowishness which will give a pleasant white but not a perfect white.

It is common place that washed linen is cleaner than used linen. One can see it immediately by its colour. So it is probably a psychological phenomenon that people associate a pure white colour with cleanness hygiene and innocence in contrast to a yellowish white which makes them think of impurity.

5.2.14.1   Optical brightening agents

Some 150 years ago, Stokes initially studied "fluorescence" as a scientific phenomenon. In 1929 Krais demonstrated that a yellowish viscose material appeared whiter after being immersed in an esculin solution–an effect we would call "brightening" today. Methods of brightening were not entirely unknown at the time–as early as the end of the 18th century.

Paper makers discovered that horse chestnut extracts enhanced the effect of bleach liquors.

However, they were unaware of the fact that esculin–a glycoside of 6,7-dihydroxy-couma-rin was responsible for this improvement. This was the first demonstration on the yellow-compensating effect of colourless but bluish fluorescing chemicals. For some 50 years, papermaker and coaters have availed themselves of optical brighteners to eliminate the yellowness in paper products.

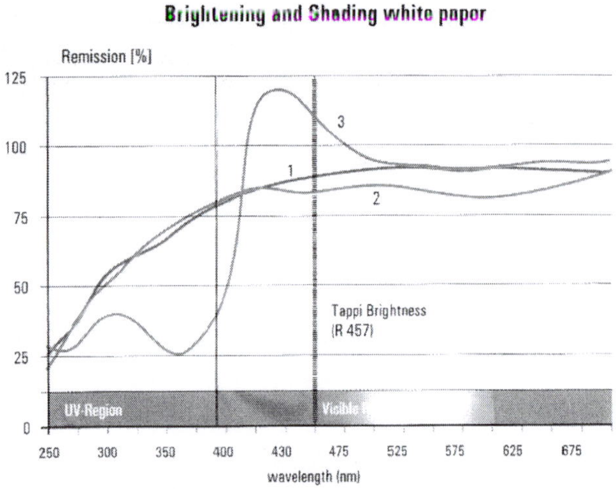

Figure 5.13. Brightening and shading white paper

Researches followed to make a perfect white or a near perfect white. This resulted in the production of optical whitening agents. Optical whitening agents act as fluorescent dyes though they are colourless. These agents absorb the UV -rays, which are invisible to the human eye and convert them into visible blue–violet rays, which result in a higher degree of whiteness. Optical brighteners resemble dyes in all respects except that they have no colour and are thus called colourless dyes. They absorb ultraviolet light (300–400 nm) from daylight and emit it in the visible range (400–500 nm) at the blue region of the spectrum.

Optical brighteners or whiteners are chemical compounds that absorb invisible ultraviolet radiation from daylight and re-emit it as visible blue fluorescent light. OBA's take up invisible UV light in the spectral band of 300–400 nm and, owing to the Stokes's shift, re-emit it in the visible higher wavelength spectrum of approx. 400–450 nm.

The green curve (1) shows the spectra reflection of the base paper, which has a "blue deficiency" so the optical brightener has to emit blue light to offset the deficiency. However, since the brightener rightly over compensates this deficiency, more light is re-emitted in the visible blue range than was originally irradiated, paper brightness is increased and the paper appears whiter to the eye (red curve 3). Alternatively, shading dyes may be added which give the converse effect. In this case the spectral reflection in the yellow and red bands is reduced as uniformly as possible to reach the blue region in order to compensate for the blue deficiency. Overall, the reflected quantity of light is decreased and the luminosity of the paper is reduced (blue curve 2).

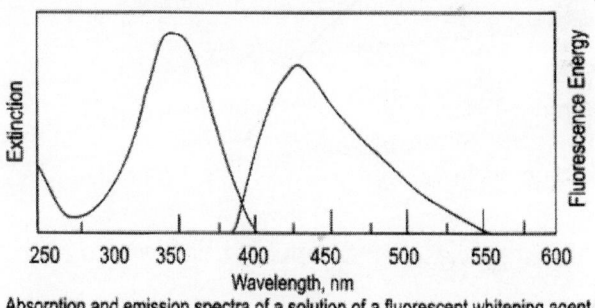

Absorption and emission spectra of a solution of a fluorescent whitening agent

Figure 5.14.

Near-white polymeric substrates ego bleached cloth, possess yellowness caused by absorption in the blue region when an OBA is applied, the blue fluorescence complements the yellowness and adds a bluish hue to the substrate, which the eye appreciates as a brilliant white. The whiteness thus obtained is much superior to that achieved by chemical bleaching even if the latter is carried out to the limit where the cloth is weakened by chemical degradation or if chemical bleaching is supplemented by blue tinting. A well-bleached substrate is essential for optical whitening to be effective. OBAs are used in washing powders and on paper for about 75% of total. Their use in the finishing of white textile goods, particularly cotton is very important, it is about 20% of total. It is also used in mass whitening of synthetic fibres during manufacture or during melt or solvent spinning.

5.2.14.2   What is fluorescence?

When a dye molecule absorbs a photon of light, an electron is excited from the singlet ground state, S_0 to a higher electronic energy level, particularly the first level, S_1. Then the molecule is thus in the vibrationally excited state. However, vibrational relaxation takes place very rapidly (in about 10^{-12} s) and the molecule settles into the vibrational ground state of S_1 losing thermal energy to its surroundings.

Absorption (A) of light quanta by the brightener molecules induces transitions from the singlet ground state S to vibrational levels of the electronically excited singlet states (S_1). Several routes deactivate brighteners in the S_1 state. Fluorescence results from radiative transitions to vibrational levels of the ground state (F). Deactivation processes competing with fluorescence are mainly non-radiative deactivation to the S_0 state (IC) and non-radiative transition to the triplet state (inter system crossing, ISC). The efficiency of fluorescence is measured by the quantum yield:

$$\Phi = \frac{\text{Number of quanta emitted}}{\text{Number of quanta absorbed.}}$$

It is determined by the relative rates of fluorescence emission and the competing processes. When fixed in solid substrates, brighteners fluoresce with high quantum yields.

5.2.15   Energy diagram of optical brighteners and transitions

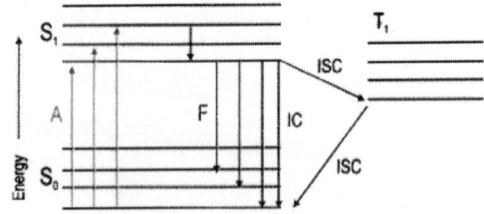

A–Absorption, F–Fluorescence, IC–Internal Conversion, ISC–Inter system Crossing, S–Singlet State, T–Triplet State.

Figure 5.15. Energy diagram of OBA

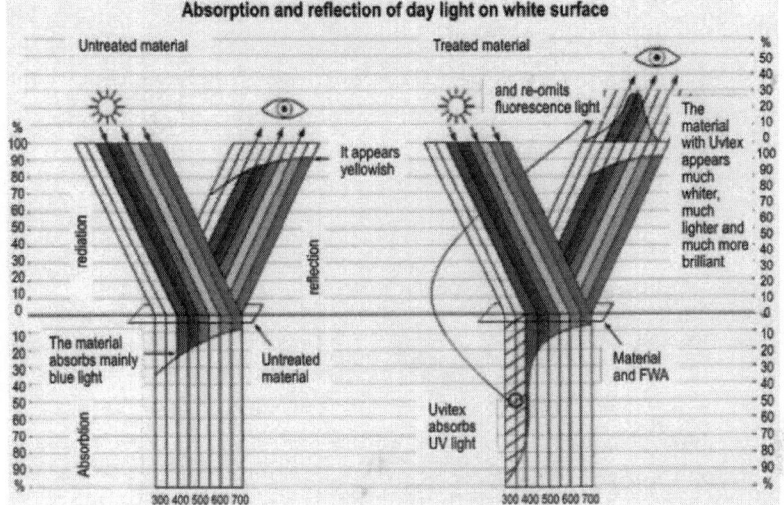

Figure 5.16. Absorption and reflection of daylight on white surface

5.2.15.1   Different groups of optical brightening agents

- Stilbene
- Benzothiazole
- Benzoaminoazole

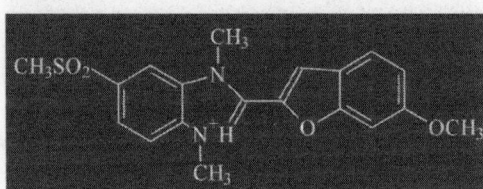

Figure 5.17. Benzoaminoazole

- Benzoxazole
- Coumarin

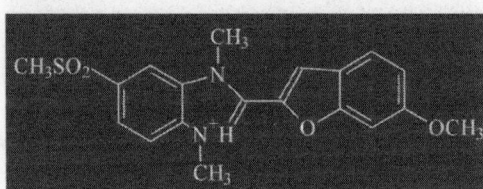

Figure 5.18. Coumarin derivatives

- Pyrazolin

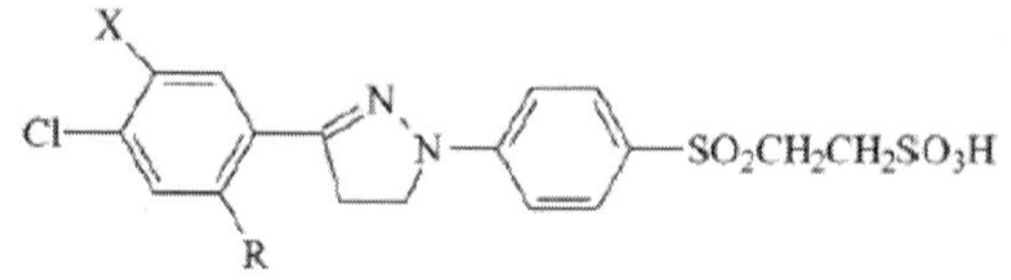

X = H, R = H or X=Cl, R = Ch,

Figure 5.19. Pyrazoline compound

5.2.15.2   General molecular structure OBA

- An extended conjugated system of double bonds (=) should be present in one plane, in the molecular structure of optical brightening agent.
- Such structure is required for excitation from ó to ó* i.e., for fluorescence to occur.
- Planarity of molecule ensures fluorescent efficiency and high affinity to the substrate.

Owing to their molecular structure, all optical brighteners have a special feature that must be taken into account for both internal and surface application: they will only give maximum performance if they are anchored to an appropriate medium. The reason is that the brightening molecule only develops its maximum fluorescence if it is retained in a plane. As long as the molecule remains in its transformed state, parts of the molecule are free to oscillate and tend to turn out of the plane. The mesmeric system of π-electrons is in this case prevented from optimally spreading across the entire molecule so that fluorescence losses are experienced.

It is possible to anchor the OBA because of its affinity to cellulose fibres. The term "substantivity" in this context describes the capability of an anionic molecule, such as a substantive dye or an optical brightener, to attach to anionic cellulose fibres. The driving force behind this mechanism is the interactions between the π-electrons of the brightener molecule and the hydroxyl groups of the cellulose or hemi-cellulose. Similar effects may be observed with –OH groups of starches, CMC (Carboxy Methyl Cellulose), PVOH (Polyvinyl Alcohol) or the methylol groups of MF (MelamineFormaldehyde) resins. These effects produce the so called carrier properties of the products concerned.

Cis-trans isomerism

All commercial optical brighteners are available in the low energy *trans*-form. The only form in which an optical brightener is active. When being raised to a higher energetic level by intensive day light irradiation, the molecule is eventually irreversibly transformed to the optically inactive *cis*-form.

5.2.15.3   General requirements of an OBA for a textile processor

It should have high stability and fastness. Also it should have affinity towards the fibre for which it is meant. After treatment it should give good whiteness index. It should not be yellow on keeping or in contact with any chemical (phenolic yellowing).

High affinity type of whiteners exhaust to about 80% on the fibre at equilibrium from exhaustion bath. Affinity can be increased by adding a salt to the treatment bath. It is effective in the pH range of 5–11. It is particularly suited for pad-batch application, together with a resin treatment or peroxide bleach. These are mainly used for whitening regenerated cellulose, which requires *OBAs* of higher affinity than for cotton.

Introducing an extra two sulphonate groups into the molecule of high affinity *OBA,* solubility can be increased. Thereby it reduces the affinity for cotton, increases the substantivity. With low-medium affinity whiteners good levelling can be achieved. They are very sensitive to salt concentration and in the presence of 0.5% w/v sodium sulphate, exhaustion of greater than 95% is achieved. They are ideally suited to a padding process.

A third type of OBA has low-affinity even in the presence of salts. This type includes the DASC (Diaminostilbenedisulphonic acid-cyanuric chloride) products containing 6 sulphonic groups. Due to their high stability to acids, these products are particularly suitable for use in conjunction with resin treatments involving highly acidic conditions.

Generally anionic whiteners are used for cotton, wool, and polyamides, cationic whiteners are used mostly for polyacrylonitrile fibres and non-ionic whiteners are used for polyester, acetate, acrylic and polyamide fibres.

Most customers require high light fastness, as we often hear that optical brighteners disapppear, fade or yellow. This is not really true–they are just deactivated when exposed to daylight. As with dyes, the light fastness of a particular OBA depends on the substrate and cannot be stated generally, for example:

- In highly diluted aqueous solutions, for instance, optical brighteners very quickly lose most of their optical activity when exposed to daylight. Their light fastness is below 1 of the blue wool scale.

- On exposure to light, OBA's used in paper coatings usually exhibit a restricted light fastness rating of around 1.5 of the blue wool scale.

- When internally applied, OBA's perform better in terms of light fastness, coming in the range of ratings 1–3 of the blue wool scale, depending on their type and concentration.

- When used in paper laminates that are impregnated and pressed with melamine resins, OBA's attain light fastness levels >6. However, the titanium dioxide required to give the desired opacity of the laminate seriously impairs the performance of optical brighteners.

5.2.15.4   Saturation limit or greying or greening

Generally speaking, all OBA's reach a maximum whiteness level. In terms of the maximum attainable brightness, however, the various types of brighteners behave differently. The saturation limit may be defined as the dosage of optical brightener beyond which no further brightness gain is perceptible by the human eye. Instead, greying of the paper may be experienced because the fibres to be brightened are saturated. Under acid sized conditions greening can occur. It is generally believed that for OBA's to function efficiently, they must be present on the substrate in a monomolecular layer. As soon as enough OBA is added so that 2two molecules, or more, of OBA are present together forming a particle, the fluorescence is impaired and the shade becomes greener! yellower. This indicates the start of the saturation point.

5.2.15.5   OBAs for cotton and viscose

Most of the OBAs used for cotton and regenerated cellulose are of the DASC (Diaminostilbenedisulphonic acid-cyanuric chloride) type. Preparation of DASC type OBA they are prepared from cyanuric chloride (CC) which is condensed with 4,4'-diaminostilbene-2,2'-disulphonic acid (DAS).

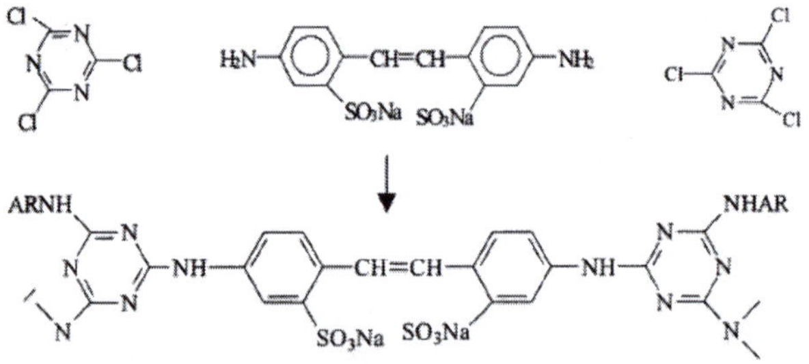

Figure 5.20. Diaminostilbenedisulphonic acid-cyanuric chloride

By varying the number of sulphonic acid groups in the aromatic amine $ArNH_2$ from none to two, an *OBA* may be synthesised containing from two to six sulphonic groups. The number of hydroxyl groups in the aliphatic amine can also be change. Thus the polarity of a DASC whitener and its solubility in water can be readily varied within a wide range.

5.2.15.6   OBAs for polyamide

These include DASC whiteners of both high and medium affinity as well as triazolestilbene disulphonic acids and DSBP whiteners. Pyrazoline compounds as is naphthotriazol stilbene derivatives are also effective for polyamide. Coumarin whiteners have poor light fastness. Benzoxazole derivatives are also used for the whitening of polyamide textiles.

Application
1. Exhaustion technique
Typical processing sequence for polyamide fibres.
Reduction bleach bath

Recipe

Quantity	Unit	Chemicals
0.5–1.0	%	Suitable OBA
2–5	g/l	Sodium hydrosulphite
0.4–1.0	g/l	Sodium tripolyphosphate
		pH 9–10

Enter goods at 50°C. Raise to boil in 30 min and continue for around 30 min. OBA for polyamide can be incorporated into a semi-continuous or continuous padding process. In one such process, the cloth is padded with OBA, sodium dithionite and detergent at 90°C and then heated at 120–130°C for 15–20 min while bleaching takes place, the OBA diffuses into the fibre and the optical whiteness is developed. In the pad-roll process, the impregnated cloth is held for 1–2 h at 90°C. In continuous pad-steam process, steaming is for 3 min at 105°C and for shorter periods at higher temperature–high-affinity OBAs are used under slightly acid condition.

Acid-shock treatment
The open-width cloth is impregnated with scouring detergent and OBA at above 50°C. A high concentration (15 g/l) of high affinity OBA is used. The cloth then, asses through a series of hot washes at 90°C, acidified to pH 3.5–4.0, which may also contain OBA in low-concentration. The process is more suitable for nylon 6. The treatment is usually followed by heat setting.

Pad-bake (thermosol) treatment
After padding during heat treatment OBA diffuses into the fibre forming a solid solution. Heating is for 10–30 s at 190°C for nylon 6 and at 200°C for nylon 66. Thermosol whitening is less suited for polyamide than for polyester, because slower penetration into the fibre.

5.2.15.7   OBA for polyacrylonitrile

Acrylic fibres are whitened with non-ionic disperse whiteners. However, modacrylic fibres can be can be whitened with some cationic OBAs, they are fixed on the fibres by diffusing inside and forming a salt with the acid groups.

Water soluble cationic whiteners are the pyrazoline, the bis–benzoxazole, naphthalimide and group of mixed benzimidazole-benzofuran salt, etc.

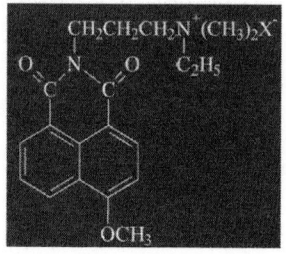
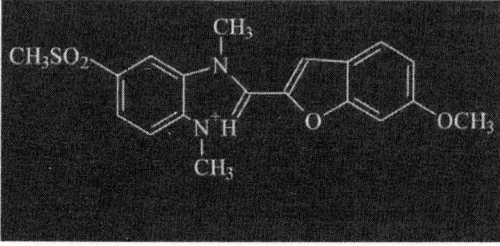

Naphthalimide   Group of mixed benzimidazole-benzofuran salt

Figure 5.21. OBA for polyacrilonitrile

Application

Many acyilonitrile fibres have high natural whiteness and do not need chemical bleaching. OBAs can be applied in a bath together with a softening agent at pH 4–6. A rapid, one stage bleaching, whitening softening and differential shrinking process for staple yarn can be done.

Enter the fabric in a solution containing the OBA, defoaming agent, formic acid to adjust pH 3–3.5, sodium chlorite 1–2 g/l, sodium nitrite 1–2 g/l, and at 40–50°C and heat. Heat from 70°C at 0.5–1.0°C per min to boil. Treat for 30 min at boil, cool, drain and rinse.

5.2.15.8   OBAs for polyester

In case of these non-polar fibres, like polyester the OBA diffuses into the polymer substrate at high temperature, forming a solid solution and for this purpose non-polar whiteners are needed. These include the styrylstilbene compound, bisbenzoxazoles, 4 phenylbenzoxazyl stilbenes, naphthotriazolyl-stilbenes and naphthalimides. Non-ionic optical brighteners made in the form of disperse dyes are used for polyester, Acetate fibres and in some cases polyamide also for fluorescence brightening. Hence generally the dyeing method will be same as that of disperse dyes.

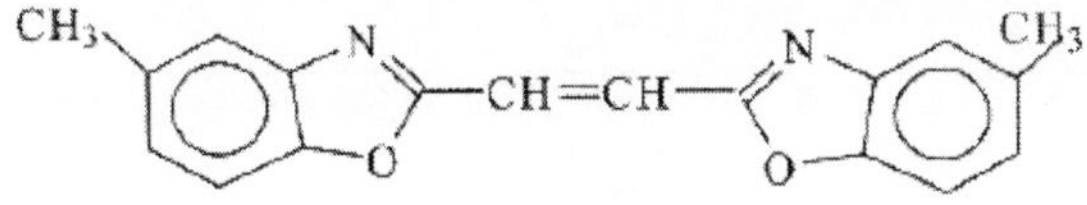

Bisbenzoxazoles

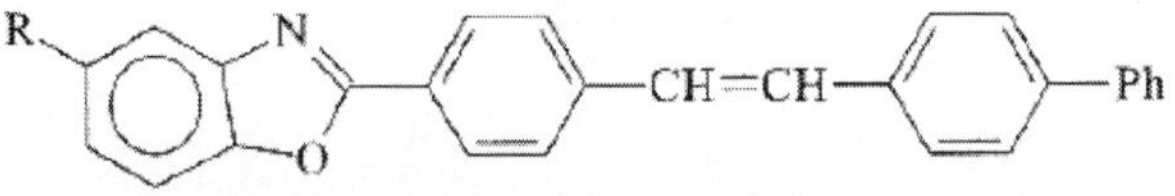

4-phenylbenzoxazyl stilbenes

Figure 5.22. OBAs for polyester

Some of these OBAs are not stable to bleaching by sodium chlorite either in the same bath or where the bleaching is carried out after applying them. Hence in these cases sodium chlorite bleaching is carried out prior to applying them. It is stable to bleaching by hydrogen peroxide either in the same bath or as an after-treatment. If the combined treatment is carried out under high temperature conditions, i.e., >110°C (230°F), the yield obtained from the OBA may be reduced.

Application of 100% polyester and PE/cellulosic blends
These OBAs can be applied to 100% polyester or polyester blends by padding or batch wise methods, i.e., high temperature or carrier dyeing.
100% polyester exhaust method
Exhaust applications can be carried out either under high temperature conditions at 120–130°C (250–265°F) or at atmospheric temperature with the addition of a carrier. Most carriers are suitable for this application with the exception of *ortho* phenyl-phenol types (OPP) which give a poor yield. It is essential to remove the carrier after dyeing in order to avoid subsequent yellowing on exposure to light. In certain instances it may not be desirable to use a carrier. In such circumstances a reduced white will be obtained, which may be improved by prolonging the dyeing time. It should be noted that no improvement will be achieved by increasing the amount of OBA than recommended.

High temperature dyeing will produce the best result, with carrier dyeing being slightly inferior. Dyeing without carrier is a little inferior to carrier dyeing. Exhaust processes are used mainly on yarn and knitted piece goods where a padding operation is not always practicable.

5.2.15.9   *Typical recipes*
Carrier dyeing

Recipe

0.1–0.4%	OBA on weight of goods
1.0 g/l	Carrier
0.25 g/l	Levelling agent
	pH adjusted to 4–5

The bath is heated to the boil over 30 min, with the carrier being added at 75–80°C (165–175°F). The temperature is maintained at the boil for 45–60 min. The goods are rinsed hot, cold, dried and heat-set. If the goods are already pre-heat set it is necessary to soap-off the fabric at 70–75°C (160–165°F) in order to remove residual carrier.

Without carrier

Recipe

0.1–0.4%	OBA on weight of goods
0.25 g/l	Sequestering agent
	pH adjusted to 4–5

The goods are entered in the above bath and the temperature is raised to the boil over 30 min and maintained at the boil for 60–90 min. The goods are finally rinsed, dried and heat-set.

High temperature dyeing
The goods are entered into a bath containing:

Recipe

0.1–0.4%	OBA on weight of goods
0.25 g/l	Sequestering agent
	pH adjusted to 4–5

The bath is heated to 120°C (250°F) over 30 min and maintained at this temperature for 30–45 min, then cooled. The goods are finally rinsed, dried and heat-set.

Padding method (Thermosol or high temperature steaming)
This is the most economical process for bulk production. The heat setting, which is necessary for the fabric, can be utilised to develop the OBA. The largest outlet for white on 100% polyester is as a background white for prints and, if necessary, the steaming required to fix the print will also give full development of the OBA. The heat required to ensure full development OBA will vary with

different types of fabrics. As a general rule, to obtain full development in 30 s requires a minimum temperature of 185–190°C (365–375°F). However, it is not recommended to heat certain fabrics to this temperature. In these cases a partial development will be obtained and in the case of printing, the full white will be developed during steaming. If full development is required the time of heat setting may be increased to 60–70 s. Times and temperatures will vary with different fabrics and stenters. The times quoted are for pre-dried fabric.

Typical recipe (50% mangle expression)
The fabric is padded through 2–8 g/l OBA.
        dried at 110–120°C (230–250°F).
        heat-set for 30 s at 190°C (375°F).

5.2.16   Polyester/cellulosic blends

Polyester/cellulose is mainly processed in woven piece form and a continuous process is normally required. The method employed is pad+heat-set where the heat-setting necessary for the fabric will also develop the OBA. There are however, instances where a padding application is not feasible, e.g., yarns and knitted goods

5.2.16.1   Pad + heat-set

Where optimum whites are required, e.g., high class shirtings, it is necessary to carry out a three-stage process as follows:
    (a) Apply OBA by pad + heat-set.
    (b) Peroxide bleach (A fluorescent brightening agent for cellulose can be added at this stage if required).
    (c) Apply finish–usually resin plus a fluorescent brightening agent for the cellulose.
    The reason for adopting this processing route is that OBA for PE does not whiten the cellulose and will, in fact, leave a slight discolouration; also the heat setting temperatures required, 180–200°C (355–390°F), will cause discolouration of the cellulosic portion of the blend. The peroxide bleaching, therefore, is carried out to clear the cellulosic portion of this discolouration. The cellulosic fluorescent brightening agent can be applied either during peroxide bleaching, if no further processing is to be given, or in the final padding bath. Care must be taken with the latter method as an excess application will cause greening on the polyester portion giving a final duller white.

Typical process for high whiteness

Singe → Desize → Scour → Bleach (chlorite or peroxide) → Dry → Pad → Dry → Heat-set (for 30 s at 190–195°C (375–385°F) → (Singe and mercerise if necessary) → Peroxide bleach (Cellulosic OBA may be added at this stage if no further finish is required.) → Dry → Finish (Padding type OBA at 3 g/l may be added to the final finish; tinting dye may also be added at this stage.)

5.2.16.2   OBAs for wool

Wool contains basic groups and is usually whitened by acid dye type OBAs, similar to those used for polyamide. In case of wool photo yellowing is a common phenomenon. OBAs sensitise the photo chemical degradation and yellowing of wool particularly the peroxide-bleached wool in wet state. Of the commercially available OBAs, those of the DASC type are said to cause the least photo yellowing. There is also process to reduce the photo yellowing of the optically whitened wool by treatment with thiourea formaldehyde resin.

Application
Optical whitening is carried out together with a reductive bleaching agent in a weakly acid medium. The acidity (pH 5–5.5) is provided by the hydrosulphite preparations used for reductive bleaching. With medium affinity whiteners the pH is adjusted to 4 by using acetic acid.

Goods are pre-bleached with hydrogen pperoxide are treated at temperature 50–60°C for 30 min in a bath containing suitable OBA, sodium hydrosulphite (2–5 g/l), sodium tripolyphosphate (STPP) (0.4–1.0 g/l) adjusted to pH 5–5.5.

Precautions to be taken to get better, bright, optically brightened cloth and avoid yellowing when curing the fabric treated with synthetic resins. It is always to do curing at lower temperature for longer time. Core alkali of the fabric is to be neutralised to avoid yellowing during the drying or curing. The use of cationic yellowing softeners, additives, filling agents, etc., has to be completely avoided and as far as possible non-yellowing softeners or non-ionic softeners has to be used.

The base goods should be properly bleached with minimum whiteness index and absorbent to achieve good whiteness on OBA treatment. Machine used while exhaust application of OBA should be free from any dye residues which may cause undesired colouration and hamper the final whiteness. Fabric should be free from metal traces, soft water should be used for OBA application. One should check the compatibility of the finishing agents or any other chemicals used either in the OBA bath or in further treatments like finishing.

The finished material should not be stored in humid or damp condition to avoid any bacteria growth. Yellowing can happen during storage of the made-up goods due to antioxidants from packaging materials hence storage in such packaging material should be strictly avoided.

5.2.17   Preparation of elastane fabric

Elastane/spandex (the commonly used generic name is spandex or elastane) fibre was first developed to replace rubber to use in fashionable garments, waistbands, etc. Du Pont Corporation first introduced these in market in 1959 with brand name "Lycra" and later on Dunlop Rubber Co. under the brand name "Vyrene". These are elastomeric man-made fibres mainly used for its property of stretchability allowing the garment to move as the body moves, providing the wearer comfort and also the fibres shape retention property. These fibres are long chain block copolymer comprising of 85% of segmented polyurethane. Though these fibres are used in all types of textiles but it is still more used in few areas such as sportswear, swimwear, undergarment, etc.

Spandex/elastin fibres are not used alone. These are being used in combination with any natural or synthetic fibre and amount of spandex fibre used depends on its end use application.

The lycra is a polymer consists of uniformly arranged soft and hard segments. The soft unarranged segment gives the property of rubber like elasticity and the hard short crystalline areas causes the resetting after deformation.

Elastane is used in knitted and woven goods. Elastane yarn is incorporated in to the textile constructions where their high elastic extension and recovery are advantageous and increases property of the garments in the garments in which the fabric is used. Elastane yarns are never used alone in fabrics, but rather in combination with other yarns. Blended with other fibres like Nylon, Polyester, Cotton, Silk, Wool and other fibres either as bare yarns or as covered elastic yarns: Co-twisted, core spun, covered or co-tangled.

Typical elastic percentages in the elastane fabrics:

Woven fabrics	2–8%
Swim/sportswear	12–20%
Corsetry	10–45%
Under wear	2–5%
Ladies hosiery	2–12%
Medical hosiery	35–50%

Fabrics containing elastane are sensitive to unsaturated oils, grease, fatty acids and their derivatives which tend to decolourise and degrade the elastane. Chlorine liberating chemicals will discolour and degrade the elastane fibres. Fumes, atmospheric pollutants or a long-term exposure to ultra violet (UV) light may decrease the initial whiteness of tricot fabrics containing elastane.

The problem when finishing elastic articles is that maximum stretchability and elastic properties must be achieved while maintaining other technological features (final width, weight per square metre, dimensional stability, etc.). Elastic articles are very sensitive to marks and should in the majority of cases be processed open/width. Low-tension treatment of the fabric is also particularly important. The stretchability and elastic recovery are primarily pre-defined by the elastane content and by the weaving or knitting structure. An incorrectly performed finishing process can, however, permanently reduce the stretchability and elastic recovery. Damage to the elastane can, for instance, be caused by incorrect storage (folded instead of in rolls, excessively long storage), the action of light during storage and even by knitting oils and other oils. In such cases there is no way back for the finisher. There is practically no means of correcting this situation. Chemical damage to the elastane fibres can be caused by excessively high quantities of alkaline, acidic, oxidation or reducing agents. However, given correct formulation and the right process sequence the elastane fibre is not at risk. During the heat setting process, the elastane can also be damaged by an excessively high setting temperature and an excessively long setting time. These problems are well-characterised and experienced users know how to prevent them. Today these problems no longer need to be the reason for inadequate stretchability and elastic recovery.

If the shrinkage properties of a woven fabric and/or knitwear are greater than that necessary for achieving the weight per square metre and width, the fabric must be heat set using hot air. Heat set fabrics normally have a greater dimensional stability and the fabric edges have less tendency to curl, considerably easing further processing.

A typical process route for elastane fabrics:

Knitted fabrics
Relaxation → Heat setting → Scouring/bleaching → Dyeing/Printing/Whitening → Finishing

Woven fabrics
Relaxation→Drying/Heat setting →Singeing → Cold Pad-batch Bleaching → Washing → Drying → Finishing

Blends with elastane should be treated in grey stage in a relaxed condition with steam or hot water prior to heat setting or/and pre-treatment steps. Relaxation releases the residual stress built-up by tensioning of the yarn during knitting/weaving.

Relaxation is carried out by

• Hot water treatment
• Steam frame
• Steam table.

5.2.17.1   Storage of grey/partly processed fabrics

Any fabric in the grey form has to be stored till it is taken for processing. Lycra fabric has to be carefully stored with care as a good storage can help and bad storage can reduce the quality of the finished fabric.

- Stacking of elastane fabric is undesirable as it will obstruct the handling of the bottom pieces that were stored first. The weight and pressure of the stack can cause undesired moiré, folds and creases which sometimes cannot be cleared with any further processes (permanent). It is best to keep the rolls of grey suspended horizontally, one by one, on metal axle-tubes. Light pieces can be stored individually in boxes. Storage of grey fabrics must be short to prevent yellowing or decay of elastane by hard yarn finish or machine lubricants that might cause unsaturated fatty acids or fatty esters. The knitter, weaver, fibre supplier or coning oil producer should make sure by tests that the lubricants used do not discolour or degrade elastane during the normal course of processing. Before grey goods are put in stock for an extended period, they should be relaxed, thoroughly scoured and batched-up again at controlled low-tension (10–20% stretch over relaxed fabric) in order to remove the winding tension which might cold-set the grey fabric, and to prevent pleats and creases from setting and becoming permanent. If lag time is expected between operations, the fabric should be wrapped in airtight, chemically inert covers, ideally black, to protect it from discolouration. This discolouration mostly affects batch edges and the external upper layers of unprotected fabric.
- Excessive coning oils should be scoured and goods dried before storage. Oils which contain glycerine or certain cationic anti-static should not be used.
- Clean dry goods should be stored in polythene bags shaded from direct sunlight. This will help to protect the goods from fumes and stray chemicals in the form of dust and vapours.
- Periods of wet storage greater than 24 h are not recommended.
- Socks and half hose, which are scoured in large batches and held for colour orders before dyeing should be tumble dried before storage. Otherwise modulus loss may occur during several days.
- If tumble drying is not available, shelf-drying is done at sufficient temperature to dry the goods overnight (60–70°C). Ventilation is necessary.
- Storage after drying should be kept for a min. to preserve the inherent properties of the elastic yarn. No strong acid or alkali should remain on the fabric since they could cause subsequent hydrolysis

degradation to ensure that no alkali is present the last wet operation should be at pH 5–6.

Finally, it is recommended to avoid storage of grey fabrics for more than two months and to always dispatch the oldest fabric first, so that to ensure the first fabrics into the grey room are first out to the dye house.

The elastane fibre is resistant to most of the chemicals used normally in the processing most of the fabrics. Elastane fibre does not lose its elasticity or other properties even in severe processes like mercerising, carbonising, scouring at 95–102°C, light hypochlorite bleaching. It is advisable to avoid hypochlorite bleaching, peroxide bleaching, etc. Chemicals which have to be avoided are unsaturated oils, greases fatty acids and their derivatives which can decolourise of affect the elasticity. Pine oil, used in printing, cutting. or boarding, can also affect stretch fabrics. It is essential to make sure that these lubricants, if present, will not discolour or degrade elastane during the usual course of processing. Greys containing elastane which need storage for an extended period of time should be thoroughly scoured and dried to prevent any discolouration or degradation by an unsaturated finish. Further, the fabrics should be protected from any contact with grease during processing.

Since the normal processing of most of the fibres are safe on elastane but longer treatment times than normally required should be avoided. Fumes, ultraviolet light, other atmospheric pollutants may decrease the initial whiteness of fabric containing elastane.

5.2.17.2   Relaxation

The grey goods, consisting of EL-fibre blends, must first be treated in the relaxed state with hot water, steam, or warm organic solvent (solvent wash). Since the grey goods are usually wider than the finished goods the width must be reduced as necessary by relaxing before thermofixing. In exceptional cases, where the starting material width is less than the finished article, thermofixing can take place directly under tension.

Regardless of end use, the fabric should be relaxed before dyeing to reduce residual stresses caused by tensioning of the elastic yarns during knitting. Heat setting will also prevent cockling or creeping of the yarns, distortion of the structure, and uneven dyeing and finishing. (2) Allowing the fabric to relax from inherent construction and processing stresses will result in improved dimensional stability.

Relaxation as first step eliminates tensions caused during the knitting or weaving of the yarn. If this is not done the finished articles may suffer from waviness, curvatures distorting the patterns, etc. These defects are particularly obvious in articles with smooth surfaces or knitted or woven goods with uniform patterns. The potential shrinkage in the grey fabric should also be removed to obtain an evenly dyed and uniformly finished product.

Knitted fabrics
Relaxation can be achieved by

1. Passing over a steam table
Steam table relaxation is preferred because complete relaxation can be achieved. The time of relaxation will depend upon fabric construction and rate of steam generation. It is also possible to relax the goods by passing them with overfeed through a steam chest on a pin chain, followed by drying on a stenter. This procedure is followed by the thermofixing step. This "two-step" approach (relaxation followed by thermofixing) gives a more uniform result than a combined relaxing/thermofixing process.

2. Steam framing
Steam framing consists in overfeeding a fabric over a steam box fixed at the sterner inlet. The sterner should then only dry the steam relaxed fabric. Combined steam relaxation and heat setting gives less uniform results than relaxation and heat setting in separate steps.

3. Continuous solvents relaxation
Continuous solvent scouring serves to relax and dry clean fabrics at the same time. It provides a good width wise relaxation and a controlled lengthwise relaxation of treated fabrics.

4. Relaxation by hot water
Relaxation by hot water occurs when a non-heat-set grey fabric is washed or dyed without any stress. This is an efficient way to relax (the fabric) but it may leave permanent creases or interfere with the subsequent heat setting of the relaxed.

Fully relaxed fabrics are stable to washing, but in most cases they are too narrow, too heavy, and prone to creasing.

Tubular fabrics
Tubular knit fabrics can be relaxed by: (a) Sending the fabric through a tensionless steam calendar. (b) Steaming in an autoclave in a batched roll.

When collapsed, tubular fabrics are slit along one edge, the remaining edge becomes the centre line in the fabric after the slit tube is opened. This line, if not removed before heat-setting, will become a permanent feature in the fabric. A few means for avoiding or removing this edge line are detailed below:

1. Take-up the tubular fabric into a truck directly under the take-down roll on the knitting machine.

2. Slit the fabric immediately off the knitting machine; pad the slit fabric through a trough containing hot water and a wetting agent; then open and heat-set the fabric on a stenter frame.

3. Slit the fabric directly on the knitting machine and roll-up in open-width form on the machine.

Woven fabrics

Articles with 6% and more elastane are relaxed grey and washed. An enzyme additive (1–2 ml/kg) in the washing liquor already facilitates light de-sizing. Greases and oils are removed at the same time to the extent that a subsequent drying and heat setting process is possible without excessive production of

1. Relaxation (washing)

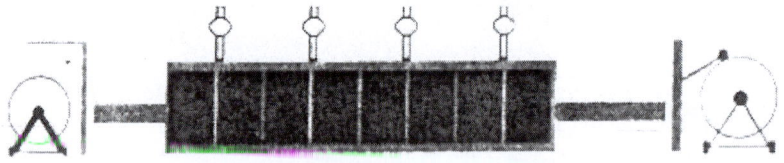

2. Drying / Heat setting (Stenter)

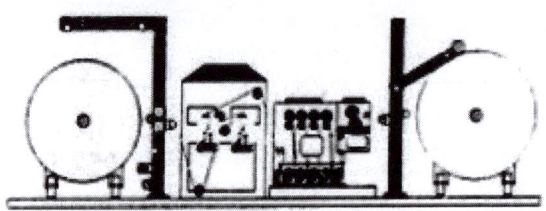

3. Singeing + Cold PB Bleaching

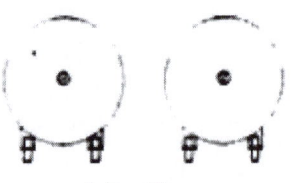

4. Dwelling

5. Washing-off

Figure 5.23. Preparation of elastane fabrics

smoke or burning in of the preparations. The dry fabric can then be singed and impregnated using a classic roller vat. Using extraction cold pad-batch bleaching, de-sizing and bleaching can be combined. After the dwelling time the subsequent washing-off takes place on a low-tension drum washing machine.

Alternate route 1 for woven

Singeing (drying) → cold pad-batch bleaching → washing →heat setting.

With this alternative process route, after singeing the fabric is impregnated at low-tension, e.g., on a padding station using cold pad-batch bleaching. As the impregnation liquor is cold, the relaxation is started but not completed. Here special attention is to be paid to ensuring that the winding process is performed with very low tension and as far as possible without edge curling. The most suitable setup here is specially arranged transfer rollers that positively feed the fabric to the A-frame. The A-frame is fixed to a moving platform. This platform then moves the A-frame automatically with increasing roll diameter. The contact pressure can be pre-selected and in this way the batch hardness controlled.

Alternate route 2 woven

Singeing → continuous de-sizing/relaxation → continuous scouring bleaching →Drying/heat setting.

The third alternative is certainly the easiest from a logistics point of view; however it requires an elaborate set of machines with correspondingly large production quantities. As both the de-sizing and the combined scouring/ bleaching process take place on storage systems, there are limits here on the treatment of articles with curling fabric edges. The edges curl particularly on heavy cotton twills with thick fabric edges (tucked in seldvedge). In such cases the de-sizing must be performed using the pad-batch process. Shock bleaching in the combined steamer can be used as well. On articles without curling edges, these rollers are swivelled out of the way. In this respect, it should also be mentioned that on un-fixed cotton/elastane fabrics, there is a risk of marks due to lay marks or crow feet with the extended bleaching process. This aspect must be checked on a case-by-case basis.

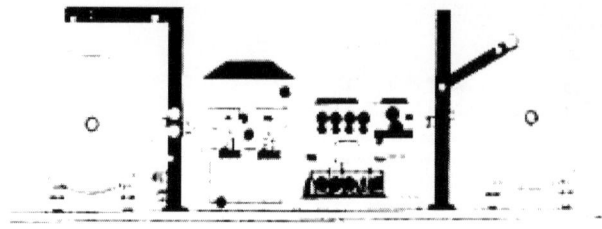

1. Singeing (and Desizing)

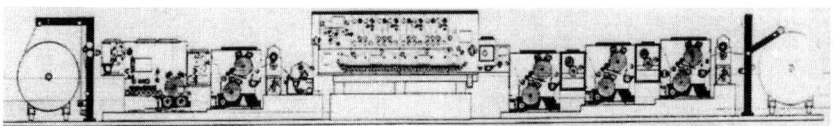

2. Relaxation (Desizing)/Bleaching

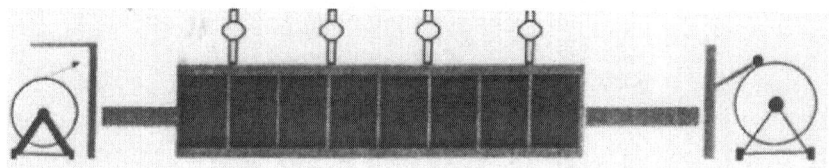

3. Drying / Heat setting

Figure 5.24.

5.2.17.3   Heat setting

It is necessary to Hheat set a fabric containing fine elastane/spandex to achieve soft dimensionally stable final fabric. Heat setting also helps to get a flat fabric with required weight and standard width.

A grey fabric from knitting machine or loom normally will have higher width than the final required width. As explained earlier in the relaxation step the fabric shrinks to a lower width than required, provided the knitting weaving parameters are maintained well. These fabrics can be heat set stretched. Heat setting has to be done before any hot wet processing to avoid curling of the selvedge and discolouration (patches) during dyeing. Heat setting after wet processing may not give the required final fabric and will have problems of wet creases not cleared, less stable, less white than the pre-heat set fabric due to less smooth, flat surface of the finished fabric.

Once the fabric is adequately relaxed, it can be heat-set. Heat setting can be done before or after the fabric is scoured, bleached, or dyed; however, some shades of colour, including white, may yellow when subjected to high heat setting temperatures. This is especially true if greige fabric is heat-set. If heat setting is performed on greige goods, spinning oils, waxes, and knitting oils may cause discoloration or yellowing that cannot be removed in subsequent scouring and bleaching processes. Greige goods can be padded in hot water with a wetting agent before heat-setting to achieve the relaxed width. However, over the time it takes to process the fabric, the padding trough can become very contaminated with the spinning and knitting lubricants as some are removed from the fabric. These contaminants can redeposit onto the fabric and result in uneven discoloration of the fabric during heat-setting. Therefore, products for spinning and knitting lubricants should be carefully chosen for potential discoloration in greige heat setting. In most cases, the yellowing of greige fabric during heat-setting can be removed with normal cotton bleaching procedures.

There is little that can be done about the discolouration if heat setting is performed following the dye process. Also, heat setting after dyeing may leave fabrics with poor stretch uniformity, variations in width, stitch distortion, or pattern distortion. (3) When heat setting is performed after proper fabric relaxation, it also helps to prevent crease, rope, and crack marks from developing in subsequent wet-processing operations. However, if the fabric is relaxed in rope form, care should be taken not to generate creases that could permanently remain in the fabric.

5.2.17.4   Heat setting of tubular fabrics

The heat setting of tubular materials can be can be heat set by cycles of high pressure steam with vacuum in an autoclave. It should be noted that because the knitted tube is rolled in a flattened configuration, the edges of the tubular goods might be permanently set during this process. It is recommended that entire dye lots be autoclaved together to avoid any dyeing anomalies that may result from non-uniform autoclave conditions. However, the heat setting stage can be omitted if the fabric is designed to optimise contraction and power. When fabrics are unrolled from the greige condition, they are usually too narrow, too heavy, and for most apparel applications, they have too much power. Therefore, heat setting is a must. After any heat setting process, the fabric should be cooled before going to any preparation, dyeing, or finishing process. This is especially true if it is going to be pad-batch dyed or printed directly after heat setting, as residual heat in the fabric could affect the resulting shade.

5.2.17.5   Heat setting of open -width material

Heat setting is usually carried out in a flat stenters with controlled heating, homogeneous heating throughout the length and width of the chamber to ensure consistent setting. It should maintain heat-up to 200°C. Stenters must comprise a large span overfeed/underfeed device, preferably coupled with an automatic weight control, to adjust fabric weight and stretch as required. In the inlet section, the stenter should have an efficient steam box. An efficient fabric cooler at the stenter outlet should fix the set.

Heat setting temperatures range from 182–196°C. A fabric heat setting temperature of 182°C is used if the desired effect is to maintain fabric weight while retaining good stretch and growth properties. However, a fabric temperature of 196°C is recommended when a sheer look with reduced stretch is desired. Temperatures above 196°C will cause the fabric to lose "power" due to the denier reduction of the spandex. In this discussion, the ability of a fabric to stretch and then recover from that stretch is referred to as power. For knit constructions that have a tendency to curl, such as jerseys and tricots, a low heat setting history may demonstrate a higher curl potential.

When open-width fabrics are heat-set on a pin stenter frame, care must be given to width control and heat distribution. The amount of overfeed needed and the frame width used will depend on the fabric weight, width, and stretch level that are desired. A fabric that is wet from a previous process (e.g., prescour or relaxation) should be rewet and padded at the entrance to the stenter to ensure uniform moisture content throughout the fabric. Often there is a delay in going from a wet process to drying. If the wet fabrics are stored in a truck or on an A-frame while awaiting drying, there may be partial or complete drying of the top layers and the edges of the stored fabric. If these goods are not rewet and brought to uniform moisture content, then the drier areas will get hot faster and have a different heat history. This will result in uneven heat setting and thus uneven dyeing, shrinkage, stretch, and recovery. Therefore, uniform moisture content in the fabric is essential for uniform drying and heat setting.

Selection of heat setting conditions is decided based on many factors chosen for a required final fabric and in most cases may not depend on the lycra used.

Some major factors considered for deciding heat setting conditions are as follows:

- Required weight width and structure of the fabric
- Type of garments to be made
- Required appearance, flatness, whiteness, stability of the finished fabric
- Colour of the fabric
- Type and count of the hard fibre used along with lycra
- Lubricant of the hard fibre and its sensitivity to heat
- Make, heat source, size and set up of the stenter used
- The type of lycra used
- The lycra content in the fabric.

The following procedure is recommended for developing new fabrics:

Many wide fabrics containing elastane/spandex are constructed so that heat setting is required to produce a satisfactory fabric with acceptable dimension stability. They are as follows:

a. Fabrics showing excessive shrinkage as determined in a preliminary test in boiling water. Take a one yard of grey sample and boil it 10 min in 0.50% (OWG) synthetic detergent and rinse, then air dry at 82°C (180°F). Measure the max hand stretch.

b. Knit structures containing a high percentage of spandex.

c. Woven structures in which the required finish width is close to the reed width.

d. Fabrics with a high stretch potential that require low-stretch in the finished state, such as ski pants.

e. When the desired finished width is greater than the greige or jammed width or when a greater fabric yield is desired. When the off-loom width is greater than the finished width, use a hot wet pre-treatment to reduce the width to manageable dimensions for heat setting. Care should be taken to avoid any folds/creases setting on the fabric.

Normally heat setting efficiency will be less than 100%. This means the heat set fabric will contract in subsequent operation. Therefore, heat setting should be followed by a wet processing to permit the residual shrinkage and relaxation to occur and to develop the full stretch potential. This usually happen in dyeing operation.

To determine the width at which the fabric to be heat set under given conditions of time and temperature. A sample of grey fabric or partially relaxed fabric sample should be heat set 10–15% above the required finished width. After heat setting the sample is treated in boiling water for 5–10 min and relax dried. Heat setting efficiency or percent retension of heat set width is calculated by dividing the finished width of the sample by the heat set width.

5.2.17.6   Effect of heat setting temperature and time on the fabric

Table 5.6. Effect heatsetting on the fabric

Property	Temperature increase	Time increase
Stability	Higher	Higher
Shrinkage	Lower	Lower
Whiteness	Lower	Lower
Flatness	Higher	Higher
Power	Lower	Lower
Width	higher	Higher

Lubricant spots	Increases	Negligible
Uniformity (PA 6)	Decreases	Negligible
Setting cost	Increases	Increases

Heat setting is usually done at 5–15% higher than the final required width because a small amount of shrinkage still remains in a fabric with elastane after heat setting. This compensates for some shrinkage expected during further wet-processing. The actual setting width can be assessed by mill testing the heat-set efficiency of a fabric (see below). Dry heat in excess of 180°C will be needed to set fabrics which contain lycra elastane. Tests show that such fabrics usually require pre-setting for 30–70 s at temperatures ranging from 185°C to 195°C. Overheating will overset the fabric and may affect the stretchability and undersetting (under heating) will cause less stability, curling, higher weight, narrower width, etc.

Saturated steam can also set fabrics with elastane. It is mainly applied to hosiery partly set by boarding, which requires milder conditions, e.g., 30–60 s at 110–120°C. Saturated steam applied for 10–15 min at 120–130°C will also set batched piece fabrics. As this technique requires vacuum in an autoclave, and does not ensure a constant fabric width, it is seldom used.

5.2.17.7   Heat setting efficiency

Heat setting efficiency can be checked as follows:
Note the width of the fabric after heat setting (HSW). Boil the sample for 5–10 min in water, then dry it relaxed and check its dry finished width (FW). The ratio of widths after and before wet relaxation (FW:HSW) indicates heat-set efficiency (HSE). This figure can then be used to estimate the heat-set width a greige fabric requires to obtain a given finished width (FW).

$$HSE = FW/HSW$$
Or $$HSW = FW/HSE$$

For example, a test sample shrinks from 160 cm HSW to 144 cm FW. The required fabric FW is 152 cm. How wide should it be set?

$$HSE = 144/160 = 0.9 \ (90\%).$$

Required HSW = 152/0.9 = 169 cm on stenter.

This figure is a close approximation but for accurate results a trial has to be taken to determine the actual results and follow the process for bulk accordingly.

5.2.17.8   Heat setting of elastane

Tricot fabric containing elastane should be heat set prior to hot wet-processing. For example, pre-treatment, dyeing, whitening to reduce shrinkage, curling and prevent decolouration.

Fabrics which are set after wet process are less stable, curls and less whiteness than pre-set fabrics. Temperature of dry heat setting has to exceed 180°C. Pre-setting is recommended for 30–70 s at 180–195°C. Oversetting causes discolouration and strength loss and undersetting causes shrinkage, curling and higher weight.

5.2.18   Pad-batch cold demineralising

Recipe

Quantity	Unit	Chemicals
3–5	g/l	Wetting agent (e.g., Ultravon CN–Ciba)
4–6	g/l	Demineralisation (e.g., Invatex SA–Ciba)

Impregnate at 20–40°C at liquor pick-up of almost 100% and batch and store for 1–3 h and hot rinse.

Fabric containing elastane yarn require careful control of processing conditions to preserve the intrinsic properties of the fibre, while obtaining the required fabric characteristics. These preparation, dyeing and finishing conditions should be selected with proper understanding because prolonged hot/ wet treatment certain chemicals, excessive tensions, high temperatures can affect the performance of elastane fibres. One should control the process parameters in such a way as to achieve the required weight, width, stability and stretch.

Tension, temperature, strength of chemicals used in processing, and duration of treatment has to be kept to a minimum to avoid any negative effect on the fibre. So, throughout the process and even storage utmost care has to be taken.

5.2.19   Scouring

Scouring is carried out to remove the lubricants and mill soil. This is comparatively simple as far as the lycra fibre is concerned. But more important is the scouring of the companion hard fibre. Hence it is advisable to follow the hard fibre scouring process which will also clean the lycra fibre. Since peracetic, sodium chlorite and sodium hypochlorite can cause yellowing or degrade the lycra fibre it is advisable to follow the hypochlorite route of scouring/bleaching.

For cotton any of the continuous or discontinuous scouring process can be followed in case of cotton/lycra fabric.

A typical procedure for scouring synthetic fabrics with lycra is given below:

Recipe

Quantity	Unit	Chemicals
0.5	g/l	Sequestering agent
1	g/l	Trisodium phosphate or 2 g/l soda ash
1–2	g/l	Non-ionic wetting agent

Set bath at 50°C with the above ingredients preferably in a long liquor to goods ratio (20:1). Circulate bath and heat it to 80°C over 15 min. Run 30–45 min at 80°C. Cool to 50°C, drop bath or overflow rinse. Rinse hot and cold and neutralise. In case of heavily soiled fabric it is advisable to pre-run in a solvent emulsion and 5 ml/l TSP (Trisodium Phosphate) solution and the scour as usual. The scouring recipes may be adopted as per the type hard fibre in the fabric.

5.2.20   Bleaching

Elastane can be bleached with hydrogen peroxide in the normal bleaching route or by stabilised reductive bleaching agent also. Sodium hypochlorite, sodium chlorite and peracetic acid should be avoided as these chemicals would degrade elastane and/or turn yellow. Elastane is often contaminated with iron and in this case it is recommended to a demineralising (details given earlier) in the beginning of the pre-treatment process.

5.2.20.1   Discontinuous one step peroxide bleach

Recipe

Quantity	Unit	Chemicals
0.5–1.5	g/l	Peroxide stabilise (e.g., Tinoclarite CS)
1.0–2.5	g/l	NaOH 100%
1.0–2.0	g/l	Wetting agent and scouring agent (e.g., Invatex DA–Ciba)
2.0–10.0	ml/l	Hydrogen peroxide (35%)

Treat 40–60 min at 98°C at an MLR of 10:1 cool the bath and drain and rinse.

Pad-batch cold peroxide bleach for cottons/elastanes

Recipe I

Recipe

Quantity	Unit	Chemicals
12–20	ml/kg	Peroxide stabilise (e.g., Tinoclarite CS)
25–30	g/kg	NaOH 100%
0–5.0	g/kg	Persulphate (preferably sodium persulphate)
40–60	ml/kg	Hydrogen peroxide (35%)

Recipe II

Recipe

Quantity	Unit	Chemicals
10–16	ml/kg	Peroxide stabilise (e.g., Tinoclarite CS)
8–15	g/kg	NaOH 100%
6–12	ml/kg	Sodium silicate (38°Bé/72°Tw)
40–60	ml/kg	Hydrogen peroxide (35%)
0–5	g/kg	Persulphate (preferably sodium persulphate)

Impregnate at room temperature at pick-up of about 90–100% and store for 16–24 h at room temperature and wash.

5.2.20.2   Discontinuous reductive bleach for cotton elastane

For a better whiteness a reductive bleach can be followed after the peroxide bleach.

Recipe

Quantity	Unit	Chemicals
0–5.0	g/l	Stabilised reducing agent
0.3–0.2	%	Optical whitening agent PDR (or a combination of OWA)

Treat at a MLR of 10:1–20:1 for 30–40 min at 95°C and hot rinse.

Mostly either bleaching or scouring can be done as per cotton fabric for cotton/elastane without much changes.

A reducing bath can increase the whiteness of Llycra and hence the following treatment can be given to the fabric if required.

Reducing scour

This method can be used to improve the natural whiteness of lycra in a grey fabric. A typical procedure is as follows:

Set bath at 50°C containing 1.2 g/l non-ionic detergent and 1 g/l trisodium phosphate or 2 g/l soda ash along with 3–5 g/l sodium dithionite (hydrosulfite). Heat to 75–80°C. Treat 45–60 min at this temperature. Cool and drop bath. Run 15 min in a fresh bath containing 0.5 m/l hydrogen peroxide 35%. Rinse cold.

Reducing bleach
This technique applies more reducing chemicals and exerts a stronger bleaching action on a fabric containing lycra than the reducing scour.

Set bath at 50°C containing 5–10 g/l sodium dithionite (hydrosulphite) and 5–10 g/l sodium metabisulfite. Heat to 80–85°C. Treat 45–60 min at this temperature. Cool, drop bath and rinse. Treat 15 min in a fresh bath with 0.5 ml/hydrogen peroxide 35%.

These processes also give additional help for full white fabric to achieve higher whiteness especially where the lycra percentage is higher.

5.2.20.3   Full white
One can follow the whitening process for the hard fibre used in the lycra fabric.

Elastane/lycra blends
The blends of viscose/elastane contain more than about 5% or 6% elastane, it will be very difficult to achieve satisfactory dimensional stability in the final garment (e.g., less than 5% shrinkage on washing) unless the fabric is pre-set before wet-processing.

For best results we recommend:
Pad the greige fabric in:
Wetting/scouring Aagent 20 g/l.
Lubricating agent 5 g/l.

Heat set for 30–60 s at 190–195°C (depending on type of elastane). After heat setting the fabric will feel very soft and bulky compared to fabric heat-set without auxiliaries, and when put into water the knitting and spinning oils and charred colour are immediately rinsed out of the fabric. When the heat set fabric is loaded into the dyeing machine, a quick hot rinse is applied, then the fabric is scoured/bleached to complete the removal of silicone oil. Because the fabric already contains Lubrifil LAF, no further addition of anti-crease agent is necessary in the scouring/bleaching bath.

5.2.20.4   Optical white for cotton/lycra and viscose/lycra
Viscose/lycra (e.g., 90.5%/9.5%) fabrics can be susceptible to damage and loss of mechanical strength when bleached by conventional high temperature peroxide bleaching processes. A batch bleaching process using TAED (tetra

acetyl ethylene diamine) at lower temperatures has been developed to provide improved whiteness and minimum strength losses, especially when the fabric is so brown after setting.

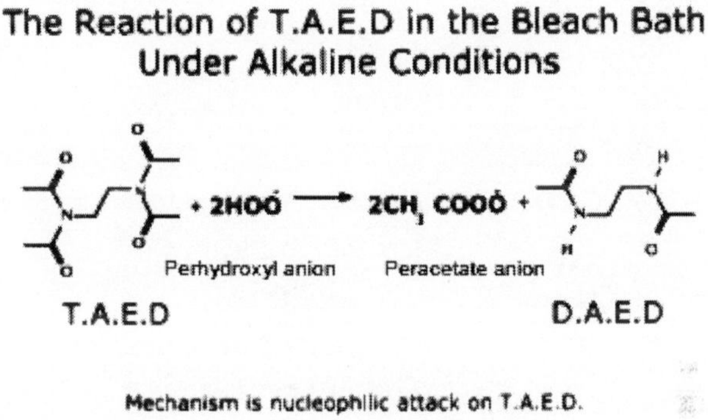

The Reaction of T.A.E.D in the Bleach Bath Under Alkaline Conditions

Mechanism is nucleophilic attack on T.A.E.D.

Figure 5.25. The reaction of T.A.E.D in the bleach bath under alkaline conditions

The bleaching process is carried out at 70°C for 20 min at near neutral pH followed by pH adjustment to higher alkalinity and a further 20–30 min bleaching.

Recipe

Quantity	Unit	Chemicals
2.0–3.0	g/l	Peroxide stabiliser
3.0–5.0	ml/l	Hydrogen peroxide 35%
2.0–2.2	g/l	TAED

Caustic soda to adjust to pH 10.5.
Run 20–30 min at 70°C.
Apply OBA in second bath.

Recipe

Quantity	Unit	Chemicals
1.0–2.0	g/l	Stabilised sodium dithionite
0.2–0.6	%	Suitable optical whitening agent

Set bath at 40–50°C. Raise to 60–80°C at 1–2°C per min. Run at 60–80°C for 20 min. Cool and drain.

5.2.21   Combined scouring/bleaching of elastane fabrics

Cellulose/elastane blend

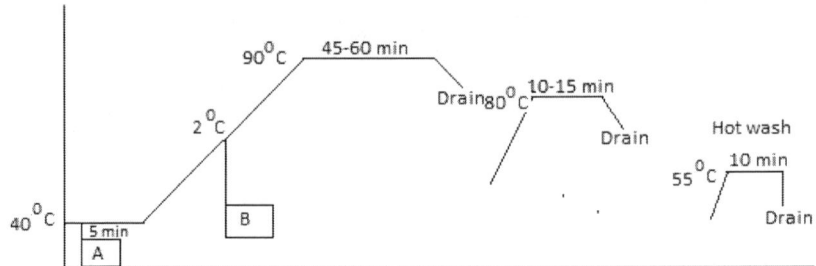

Figure 5.26. Combined scouring and bleaching process for Elastane

Addition A

2.0 g/l	Lubricating agent
1.0 g/l	Wetting and scouring agent
1.0 g/l	Dyebath conditioner
3.0 g/l	Caustic soda 100%
0.7 g/l	Peroxide stabiliser

Addition B

3–4 g/l	Hydrogen peroxide 50%

Process: Load the material. Add the chemicals mentioned in A in the above paragraph at 30–40°C. Raise the temperature to 98°C. While raising the temperature at 50–60°C, dose hydrogen peroxide. Run at 98°C for 60 min. Drop the temperature to 80°C. Drain, give hot rinse.

If full white: Set the bath at 55°C and pH 7.0 OBA: 1.0 g/l.

Run the bath for 10 min.

Drain the bath or start dyeing in the same bath.

Synthetic/spandex blend

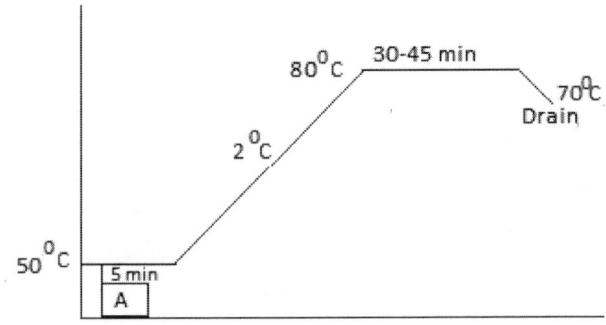

Figure 5.27.

Addition A

2.0 g/l	Anticreasing agent
1.0–2.0 g/l	Wetting and scouring agent
1.0 g/l	Dyebath conditioner
2.0 g/l	Soda ash

Process: Add the chemicals mentioned in A (as shown above) at 50°C. Raise the temperature to 80°C. Run the machine for 30–45 min. Drop the temperature to 70°C.

Drain. Give hot rinse. Cold rinse.

5.2.22   Modern continuous bleaching

The material is fed to the continuous bleaching range (CBR) desized, washed, chemical padded, steamed, washed, neutralised and dried. The process of singeing, desizing can be done by batch wise process and balance process can be done continuous. In such cases first portion of the CBR is washing unit which helps in washing-off the desized material.

The next portion is wet-on-wet padding of the bleaching chemical. In Benninger machine it is done by impacta which has following features. Other machineries will have equivalent machines.

Features

- Controlled liquor pick-up.
- Special chemical dosing.
- Exact application of chemical–Robust dosing system.
- Both high and minimum pick-up application.
- High liquor exchange–wet-on-wet padding.
- Fast penetration into fibres with optimum saturation of fabric with chemical liquor since the contact time is short and wet on wet padding.
- Controlled chemical consumption.

Automated dosing system for each chemical. There is a feed pump and induction flow meter to measure each chemical. Chemicals are pumped in ml/kg of fabric in running meter. Level of pad liquor in automatically controlled. Initial filling in litre according to the total volume and top up is done in ml/kg.

Calculation for dosing system (w.r.t Benninger CBR)

Rf = Reinforcement factor

P1 = Entry pick-up

P2 = Exit pick-up

E = Exchange factor

Example: P1 = 0.6; P2 = 1.0; E = 0.8 then, Rf = 2.2.
Calculation (Benninger)

Example:
Available data
Pick-up at inlet = 50%
Pick-up at exit = 95%
Exchange factor = 0.7

$$\text{Effective pick-up} = \frac{\text{Chemical on weight of fabric (ml/kg)} \times 100\%}{\text{Concentration in the bath ml/l}} \quad \underline{\text{or}}$$

$$= (\% \text{ Exit pick-up} - \% \text{ Entry pick-up}) + (\% \text{ Entry pick-up} \times \text{ Exchange factor})$$

In the above example

$$= \frac{(95 - 50) + (50 \times 0.7)}{100} = 0.8$$

$$\text{Initial concentration in ml/l} = \frac{\text{ml/kg}}{\text{Effective pick-up}} = \frac{35 \text{ml/kg}}{0.8} = 43.75$$

Initial volume of chemical in litres =

$$\frac{\text{Volume of impacta} \times \text{Initial concentration ml/l}}{1000}$$

$$= \frac{90 \text{ litres} \times 43.75 \text{ ml/l}}{1000}$$

$$= 3.91 \text{ litres}$$

$$\text{Exchange factor} = \frac{(\% \text{ Effective pick-up} - \% \text{ Exit pick-up} + \% \text{ Entry pick-up})}{\% \text{ Entry pick-up}}$$

$$= \frac{(80 - 100 + 50)}{50} = 0.6$$

Factors influencing exchange factor
1. Temperature: Increase in temperature of the saturator (up to 60°C) leads to a high exchange factor.
2. Fabric weight: With high fabric weight exchange value will decrease.
3. Speed: Increase of speed will reduce the exchange.
4. Fabric construction: Exchange factor is related to construction of the fabric.

5. Yarn quality: Type of yarn will influence exchange factor.
6. Type of fabric: Hydrophobic fibres reduce exchange factor.
7. Level of the saturator: Higher the level in the saturator will increase the exchange factor.

General guideline of Eexchange factor (for 100% cotton woven)

Benninger impacta

Table 5.7. Guideline Exchange factor for Impacta (Benninger)

Fabric GSM	0–30 m/m	31–60 m/m	51–80 m/m	>81 m/m
100–150	0.08	0.7	0.65	0.55
151–200	0.76	0.65	0.6	0.53
201–250	0.72	0.63	0.58	0.52
251–300	0.7	0.6	0.56	0.5
301–350	0.66	0.58	0.54	0.5
>351	0.62	0.55	0.5	0.5

Correction factor for different impregnation units.
1. Roller Vat 20 m with pressure roller e.g., DA4   +0.15
2. Roller Vat 20 m with pressure roller e.g., D4   +0.05
3. Booster e.g., AU   −0.20
4. Padder e.g., Beicoflex CU with under liquid pressure roller   −0.10
5. Padder e.g., Beicoflex CU without under liquid pressure roller −0.20
6. Kiss Roll (Bendip) liquid exchange = 0 (no liquor exchange)
7. Flexinip (Kuster)   liquid exchange = 0 (no liquor exchange)
Flow rate check for inductive flow meter and regulative values

$$\text{ml/min} = \frac{\text{Fabric weight (g/m)} \times \text{Recipe (ml/kg)} \times \text{Speed (m/m)}}{1000}$$

Rate of water addition (carried out fully automatic)

$$\text{ml/kg} = [(\% \text{ Exit pick-up} - \text{Entry pick-up}) \times 1000] - \text{Total chem consumption (ml/kg)}$$

$$\text{Water ml/min} = \frac{\text{water ml/kg} \times \text{fabric weight g/m} \times \text{speed m/min}}{1000}$$

Example
Entry pick-up = 55%
Exit pick-up = 100%
Volume of impacta = 90 litres

Fabric width = 1.8 m
GSM of fabric = 170 g/m²
Speed = 60 m/m
Exchange factor = 0.70
Effective pick-up = 83.5%

Guideline recipe

S.No.	Chemical	ml/kg	ml/l	Initial fill-up	Dosing of chemical ml/min
1	NaOH 48°Bé	25	29.94	2.69	612
2	H_2O_2 50%	40	47.9	4.31	979.2
3	Stabiliser	10	11.8	1.08	244.8
4	Scouring agent	4	4.79	0.43	97.9
	Total amount			8.51	1933.9
	Water	Auto dosing		81.49	6328.1

There are two types of automatic dosing system. In direct dosing system pure chemicals are directly pumped to the impregnation bath and water is directly added to the bath and circulated. In indirect dosing chemicals are diluted with water and pumped to the impregnation bath (Benninger).

Direct dosing
1. Dosing of individual component into the liquor circulation of the impregnation compartment.
2. Water is added to the impregnation bath, circulated and level controlled.
3. Dosing product in ml/kg of fabric.
4. All are PLC (Programmable Logic Controller) controlled.

Indirect dosing
1. All feeding including water are blended in a mixture and fed into the impregnation tank.
2. Dosing product in ml/kg of fabric.
3. Metering of liquor pick-up.
4. Rapid recipe changes recipes, concentration controlled and PLC operation.

5.2.22.1   Example recipes (Benninger CBR example)

I. Enzymatic desizing and peroxide bleaching
Enzymatic desizing–injecta

Recipe

Quantity	Unit	Chemicals
4	ml/kg	Wetting agent (e.g., Sandozin MRZ liquid–Sandoz)
2	ml/kg	Suitable enzyme (e.g., Bactosol PHC–Sandoz)
0–1	ml/kg	Plexopho eco liquid (Sandoz)
0–0.5	ml/kg	Sandozin NAN liquid (Sandoz)

Few seconds at 70–80°C.
Peroxide bleaching–impacta

Recipe

Quantity	Unit	Chemicals
2	ml/kg	Wetting agent (e.g., Sandozin MRZ liquid–Sandoz)
6	ml/kg	Stabiliser (e.g., Stabiliser SIFA Sandoz)
20–20	ml/kg	NaOH 48°Bé
30	ml/kg	Peroxide 50%
0–1	ml/kg	Sirrix AK liquid (Sandoz)

Steam 15 min at 100°C.
Neutralisation and peroxide killer

Recipe

Quantity	Unit	Chemicals
2	ml/kg	Neutralising agent
1	ml/kg	Enzymatic peroxide killer

Washing at 40°C.
II. Alkaline extraction and peroxide bleaching

Alkaline extraction

Recipe

Quantity	Unit	Chemicals
4	ml/kg	Wetting agent (e.g., Sandozin MRZ liquid–Sandoz)
2	ml/kg	Sirrix DNA liquid (Sandoz)
10	ml/kg	NaOH 48°Bé
0–0.5	ml/kg	Sandozin NAN liquid (Sandoz)

Few seconds at 70°C.

Peroxide bleaching–impacta

Recipe

Quantity	Unit	Chemicals
2	ml/kg	Wetting agent
6	ml/kg	Stabiliser
15	ml/kg	NaOH 48°Bé
25	ml/kg	Peroxide 50%
0–1	ml/kg	Sirrix AK liquid (Sandoz)

Steaming at 100°C for 15 min.
Neutralisation and peroxide killing as in I.

III. Enzymatic desizing and alkaline scouring
Enzymatic desizing–injecta

Recipe

Quantity	Unit	Chemicals
4	ml/kg	Scouring cum wetting agent
2	ml/kg	Desizing agent
0–1	ml/kg	Chelating agent
0–0.5	ml/kg	Sandozin NAN liquid (Sandoz)

Few seconds at 70–80°C.
Alkaline scouring–impacta

Recipe

Quantity	Unit	Chemicals
6	ml/kg	Scouring cum wetting agent
2	ml/kg	Demineralising agent
60	ml/kg	NaOH 48°Bé (Process A)
30	ml/kg	NaOH 48°Bé (Process B)

Process A–Tight strand 2–3 min at 100°C.
Process B–Roller bed 10–15 min at 100°C.
Neutralisation only by 2 ml/l suitable neutralising agent–Washing at 40°C.

IV. Acid cracking–Peroxide bleach

Shock acid cracking–injecta

Recipe

Quantity	Unit	Chemicals
4	ml/kg	Scoring cum wetting agent
8	ml/kg	Demineralising agent

Few seconds at 70–80°C.

Peroxide bleach–impacta

Recipe

Quantity	Unit	Chemicals
4	ml/kg	Wetting agent
6	ml/kg	Stabiliser
20	ml/kg	NaOH 48°Bé
30	ml/kg	Peroxide 50%
0–1	ml/kg	Sirrix AK liquid (Sandoz)

Neutralisation and peroxide killing

Recipe

Quantity	Unit	Chemicals
2	ml/kg	Neutralising agent
1	ml/kg	Enzymatic peroxide killer

Washing at 40°C.

V. Combination of pad-batch and pad-steam

Pad-batch enzymatic desizing

Recipe

Quantity	Unit	Chemicals
6	ml/l	Wetting agent
2	ml/l	Suitable desizing agent
0–1	ml/l	Plexophor eco liquid (Sandoz)
0–0.5	ml/l	Sandozin NAN liquid (Sandoz)
0–1	ml/l	Sirrix AK liquid

Adjust the pH 6–6.5, pad-batch and rotate the batch at room temperature for 4–6 h.

Peroxide bleaching (impacta) and neutralisation (extracta) as in IV.

VI. Combination of pad-batch and pad-steam II

Cold pad-batch bleaching

Recipe

Quantity	Unit	Chemicals
6	ml/l	Wetting agent
6	ml/l	Stabiliser
30	ml/l	NaOH 48°Bé
50	ml/l	H_2O_2 50%
0–1	ml/l	Sirrix NAN liquid (Sandoz)
0–5	g/l	Sodium perborate

Storage 16–24 h at room temperature.

Pad-steam peroxide bleach–injecta

Recipe

Quantity	Unit	Chemicals
2	ml/kg	Wetting agent
5	ml/kg	Stabiliser
10	ml/kg	NaOH 48°Bé
20	ml/kg	Peroxide 50%
0–1	ml/kg	Sirrix AK liquid (Sandoz)

Steam for 10 min at 100°C.
Neutralisation and peroxide killing as usual.
Washing at 40°C.

5.2.22.2   Example recipes for preparation–Erbatech CBR

I. Enzymatic desizing

Recipe

Quantity	Unit	Chemicals
4	ml/kg	Wetting agent
2	ml/kg	Desizing agent
0–1	ml/kg	Plexophor eco liquid (Sandoz)
0–0.5	ml/kg	Sandozin NAN liquid (Sandoz)

Peroxide bleaching

Recipe

Quantity	Unit	Chemicals
2	ml/kg	Wetting agent
6	ml/kg	Stabiliser
20	ml/kg	NaOH 48°Bé
30	ml/kg	Peroxide 50%
0–1	ml/kg	Sirrix AK liquid (Sandoz)

Steam for 15 min at 100°C.

Neutralisation and peroxide killing

Recipe

Quantity	Unit	Chemicals
2	ml/kg	Neutralising agent
1	ml/kg	Enzymatic peroxide killer

Washing at 40°C.

II. Alkaline extraction and peroxide bleaching

Alkaline extraction

Recipe

Quantity	Unit	Chemicals
4	ml/kg	Wetting Agent (e.g., Sandozin MRZ liquid–Sandoz)
2	ml/kg	Sirrix DNA liquid (Sandoz)
10	ml/kg	NaOH 48°Bé
0–0.5	ml/kg	Sandozin NAN liquid (Sandoz)

Few seconds at 70°C.

Peroxide bleaching

Recipe

Quantity	Unit	Chemicals
2	ml/kg	Wetting agent
6	ml/kg	Stabiliser
15	ml/kg	NaOH 48°Bé
25	ml/kg	Peroxide 50%
0–1	ml/kg	Sirrix AK liquid (Sandoz)

Steaming at 100°C for 15 min.
Neutralisation and peroxide killing as in I.

III. Desizing and alkaline scouring

Enzymatic desizing

Recipe

Quantity	Unit	Chemicals
4	ml/kg	Scouring cum wetting agent
2	ml/kg	Desizing agent
0–1	ml/kg	Chelating agent
0–0.5	ml/kg	Sandozin NAN liquid (Sandoz)

Few seconds at 70–80°C.

Alkaline scouring

Recipe

Quantity	Unit	Chemicals
4	ml/kg	Scouring cum wetting agent
2	ml/kg	Demineralising agent
60	ml/kg	NaOH 48°Bé (Process A)
30	ml/kg	NaOH 48°Bé (Process B)

Process A–Tight strand 2–3 min at 100°C.
Process B–Roller bed 10–15 min at 100°C.
Neutralisation only by 2 ml/l suitable neutralising agent–Washing at 40°C.

IV. Combined pad-batch desizing and pad-steam bleaching

Pad-batch enzymatic desizing

Recipe

Quantity	Unit	Chemicals
6	ml/l	Wetting agent
2	ml/l	Suitable desizing agent
0–1	ml/l	Plexophor eco liquid (sandoz)
0–0.5	ml/l	Sandizin NAN liquid (Sandoz)
0–1	ml/l	Sirrix AK liquid

Adjust the pH 6–6.5, pad-batch and rotate the batch at room temperature for 4–6 h.

Peroxide bleaching

Recipe

Quantity	Unit	Chemicals
2	ml/kg	Wetting agent
6	ml/kg	Stabiliser
20	ml/kg	NaOH 48°Bé
30	ml/kg	Peroxide 50%
0–1	ml/kg	Sirrix AK liquid (Sandoz)

Steaming at 100°C for 20 min.
Neutralisation as in I.

5.2.22.3   Recipes for kusters CBR

The CBR consists of Compacta and Turbo Flush washing compartments. In Turbo wash compartments combination of surface washing and penetration counter flow liquor circulation at high circulation rates are adopted. Fabric is threaded at an angle with large circumference guide rolls arranged very close at an angle to avoid crease formation two guide rollers with water cascade trough acts as a separate sub-compartments inside each Turbo Flush compartment ensures better washing efficiency. This effects turbulent washing of surface as well as inside the fabric. Liquor exchange is done via cascades. Liquor nip by way of hydrodynamic pressure assures penetration washing.

Compacta washing is more like a traditional washing compartment. Fabric is threaded between vertical driven rolls ensures high reaction time, high degree of washing effectiveness at the same time high fabric speed with controlled fabric tension.

5.2.22.4   Example recipes

1. Pre-treatment–Enzymatic desizing–Peroxide bleach

Enzymatic desizing

Recipe

Quantity	Unit	Chemicals
4	ml/kg	Wetting agent
2	ml/kg	Desizing agent

0–1	ml/kg	Plexophor eco liquid (Sandoz)
0–0.5	ml/kg	Sandozin NAN liquid (Sandoz)

Few seconds at 70°C.

Peroxide bleach

Recipe

Quantity	Unit	Chemicals
2	ml/kg	Wetting agent
6	ml/kg	Stabiliser
20	ml/kg	NaOH 48°Bé
30	ml/kg	Peroxide 50%
0–1	ml/kg	Sirrix AK liquid (Sandoz)

Steaming at 100°C for 15 min.

Neutralisation and peroxide killing

Recipe

Quantity	Unit	Chemicals
2	ml/kg	Neutralising agent
1	ml/kg	Enzymatic peroxide killer

Washing at 40°C.

2. Alkaline extraction–Peroxide bleach

Alkaline extraction

Recipe

Quantity	Unit	Chemicals
4	ml/kg	Scouring cum wetting agent
2	ml/kg	Demineralising agent
100	ml/kg	NaOH 48°Bé (Process A)
0–0.5	ml/kg	Sandozin NAN liquid (Sandoz)

Few seconds at 70°C.

Peroxide bleaching

Recipe

Quantity	Unit	Chemicals
2	ml/kg	Wetting agent
6	ml/kg	Stabiliser

15	ml/kg	NaOH 48°Bé
25	ml/kg	Peroxide 50%
0–1	ml/kg	Sirrix AK liquid (Sandoz)

Steaming for 15 min at 100°C.
Neutralisation and peroxide killing as above.
Other recipes as given in Benninger section.

5.2.22.5   Recipes for Mezzera CBR

Same as Benninger Ppre-treatment recipes (woven).

Recipe for peroxide bleach on Mezzera CBR

Quantity	Unit	Chemicals
4–8	ml/kg	Stabiliser
15–25	ml/kg	NaOH 48°Bé
25–35	ml/kg	H_2O_2 50%
1	ml/kg	Sandozin NAN liquid (Sandoz)

Steaming for 20 min at 100°C.
Neutralisation and peroxide killing
Recipe

Quantity	Unit	Chemicals
1	%	Neutralising agent
0.35	%	Enzymatic peroxide killer

First wash box at 90°C.
Second wash box at 60°C.

PART VI

Pretreatment of substrates – Bio-preparation and other technologies

6.1   Bio-preparation

A good pre-treatment of cotton material aims at imparting a good and uniform wettability to cotton and remove all natural and added impurities to make cotton accessible to dyes and finishing chemicals without damaging the cotton fibre. Many chemicals and dyestuffs are used for this purpose are not environmental friendly. Biotechnology is one area which can help in making textile processing less harmful to environment. This is an emerging area where the use of enzymes in textiles is increasing popularly due to its harmless and eco-friendly processing applications–especially pre-treatment.

Pre-treatment involves desizing (for wovens) followed by scouring and bleaching. Desizing is carried out to remove natural waxes, pectins and added impurities such as knitting or spinning lubricants, whereas bleaching removes natural colouring matter and is usually carried out for full white or lightshades as per the required brightness. Now-a-days, there is a tendency to combine scouring and bleaching to reduce processing time and consequently the cost. Conventional scouring as well as combined scouring and bleaching use highly alkaline conditioners for treatment. Caustic soda, a cheap basic chemical along with a suitable detergent, at a high temperature of about 95–98°C is used for scouring purpose. However, caustic soda is non-specific in its action i.e., apart from removing natural impurities like pectin, it can also attack cotton fibre, resulting in oxidative damage to the fibre. This not only reduces the strength of the cotton fibre but can also lead to uneven or low-dye uptake. Since caustic soda needs to be removed completely before subsequent dyeing or finishing, the processor has to use a large quantity of water for washing-off the caustic soda. This ultimately leads to increase in the volume of effluent. Salt is generated leading to increase in total dissolved solids (TDS), which are difficult to reduce by normal effluent treatment.

From time immemorial enzymes were used in various stages of textile processing. A classic example is the use of enzyme in desizing. The application of enzymes to textile materials is done during various stages of processing. Applied research on microbial cellulases, hemi-cellulases and pectinases has not only generated significant scientific knowledge, but has also revealed their enormous potential. Advances in enzyme field have already led to the development of new products, opened new markets, speeded up production of pure products and helped to reduce the pollution load. It also offers the Textile Industry the ability to reduce costs, protect the environment, address health and safety issues and improve quality and functionality. This is because the enzymes are biodegradable and they work under mild conditions and save the precious energy. Enzymes, being biocatalyst and very specific, are used in small quantities and have a direct consequence of lesser packing material and lower transportation impact. In all, the enzymes are wonder products. Enzymes and processes have been designed for textiles wet-processing starting from desizing to finishing. In case of Denim, enzymes are extensively used for Stone wash effect, bleaching as well as finishing purpose. In the long-term, more and more of the cumbersome and polluting chemical procedures employed by the textile industry will be substituted or supported by the biotechnological processes.

The enzymatic process called bio-scouring allows cotton to be treated under very mild conditions. Due to use of less rinse water, bio-scouring process reduces both effluent as well as water consumption. In bio-scouring alkaline pectinase enzyme is used for natural cellulosic fibres such as cotton, linen, hemp and blends. It removes pectin and other impurities from the primary cell wall of the cotton fibres without degradation of cellulose and thus has no negative effect on strength properties of the fabric. Depending on the type of enzyme the application can be done in machines normally.

Enzymes are used because they accelerate reactions under mild conditions. They are effective at relatively low-temperatures and environmentally benign processing conditions, and they can be used to replace harsh solvents and other organic compounds.

Bio- scouring has following advantages

Enzymatic bio-scouring is a milder process and certainly more eco-friendly than alkaline boiling off.

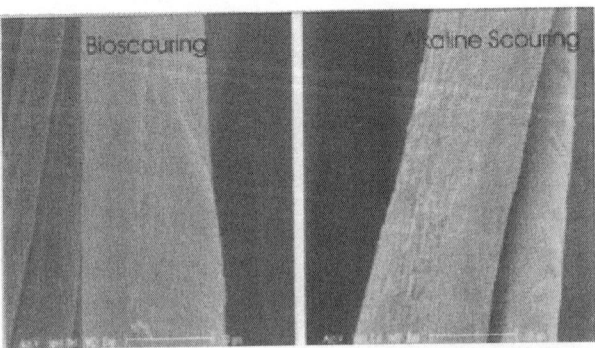

Figure 6.1. Comparison of material scoured by conventional and Bbio -scouring

- Over 50% reduction of TDS, COD and BOD in the effluent.
- Can be combined with enzymatic desizing.
- Core alkali migration no longer a problem in further processes.
- Overall savings in utilities up to 30% (less water and energy required).
- Improved strength when compared to conventional scouring.
- Lower weight loss, up to 1–2% compared to 4% of conventional alkaline scouring.
- Better softness.
- No cellulosic/fibre damage.
- Less chemicals used.
- Mild conditions allow treatment of linen and blends.

Table 6.1. Comparison of conventional scouring and bio -scouring

	Bio-scouring	Conventional scouring
pH value of process	8–9.5	13–14
Temperature	50–60	95–100
Residual pectin	10–15	10–15
Weight loss %	<2.5	3–8
Hydrophilicity (TEGEWA)	<2 s	<2 s
Fibre damage (DP component)	<0.05	<0.05
Degree of whiteness	5–10 points improved	1–2 points improved
Handle	Very soft	Harsh
Dyeability	Good	Good
Water saving	20–50%	

Enzymes are proteins, which catalyse specific chemical reactions and are known as "bio-catalysts". Chemically, an enzyme is defined as being a bio-catalyst made of a proteinic complex compound containing a sequence of approximately 200–250 amino acids and are present in all biological systems. The term "enzyme" is derived from the Greek word "enzymos" and it literally translated means "cell" or "ferments". Classification of enzymes was standardised in 1955 by the International Biochemistry Union based on the type of reaction catalysed. The conscious knowledge and knowhow of controlling enzyme processes combined with large scale of production handled to a long life products known as "industrial enzymes".

It should be emphasised that enzymes are nothing more than biological catalysts, consisting of complex, three-dimensional proteins that are composed of polypeptide chains. Most enzymes for industrial use usually have either a fungal or a bacterial source. Although these proteins greatly accelerate reactions, enzymes cannot cause a reaction to occur that ordinarily would not occur, however, in the absence of enzymes an extremely long-time on the order of months or years may be required for some reactions to occur. A great part of the specificity of an enzyme depends on the sequence of its constituent and also of its spatial configuration. Enzymes act specifically on a substrate just like chemical catalysts do. The major difference in enzymes from the chemical catalysts are that they are temperature sensitive having relatively low-energy of activation and are usually active over a narrow range of pH. Enzymes react only with a specific substrate, which fits within the active site of the enzyme molecule.

Enzymes have been used in breweries and food processing units for years. Enzymes are present in all biological systems. The traditional applications in textile wet-processing the primary use of enzymes is in the promotion of hydrolysis specific substrates, a process for converting water-insoluble material to products that dissolve in water and can be washed away. The use of the enzyme, amylase in the 1850's for removal of starch size from cotton fabric first proved the effectiveness of enzymes as positive environmental alternatives to harsh chemicals in textile wet-processing. Until quite recently, however, cellulase was the only other enzyme to find wide use in textile wet-processing. So called stone washing of Denim jeans to produce the worn look that is so popular, and bio -polishing of cotton fabric to remove surface fibres are areas in which cellulase has found very wide international use.

Enzymes are used in the textile industry because of the following reasons:
1. Act only on specific substrate.
2. Operate under mild conditions.
3. Are safe and easy to control.
4. Accelerate the reactions.
5. Can replace harsh chemicals.
6. Biologically degradable.

6.1.1   How an enzyme works

Enzymes are designed to act in living cells, hence they can work at atmospheric pressure and in mild conditions of temperature and pH. Higher reaction rates, at lower energy consumption can be achieved by enzyme catalysts as compared to traditional chemical catalysts. Enzymes can bring about hydrolysis, oxidation, reduction, coagulation and decomposition, and the most common effect of interest is hydrolysis. Enzymatic hydrolysis follows kinetics of first order reaction. Enzymes are used to alter specific substrates, such as the surface of a fabric. For an enzyme to be effective with a particular substrate, the two must be able to "fit" together. There must be a reactive site available on the substrate that is compatible with the enzyme being used. After a certain period of time, the enzyme can be removed or rendered inactive. An enzyme can be rendered inactive by changing processing conditions so that enzyme activity is no longer favoured, and is typically achieved by increasing pH (raising alkalinity), increasing temperature, or by strong ionic changes.

Enzymes are highly specific and this specificity can be explained by the lock and key theory. In order to catalyse a reaction, the enzyme molecule has to form a complex with the substrate. The binding sites of the enzyme molecule recognise the corresponding domain of the substrate molecule. After proper orientation of both the molecules, the reactive site of enzyme molecules have access to the appropriate part of the substrate molecule. After the reaction is completed, the products detach themselves from the complex. Very high concentration of the enzyme leads to crowding over the surface of the substrate, which will cause bottle neck at the binding sites and thus will lower the reaction rate.

Adverse temperature conditions lead to the denaturation of proteins and moderately high temperature results in faster movement of the enzyme. Maximum reactivity of the enzymes differs at various pH values due to the difference in proton donating or proton accepting property, which decides their ionisation. Most enzymatic activities are extremely sensitive to the pH and its variations. Enzymes, basically, are composed of proteins exhibiting zwitter ion properties. The proton donating or accepting groups in enzyme catalytic sites are at their required state of ionisation and at a selected pH of the enzymes at which its activity is optimal. Any increase or decrease in pH results in lowered reaction rates. So there is a presentation of the speed of an enzymatic reaction as a function of the pH–normally a bell like curve comprising of a phase of enzymatic activity, one maximum activity and one of declining activity, with the mid-point called the optimum pH. A variation of pH during the course of reaction may bring about an alteration in the protein structure with a denaturing effect on enzyme or the ionisation of the active site. Buffers are used so as to avoid deviations of the pH during the course of the reactions.

The Tables below show the optimum temperature and pH required for various enzymes used in the textile processing.

Table 6.2. *Process temperature for various enzymes used in textile industry*

Enzyme	Mesophile	Thermophile
α-amylase	60–70°C	
α-*Bacillus subtilis*		90°C
α-*Bacillus*		90°C
β-amylase	57–65°C	
Protease	38–50°C	

Table 6.3. Optimum pH ranges of various enzymes used in textile industry

Enzyme	Source	pH range
A-amylase	Aspergiluus niger	5
	Bacillus subtilis	5.7–6.3
	Malt	5.7–6
	Bacillus litchenformis	5–6.5
Cellulase	*Myotherium verucania*	5.5
	Pecillium persillum	5.3
	Trichoderma komigri	4–6

Six classes of enzymes are as follows:

- Oxidoreductase
- Transferases
- Hydrolases
- Lyases
- Isomerases
- Ligases.

Most of the enzymes used in the textile industry belong to class three i.e., hydrolases. The enzymes currently used in the textile industry namely, α-amylases, cellulases and proteases are included in the class of hydrolases. These enzymes provoke the hydrolysis of various types of linkages such as ester, acetyl, amide, etc., and split various compounds and become fixed on the residues. Scouring with acidic and alkaline pectinases have been tried in scouring of cotton materials, those enzymes were as efficient as that of conventional chemical process in terms of absorbency of the fabric. Bleaching can be carried out using glucose oxidase. Aeration plays an important role in the enhancement of enzyme reactions (oxidation of hydrogen peroxide is obtained by oxidation of glucose oxidase), to overcome the stabilising effect of glucose and protein in the subsequent bleaching.

Enzyme catalysed reactions occur within the confined site called "active site". Enzymes contain a true activity Centre in the form of three dimensional structures like fissures, holes, pockets, cavities or hollows. The number of active sites per molecule is very small, generally one. The overall rate of reaction depends on the time required to form the enzyme–substrate complex and the time required to form the final product. In order to catalyse a reaction, the enzyme molecule has to form a complex with the substrate. Two phases of enzyme action 11 explained by the kinetics of enzymatic-catalysis follow the Michaelis–Menten equation, which can be expressed in the simplified form as follows:

$$E + S = (E-S) \; E + P$$

Enzyme + Substrate→Enzyme – Substrate complex.

Enzyme–Substrate complex → Enzyme + Product of enzyme action.

The action of enzyme, first of all forms an enzyme–substrate complex. Direct physical contact of enzyme and substrate is required to obtain the complex. The mechanism of enzymatic hydrolysis of cellulosic materials is complicated and not yet fully understood. In order to form the E–S complex, the binding and reactive sites of enzyme recognize corresponding domains of the substrate molecules. This makes the enzyme specific towards the substrates. Finally, the complex disintegrates with the release of the reaction products and the original enzymes are once again available. The products of enzymatic size cleavage are much more soluble in water. Synergism of different components in the enzyme complex and inhibition mechanism further complicates the reaction. Enzyme diffusion plays a much more decisive role in the heterogeneous system of soluble enzyme and solid substrate. The kinetics of reactions, depends on the diffusion of the enzyme into the solid phase of the substrate and the diffusion of the reaction products out of their solid phase into liquor. For cotton, the restrictions of enzyme to the fibre surface is easily achieved, because cellulose is a highly crystalline material and possess only small amorphous areas, making the diffusion of enzymes into the interior of the fibre nearly impossible. Thus by regulating the enzyme dosage and choosing type of the enzyme, it can be confined to the surface of cotton and to the amorphous regions, leaving the fibres intact.

Enzyme hydrolysis follows the kinetics of first order equation. The active sites of the enzyme completely fit with the substrate, known as lock and key fashion. Interaction between substrate and active site takes place resulting in the formation of a number of weak bonds including hydrogen bond, Van der Waals interactions, etc. Formation of each weak interaction in the enzyme–substrate complex is associated by a small release of free energy known as binding energy. This binding energy is used as a major source of free energy by enzymes to lower the activation energy of reactions so that the reaction proceeds at a much faster rate. It is to be noted here that no enzyme alters the equilibrium constant of any catalysed reaction and it only provides lower energy reaction path so that the rate of reaction is accelerated.

6.1.2   Mechanism of enzyme action

The possible mechanisms that are responsible for the enzymatic reaction can be explained using lock and key mechanism, competitive bogie and nano competitive bogie mechanisms. All the mechanisms discussed here may act on the enzyme–substrate system simultaneously.

An enzyme has a specific three dimensional shape. This shape and location of the action site on the enzyme, control the specificity of the molecule. At the surface of the substrate, the enzymes serve to accelerate the

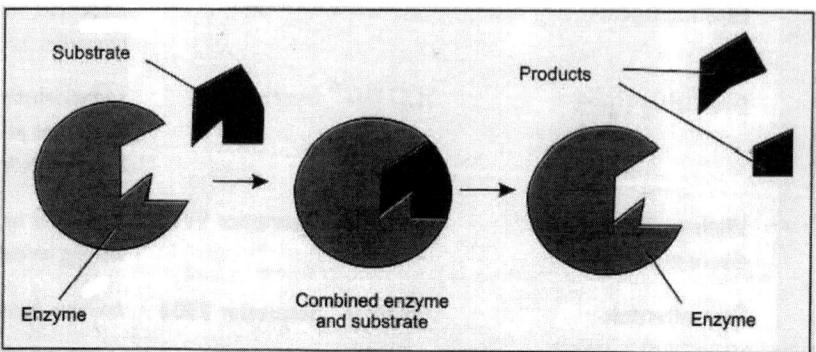

Figure 6.2. Lock and key mechanism of enzyme action

reaction of the substrate with the environment to produce reaction products. Since the enzymes are catalysts, they themselves are not changed by the reaction but the substrate undergoes changes (see Figure). After the reaction has taken place, the enzyme is released to be re-adsorbed onto another substrate surface of the substrate. The process continues until the enzyme is "poisoned" by a chemical bogie or inactivated by extreme temperature, pH or by other negative conditions in the processing environment.

6.1.3   Enzyme activators and inhibitors

Certain enzymes require metal ions as activators, especially bivalent metallic cations. These metal ions stabilise the structure of the enzyme substrate complex to the attack of the enzyme or again make the action of the ferment more efficacious. Certain metals are capable of becoming part of the enzyme and they enter into the constitution of the prosthetic group and take part in the ion exchanges. The activator ion can be considered as the prosthetic group itself. This is the case with the hydrolysis of starch by α-amylase, in the presence of chloride ion and the cation Ca^{++}. Certain wetting agents also exhibit stabilising action on enzymes. Agents which do not affect pH of the system should be selected and so the non-ionic agents are generally preferred.

Inhibitors are chemicals, such as alkalis, antiseptics and acid liberating salts that tend to inhibit the enzyme activity, which could either be reversible or irreversible. The inhibition takes place by blocking certain useful groups. Sometimes, the reaction products themselves may inhibit the enzymes. With the competitive type of inhibition, the inhibitor possesses an affinity for the enzyme with which it combines, thus actually creating competition between itself and the substrate. Action of competitive type inhibitor is reversible. In non-competitive type of inhibitions, the inhibiting substance associates itself with the ferment disturbed and without being competed with enzyme–substrate combination. These agents reduce V_{max} of the enzyme catalysts. In general, cations of heavy metals are lethal to enzymes and their effects are detrimental, for example, the salts of Pb, Hg, Cu, Fe. Reducing and oxidising agents are also known to act as inhibitors of enzymes, as their oxidising reducing properties can destroy the enzyme molecules. Sequestering agents are also lethal to enzymes of catalases or peroxydase types.

6.1.3.1   Factors influencing enzyme reactivity

The reaction depends on various independent and dependent variables involved in the formation of enzyme–substrate system and the parameters include:

- Enzyme concentration
- Concentration of the substrate
- Duration of reaction or incubation period
- Temperature of reaction
- pH of the system
- Presence of activators
- Presence of inhibitors.

Effect of temperature and pH is one of the effects changing the shape of the enzyme so that a lock and key mechanism is no longer possible. The sensitivity of enzyme to temperature and pH is advantageous in a respect that an enzyme process can be stopped at any stage by altering the temperature or addition of alkali or acid.

6.2   Cotton fibre morphology

In order to understand better the need for preparation of cotton, it is useful to review the morphology of the fibre. Cotton is a seed hair fibre, i.e., each fibre consists of a single cell that grows from the epidermis of the cotton seed. During the first 20 days of growth, the fibre's longitudinal growth continues to increase until a maximum is reached that is dependent on cotton type and growing conditions. The circular cross-section of the fibre at this time consists mainly of a lumen of cell sap contained within a primary wall composed of cellulose, pectin, protein, and wax. The next stage in the development of the fibre occurs during the next 35–50 days when the secondary wall begins to develop. This cellulose wall develops from the outside to the inside, and constitutes the bulk of the fibre. After the cotton boll is opened the lumen shrinks and the circular fibre collapses to form a more flattened cross-section. At this point the fibre consists essentially of a primary wall, a secondary wall and a lumen that contains residues of the intracellular matrix.

6.2.1   Primary wall

Since it is the primary wall surrounding the fibre that is responsible for the lack of absorbency of unprepared fibre, it is useful to examine the composition of his wall in greater detail. The primary wall consists of a matrix of cellulose in which is dispersed a mixture of pectins, proteins, wax, other organic compounds, and "ash". The approximate percentage of each of these materials is given in Table below.

6.2.1.1   Composition of the primary wall in cotton

Table 6.4. Composition of the primary wall in cotton

Component	Percentage
Cellulose	52
Pectins	12
Proteins	12
Wax	7
Other compounds	14
Ash	3

The pectins consist of polygalacturonic acids that have been converted to the form of calcium, magnesium, iron, or even aluminium salts that are not soluble in water. The building blocks of the proteins are known to consist of complex amino acids, while the wax (consists mainly of higher aliphatic alcohols, fatty acids, and their esters. Other organic compounds found in the primary wall consist of complex carbohydrates. The "ash" is composed of alkaline earth components and potassium and phosphate ions. It is likely that the pectins, proteins, wax, and other material are not uniformly distributed within the primary wall. It is reasonable to assume that the primary wall non-cellulosic components are concentrated more in the outer zone of the primary wall and decrease in concentration as the secondary wall is approached. In fact, one can argue that the so-called cuticle is not a distinct layer, but is merely the outermost region of the primary wall where the concentration of non-cellulose components is the highest.

6.2.1.2   Traditional preparation

The traditional technique for making cotton fibre absorbent in preparation for subsequent wet-processing involves scouring the cotton textile material in a hot alkaline solution of sodium hydroxide, containing wetting, dispersing, emulsifying, sequestering, or other chemical agents. Such a treatment tends to remove the non-cellulosic material, thereby interrupting the surface of the primary wall to form breaches in the matrix that permit water to penetrate the cotton fibre interior. Hot alkaline scouring can, however, result in the production of oxycellulose with associated fibre strength loss unless precautions are taken. In addition, hot alkaline scouring results in environmentally harsh chemicals being added to the textile effluent.

6.2.2   Enzymes

It should be emphasised that the enzymes are nothing more than biological catalysts, consisting of complex, three-dimensional proteins that are composed of polypeptide chains. Most enzymes for industrial use usually have either a fungal or a bacterial source. Although these proteins greatly accelerate reactions, enzymes cannot cause a reaction to occur that ordinarily would not occur, however, in the absence of enzymes an extremely long time on the order of months or years may be required for some reactions to occur.

In textile wet-processing, the primary use of enzymes is in the promotion of hydrolysis of specific substrates, a process for converting water-insoluble material to products that dissolve in water and can be washed away. The use of the enzyme, amylase, in the 1850's for removal of starch size from cotton fabric first proved the effectiveness of enzymes as positive environmental alternatives to harsh chemicals in textile wet-processing. Until quite recently, however, cellulase was the only other enzyme to find wide use in textile wet-processing. So called stone washing of Denim jeans to produce the worn look that is so popular, and bio -polishing of cotton fabric to remove surface fibres are areas in which cellulase has found very wide international use. Enzymes are used in many textile process and scope is increasing day-by-day.

Table 6.5. Uses of various enzymes in textile processing

Enzymes	Textile uses
Cellulases	Bio-finishing, bio-polishing, anti-pilling, softness, smoothness, lustre improvement and stone-washed effects on denim

Amylases	Used to hydrolyse starch in desizing
Proteases	In household washing agents better removal of protein containing soil or stains. Anti-felting of wool, accompanied by high loss of weight, tear strength and of the typical handle 2, 3 degumming of silk with the problem of silk fibroin damage
Lipases	In detergents for the hydrolysis of lipids
Pectinases	Hydrolysis of pectins, for example, in cotton preparation and retting of flax and hemp before reactive dying or printing of peroxide bleached fabrics and yarn
Catalases	Catalyse the decomposition of hydrogen peroxide, important before reactive dying or printing of peroxide bleached fabrics and yarn waste water and potentially toxic decomposition compounds (aromatic nitro-compounds)
Peroxidases	Used as an enzymatic rinse process after reactive dyeing oxidative splitting of hydrolysed reactive dyes on the fibre and in the liquor, providing better wet fastness, decolourised
Ligninases	Removal of burrs and other plant compounds from raw wool
Collagenases	Removal of residual skin parts in wool
Esterases	In development: polyester finish, removal of oligomers
Nitrilases	In development: polyacrylonitrile preparation for better colouration
Laccases	Used to decolourise indigo

6.3   Bio-preparation

Modern research in the area of stone washing or bio-polishing has revealed that cotton fabric that is treated with cellulase in concentrations high enough to achieve a positive response in these two areas, results in a fabric that is highly water absorbent. Such high concentrations of cellulase, however, bring about too much fibre weight loss or destruction to be used in routine fabric preparation. When used much lower concentrations of cellulase in attempts to make cotton fibre absorbent and were pleased to find that such treatment did definitely have the ability to make cotton absorbent. In fact, fabric treated with cellulase was more absorbent, but the cellulase enzymes used in various investigations were by no means pure. Now it is known that it is the pectinase impurities present in cellulase that is responsible for the greatly improved water absorbency of cotton that is found to occur when cotton is subjected to a mild cellulase treatment.

The pectin which is present in the cotton fibre primary wall is powerful biological glue. As noted previously, this biological glue consists of polygalacturonic acids that to a great extent have been convened to calcium, magnesium, iron, or other salts during fibre growth. These mostly water insoluble pectin salts serve to bind the waxes and proteins together in the primary wall matrix to form the fibre's protective barrier during growth. Hydrolysis of the pectin in the primary wall matrix therefore promotes an efficient interruption of the matrix RO (reverse osmosis) achieve superior water absorbency without the negative side effect of cellulose destruction.

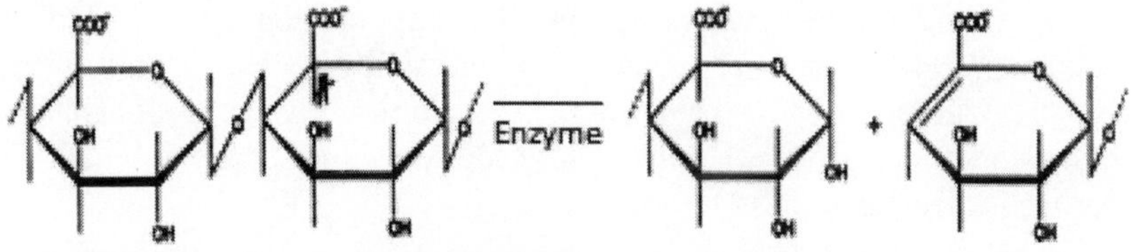

Figure 6.3. Hydrolysation of pectin

However, not all pectinase enzymes are equally effective in promoting hydrolysis of the pectin. The best pectinases seem to be those that can function under slightly alkaline conditions even in the presence of chelating agent. Alkaline pectinase functions well at moderate temperatures under buffered, mildly alkaline conditions–even when chelating agents and selected wetting agents are included in the preparation bath.

In a practical enzymatic scouring operation, lipase and protease enzymes are used. Stock solution of 18% weight by volume solution of sodium chloride is used to reduce the activity to an appropriate level. But at high temperatures, certain metal ions are detrimental to the enzymes, including Cu, Cd, Zn, Hg and oxidizing agents such as polyphosphates. EDTA, and certain anionic surfactants are quite important for the enzyme performance–amount of woven cotton is sized with a mixture of starch based polymers and tallow based lubricants. Starch and triglycerides are known to form insoluble complexes. The lipase is expected to break these complexes thus making the removal of starch much easier. Being triglycerides, these lubricants can be, almost, completely hydrolysed by lipases, yielding glycerol, fatty acids, mono and diglycerides as reaction products. Glycerol is completely water soluble, and fatty acids are easily removed during the subsequent or simultaneous scouring operation monodiglycerides act as efficient surfactants or emulsifiers. Addition of lipase is not only beneficial to desizing but also in the scouring and the whiteness, i.e., total preparation. Complementary addition of these two enzymes at a pH of 7 and temperature of 60–80°C, produces higher uniformity in the desized fabrics than that attained using amylase alone.

6.3.1   Properties of enzymatic scoured fabrics

Enzymatically desized and scoured cotton materials have comparable nitrogen content, protein and fat contents as that of conventionally treated fabrics and the fabrics showed similar absorbency in the fabric. When the enzymes are used for treatment of polyester and the polyester blended fabrics, the whiteness of the fabrics after washing is improved by using detergent containing pectinase along with other ingredients. Pectinases and combinations of pectinase cellulose were more effective.

Depending on the type of enzyme the application can be done in machines normally used for textile wet processing e.g., Pad, J box, Jiggers and Jets. A dosage of 0.4–0.5% (owf) is recommended The treatment time will depend on the process applied. Bio- scouring can also be combined with other wet processes such as follows:

- Desizing and Bio -scouring
- Bio -scouring and dyeing
- Bio- scouring and Bbio- polishing
- Bio- scouring, Bbio- polishing and dyeing

Combining these processes offer considerable time, water and energy savings.

6.3.1.1   pH control

pH value is an important parameter in this application. Generally alkaline pectinase work in abroad pH range of 7.5–9.5. The use of calcium sequestering agents should be done carefully so as not to in activate the enzyme.

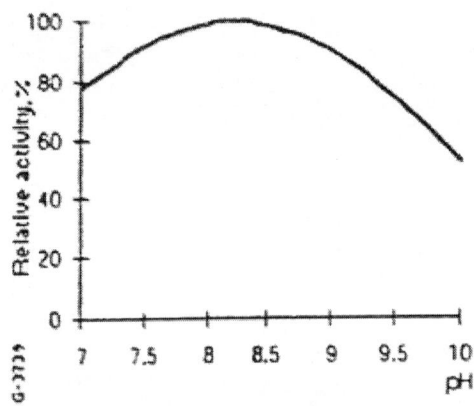

Figure 6.4. Effect of pH on pectinase at 55°C
(tested on substrate: woven desized cotton twill)

6.3.1.2   Temperature

The right temperature of the process is important for optimum performance of the enzyme. The temperature only affects the speed of the process not the efficiency of pectin degradation. Increasing the temperature above 60°C helps in improved wax removal, although the enzyme gets unstable. Many times better absorbance of the enzyme is observed at lower temperatures.

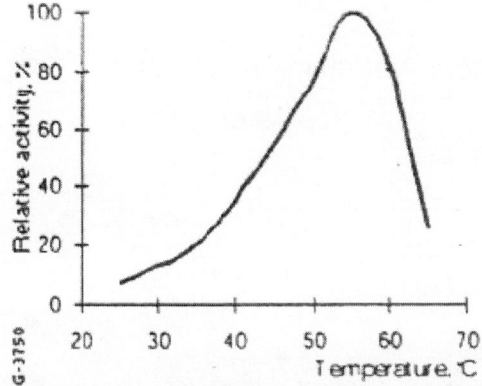

Figure 6.5. Effect of temperature on pectinase at pH 8.2
(tested on substrate: woven desized cotton twill)

Hence better performance of the process, therefore it is recommended to start the impregnation at temperature around 30–40°C and raise to the application temperature. As these enzymes are not stable for a very long time at elevated temperature, it is advisable not to keep/store the enzyme in a heated pre-mixing tank.

6.3.1.3   Wetting agents/detergents

Wetting agents are important to ensure good contact between enzyme and cotton substrate as well as for emulsification. Non-ionic surfactants do not significantly reduce enzyme activity up to dosages of 1 gpl and temperature30–60°C (e.g., Sandoclean PCJ/HPJ/NlF liq–Clarient). Anionic agents are best penetrating due to their emulsifying property. But anionic surfactants reduce the enzyme activity over a time. Low dosages of anionic surfactant may be used but it should not be mixed with enzyme before use. In general a mix of anionic and non-ionic surfactant is best recommendation.

6.3.1.4   Sequestering agents

The pectin removal capacity of pectinase is increased by using calcium binding buffers like phosphate, silicate or carbonate. Sequestrates like EDTA will also enhance pectin removal but should be applied with caution as any surplus over the metal ions present will inactivate the enzyme.

Removal of calcium is a key factor for successful scouring. Ca, Mg divalent ions must be removed from the system in order to aid wax removal and avoid bleach damage during subsequent bleaching treatment. Generally adding sequestering agents together with the enzyme is not recommended where addition is unavoidable (mild chelating agents).

6.3.2   Bio-preparation on Jet dyeing

6.3.2.1   Bio-preparation process

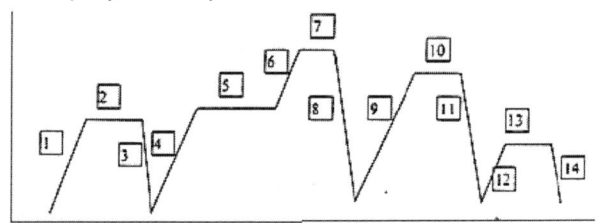

Figure 6.6. A typical bio-preparation procedure for jet dyer

Pre-rinsing (1–3) → Bio-scouring (4–5) → Emulsification (6–8) → Hot rinsing (9–11) → Cold rinsing (12–14).

A) Pre-rinsing
Pre-rinsing is a treatment which thereby wets fabric and removes some impurities including chemical additives such as knitting oil, etc. This process is optional and should be evaluated on a case by case basis. Wetting agent of about 0.5 g/l is often used. Lubricants (0.75 g/l) may be used for some machines, depending on the configuration of the equipment. A good wetting agent should wet the fabric almost immediately. The step is often carried out at a liquor/fabric ratio of 8:1–30:1 and 50°C for 10 min in order to secure adequate wetting. The water is drained at the end of this process and the fabric is ready for Bbio -preparation.

B) Bio-scouring
Bio-preparation with pectinase is a process in which pectic substrates in cotton cell wall are hydrolysed. This is carried out by an alkaline pectinase under the following typical conditions:
Enzyme dosage: Typically 0.02–0.07% owf*
Liquor ratio: 5:1–30:1
pH: 8.0–9.5
Temperature: 50–65°C
Time: 5–30 min
Emulsifier: Typically 1.0 g/l non-ionic or blend

Buffer: Sodium borate (2 mM), sodium carbonate or other typical buffer. *Assumes a product strength of 3000 APSU/g owf is defined as the amount of enzyme solution added per weight of the fabric.

C) Emulsification

For instance, 93°C and 5 min emulsification is often used for dark dye shade fabrics where 0.05 g/l EDTA and 1.0 g/l of an appropriate emulsifier is added at 10:1 liquor ratio (LR).

D) Hot rinsing

This process can be carried out at temperature ranges of 65–75°C for 5–10 min. Depending on machine, this process can be carried out in an overflow rinsing manner. A critical difference between pectinase and alkaline rinsing is the extent of amount of water needed. The amount of rinsing is greatly reduced compared to NaOH scouring where reducing alkalinity is the main purpose of hot rinsing.

E) Cold rinsing

At the end of bio-preparation, a water rinsing is often needed to lower the fabric temperature. Room temperature water is often used and rinsing is performed for about 5–10 min. It is recommend to lower the temperature below 35°C in order to conveniently carry out the next bleaching or dyeing process. For best results, it is better to carry out an immediate bleaching or dyeing after bio-preparation.

In most cases, the hot rinsing can be significantly reduced from traditional caustic based preparation. Bio-preparation has limited impact on motes as well as fabric whiteness which is not a significant consideration for dark shades. For medium shade to full white products, a further bleaching process is required to achieve satisfactory whiteness and mote removal.

6.3.2.2   All-in-one process

With more efficient enzymes we can have all-in-one process.

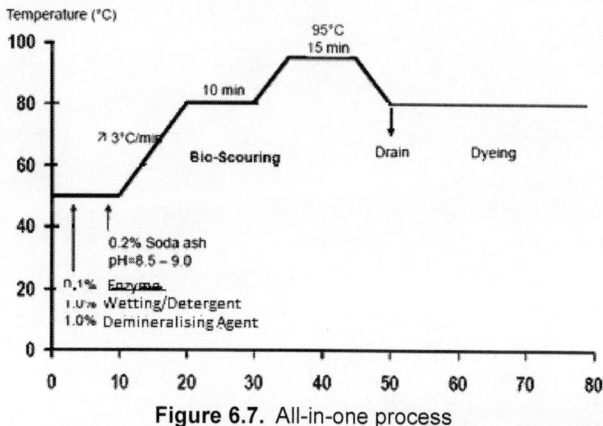

Figure 6.7. All-in-one process

6.3.3   General yarn Bio-preparation procedure

Bio-Preparation (1–2) → Emulsification and Chelation (3–4) → Hot rinsing (7).

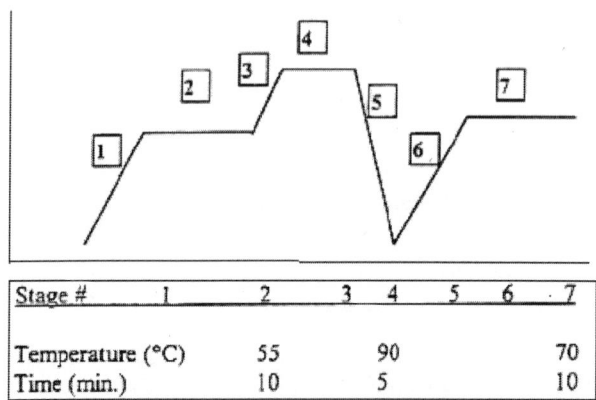

Stage #	1	2	3	4	5	6	7
Temperature (°C)	55		90				70
Time (min.)	10		5				10

Figure 6.8. Bio-preparation

Guideline recipe and process

Quantity	Unit	Chemicals
0.02–0.07	owf	Pectinase enzyme
pH 0.8–9.5		Tetra borate, carbonate (buffer)
0.05	g/l	EDTA, sodium tetrapolyphosphate (chelator)
1	g/l	Non-ionic emulsifier
0.5–0.75	ml/l	Wetting (non-ionic or anionic)
pH 0.8–9.5		pH range
5:1–15:1		Liquor ratio
55°C		Temperature
10–30	Min	Time

Refer process graph:
1. Add enzyme, wetting agent, buffer, emulsifier.
2. Heat to 55°C and hold 10 min.
3. Do not drain, add EDTA (limited quantity) heat to 90°C.
4. Hold 5 min.
5–7. Over flow rinse or normal rinse to pH 7.

(Conventional process is identical, except stage 2 is typically 30 min, add 1.0–4.0%. Caustic (owg) and the temperature is 100°C).

6.3.4   Simultaneous yarn Bio-preparation and dyeing procedure

Bio-Preparation and dyeing (1–5) → Hot rinsing (6–7).

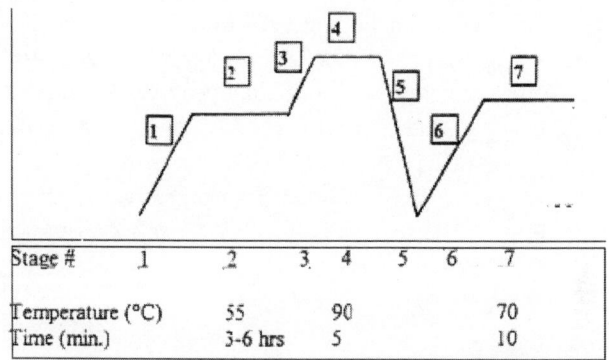

Figure 6.9. Bio-preparartion and dyeing process

Refer dyeing Graph:

1. Add dye, salt, sequestering agent, emulsifier, dyeing auxiliaries, pH to be adjusted to 8–9.5.

2. Add enzyme and then add the buffer slowly.

3. pH is increased to dyeing requirement, raised the temperature if necessary as per dyeing protocold and dye (4).

5. Overflow rinse or normal rinse (5–7).

This is a developing procedure and has the advantages of process time and water savings. Factors to consider are tolerance to all dye auxiliaries. In general, pectinase is tolerant to -80 g/l sodium chloride or - 100 g/l sodium sulphate. Use preferably non-ionic surfactants of the range 0.2–1.0% on the weight of the fabric. No chelating compound to be added in pectinase active stage. That is the sequestering agent to be added after the actual bio-preparation is over.

In general, reactive dyes will tend to have premature dye fixation at pH 9.0 (a typical Bbio -preparation pH). Therefore, the starting pH should be about neutral and slowly raised to pH 9.0. This will avoid premature dye strike and will simultaneously activate the pectinase. The stability/compatibility of dyestuffs and typical dye protocols must be evaluated on a case by case basis to ensure that pectinase remains active.

6.3.5   Pad-roll Bio-preparation procedure

The aim of the pectinase scouring is to remove pectin from the cotton. Fibre surface, thereby enabling a much more efficient emulsification of the cotton waxes, thus creating an adequate wettability. The target of the process is to remove at least 65%, preferably 70+ % of the pectin. Pad-roll systems offer good pH and temperature control, and sufficient reaction time. The process can also be applied in continuous pad systems but will demand that the temperature be controlled below 60–65°C while maintaining the relative humidity at 100% and the reaction time extended to at least 5 min. The figure below shows a typical pectinase dosage/response at conditions feasible in a pad-roll run.

Influence of enzyme dosage on % residual pectin in woven desized cotton twill at 55°C, 20 mM phosphate buffer at pH 8.2 and 120 min reaction time.

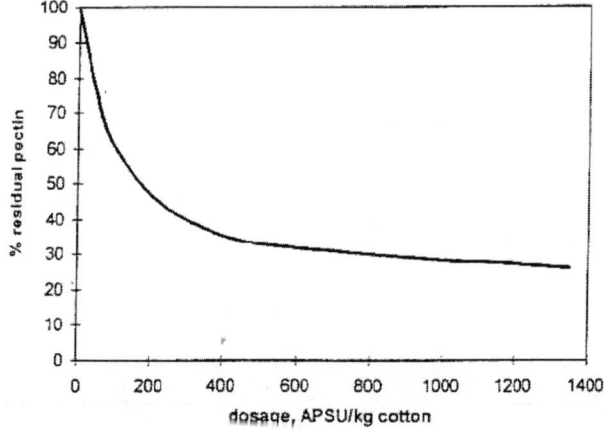

Figure 6.10. Pad roll process

Influence of enzyme dosage on % residual pectin in woven desized cotton twill at 55°C, 20 mM phosphate buffer at pH 8.2 and 120 min reaction time
1–2 Impregnation (pad-roll) → Dwelling (rotating station) → 4–5 Hot wash → 6–7 Hot rinse → 8–9 Warm rinse → 10–11 Cold rinse.

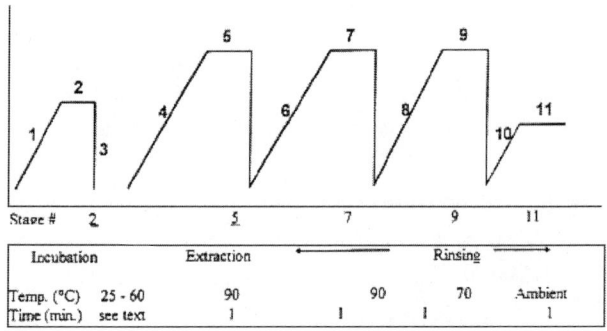

Figure 6.11. Pad-roll process

Time/temperature profile for scouring or combined desizing/scouring in pad-roll

Table 6.6.

Process	Quantity	Unit	Recipes
Impregnation	0.02–0.07	% owf	Pectinase (3000 APSU/g)
	10–20	mM	Buffer phosphate or carbonate–Start pH 9 (but can drop during reaction to 7)
	0.2–0.5	%	Non-ionic wetting and emulsifier
	70–100	%	Wet pick-up

Impregnation with recipe given, dwell for 2 – 24 hours (temp 45 – 55 °C). During washing stage add 0.2 % EDTA or 0.5 % polyphosphate 92 – 3 times) afterwards cold rinse and dry.

The above process makes the fabric ready for medium to darker shades. For lighter bright shades a mild bleach can follow if necessary, a traditional bleaching process can be used

Notes
1. Impregnation bath temperature with enzyme should be kept at 25°C and heat the solution to required temperature just before padding. Heated solution should be consumed immediately.
2. EDTA should not be used at impregnation stage.
3. In impregnation bath, one can add 0.1–0.2% anionic wetting agent, if required.
4. Desizing can be combined with bio -preparation by adding a suitable desizing enzyme which is active at the same pH and temperature.
5. Dwelling temperature: The impregnated fabric will be between ambient and 45°C. If external heating is available, the roll can be incubated at 55°C.
6. Dwelling time: Reaction time is dependent on dosage and temperature and will typically be in the range of 2–5 h (higher temperatures) to overnight (ambient temperature).
7. Washing-off: The purpose of wash-off stage is to remove the hydrolysed pectin (and starch) and further to emulsify the wax having been made accessible by the process. In this stage it is recommended that surfactant (emulsifier) and a chelating agent (e.g., 2 g/l EDTA or 5 g/l polyphosphate) is added to the water. It is done in the first wash bath of the pad system.
8. Rinsing: Rinsing to be carried out as shown in the graph above with normal water. Because enzyme preparation conditions are only mildly alkaline, extensive rinsing is not as necessary as in NaOH

scouring where reducing alkalinity is the main purpose of hot rinsing.

9. Care should be taken to avoid over drying as wettability may decrease due to melting of residual fabric.

6.3.6   Other bio-treatments–Wool

Wool is made of protein and bio-treatment features a protease that modifies the wool fibres. "Facing up" is the trade term for the ruffling up of the surface of wool garments by abrasive action during dyeing. Enzymatic treatment reduces facing up, which significantly improves the pilling performance of woollen garments and increases softness. Traditionally softening of coarser wool was carried out using various chemicals such as silicones, dichloro isocyanuric acid, tristearin and so on. Currently the problem associated with the fuzz can be eliminated by removing the fuzz using enzymatic treatment.

6.3.6.1   Enzymatic softening of wool

It was observed that the enzymatic treatment results in the removal of the protruding fibres, which are present on the surface and give a fuzzy appearance. During the enzymatic treatment, the scales present on the surface were chemically modified and due to this, a silky sheen is imparted to the surface of wool. Thus, their ability to reflect light from their surface increases, which improves lustre of the wool after enzymatic treatment. It results in the dissolution of the surface fibres, which are stiff and coarser. Thus, smooth surface is obtained after the enzymatic treatment with extra softness. Since the enzymes dissolve the protruding surface fibres completely, the softening effect achieved is also permanent.

6.3.6.2   Enzyme treatment for obtaining non-shrink wool

The shrink-resistance in wet mechanical treatment of wool is traditionally achieved by chlorine treatments. Such effect can also be achieved by treatment with proteases. Partial non-shrink action has been observed on wool when treated with *Streptomyces friedai* protease. It has been concluded that the protease enzyme did not show any effect on wool surface but they did enhance the action of a chemical treatment and a lipoprotein, lipase, as well as reducing the shrinking tendency of wool. It is essential that the enzyme treatment be followed by an application of Lanaperm resins so as to reach a satisfactory non-shrink effect.

6.3.6.3   Enzyme-enhanced bleaching of wool

Bio-bleaching of wool under both oxidative and reductive conditions under the action of protease has been of recent interest. It has been found that hydrogen peroxide bleaching in the presence of the protease considerably improved the whiteness and hydrophilicity. Enhanced bleaching has also been observed in reductive conditions by treating wool with protease papain, applied in the presence of sodium metabisulphite and sulphite. It has been pointed out that serine protease, stable to hydrogen peroxide, is active in alkaline medium and its activity is increasing with increasing hydrogen peroxide levels. In another investigation, it has been found that the whiteness obtained in wool bleaching is enhanced by the presence of a protease, with both peroxide bleaching and when sodium dithionite or bisulphite is used.

6.3.6.4   Anti-felting finishing of wool

Chlorine treatment and resin applications have been successfully and widely used in the anti-felting treatment of wool but the presence of organic halogens (AOX) is environmentally unfriendly and alternative treatments are more acceptable. With a rapid development of bio -techniques, enzymatic treatments of wool have been considered as one of the most possible alternatives to the chlorine treatment. Protease SZ conferred wool with expected area shrinkage reduction under the washing conditions, since the protease SZ has remarkable anti-felting effect on wool by Perzym process.

6.3.7   Other bio-treatments–Silk

6.3.7.1   Degummase enzymes in the processing of silk

The desizing of silk by enzyme is a well-known process which requires previous swelling of silk sericin by a selected surfactant. Application of serine protease in this context has been reported. However all the parameters of the process must be carefully controlled so that silk itself is not attacked. The raw silk thread consists of 70–80% fibroin and 20–28% of sericin. Sericin is chemically a non-filamentous protein. The process of eliminating sericin or gum from raw silk is known as degumming of silk. The purpose of boiling-off silk is three fold;

- To remove the sericin;
- To remove the reagents added by the throwster in soaking raw silk;
- To remove any dirt picked up in any of the operations of reeling, throwing or knitting.

Since all the natural and acquired impurities except sericin, constitute only a small fraction and are comparatively easily removed, the degumming process may be considered primarily as the process of cleavage of peptide bonds of sericin either by hydrolytic or enzymatic methods and its subsequent removal from fibroin by solubilisation or dispersion in water.

Degumming of silk with enzymes is based on the specificity of the attack of enzymes on certain amino acid adducts. Since sericin and fibroin contain different proportions of certain important amino acids, enzyme attacks them to different extent. Degumming with commercially available bacterial proteases is more effective than using trypsin and papain. Proteases can also be used to alter the silk fibroin surface to give an aged looking a similar way to enzyme washing of denim garments. It has also been recently reported that proteases may provide improved softness and wettability of silk fabrics. In silk, sericin and fibroin are proteins but due to specificity of enzyme, it only degrades sericin without damaging fibroin, which is not possible otherwise with conventional soda treatment. Hence, it improves degumming efficiency, imparts soft feel and ensures maximum weight reduction without affecting the fibre strength and hairiness.

Advantages of enzymatic degumming of silk are as follows:

- Removes sericin without damaging the fibroin.
- Minimum weight loss with excellent handle.
- Prevents strength loss as only acts on sericin.
- Prevents discolouration of silk.
- Reduces energy costs as application–temperature is 55–60°C.
- No haziness or unevenness as in case of metallic soaps.
- Improves lustre, hand and dye uptake properties.
- Reduction in environmental load.

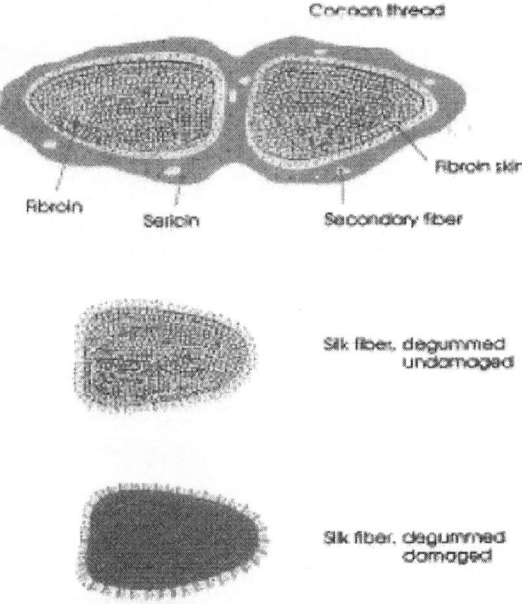

Figure 6.12. Bio-degummed silk

6.3.8   Other bio-treatments–Polyester

Enzymes are used in improving absorbency and wetting of polyester fabrics. The major drawback of alkaline hydrolysis includes damage to fabric properties such as strength and caustic discharge. Enzymes are protein molecules capable of catalysing specific chemical reaction of organic material. Hydrolases are capable of hydrolysing fatty acid or carboxylic ester, and in principle, can hydrolyse the ester linkage in polyester. Nowadays, a polyesterase enzyme is used on the polyester article under suitable conditions of concentration and time in order to obtain surface modification which results in increasing the hydrophilicity of polyester, improved stain resistance, wettability, or dyeability.

6.3.9   Other bio-treatments

6.3.9.1   *Bio-polishing*

Bio -polishing or bio -finishing with cellulase is carried out both on pieces as well as made-up garments and it can be compared to an enzymatic singeing process, carried out before dyeing or after dyeing also to give a sort of stone wash effect. It is widely used on made-up garments, without technological alternative, while on a piece traditional singeing can be used. Nevertheless, in this latter case, that particular soft hand effect typical of cellulase cannot be obtained.

Most cellulases have two functionally distinct domains: a catalytic domain and a substrate binding domain or cellulose binding domain (CBD). Both these domains are linked by a glycosylated linker–peptide. The catalytic domain have an active site in the shape of the tunnel or a cleft where the hydrolytic reaction takes place. The CBD-peptide varies from 33–240 amino acids in length.

Bio -polishing treatments permit:

- Elimination of dead or immature cotton, of neps and surface hairiness.
- "Natural" softening with an improvement of hand and drapability.
- Permanent prevention of reiterated fibrillation and pilling.
- Increase in hydrophilic properties, particularly in the case of terry fabrics.
- Better cleanliness and brightness, as well as uniformity of dye.
- Better overall quality of the material.
- The possibility of creating finishings to suit new and original fashion effects.
- Use of an entirely environmentally friendly process.

6.3.10   Other bio-treatments–Retting of flax

Retting which is the first step in processing of bast fibres, is generally a microbial process in which stem pectins are degraded, thereby freeing fibre bundles from the epidermis/cuticle barrier and from the woody shive. The practice of water retting, which produces high quality fibre, has been abandoned in western countries because of environmental pollution resulting from fermentation products of anaerobic bacteria and the high labour costs of drying.

Disadvantages of dew retting are inconsistent and often low-fibre quality. Commercial enzyme mixtures such as Flaxzym and Ultrazym are effective in separating fibre bundles from the epidermis/cuticle layer and from the shive material and also in reducing the size of bundles. Inclusion of chelators such as ethylene diamine tetra acetic acid (EDTA) or oxalic acid with the enzymes, along with mechanical disruption of stems, improved the efficiency of enzymatic retting. Properties of these enzymatically retted flax fibres compare well with those on dew-retting "cottonised" flax and good quality cotton giving slightly higher micronaire values but with greater strength.

6.3.11   Surface modification of cashmere fibres

Cashmere fibre is the fine under hair of goats produced in the extreme climates of northern China, Afghanistan and Iran. A number of organizations have examined chemical processing to improve the properties of cashmere such as shrink resistance, water repellency and dyeability. However chemical finishing processes, although giving the desired improvements in characteristics, affect the inherent desirable properties of softness and fineness. Enzymes used in the textile industry are usually hydrolases that break chemical bonds at mild conditions of temperatures and pH. Under these same conditions, it is now possible to reverse some of these reactions to create bonds. This possibility suggests that new and cost-effective methods for the addition of new finishes to textiles may be practical, producing long-lasting effects. One class of hydrolytic enzyme whose activity can now reversed successfully are proteases that conventionally are used to produce amino acids from proteins. Proteases catalyse the cleavage and synthesise peptide bonds. In contrast to cleavage of peptide bonds, the bond formation is an endothermic reaction which can be achieved by altering the concentration of the product, changing the kinetic parameters and influencing the equilibrium. A protease accelerates the attainment of the equilibrium under suitable mild conditions.

There are two possible mechanisms to produce a variety of trendy garments–Bio-washing, Bio-Ppolishing.

6.3.11.1   Bio -washing

The enzyme used for bio-washing is neutral and stable cellulase. These enzymes are produced from *Humicola isolans* which cause controlled superficial attack on the fibre leading to decolourisation without bringing about excessive loss in fabric strength. Cellulase acts on

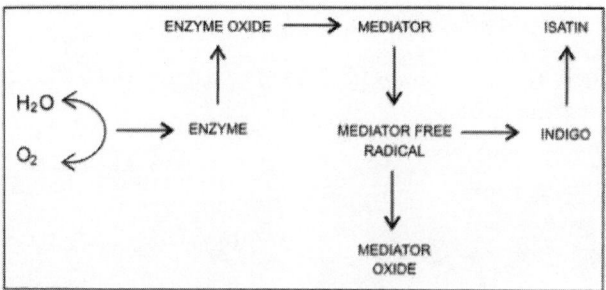

Figure 6.13. Mechanism of decolourisation

1,4 β-glucoside linkage. Here neutral stable (Bactosol IN) enzymes are mostly used, which acts at the pH of 6 and some of the cellulases where multi-component are represented as the whole cellulose complex, or in other words represented as the combination of different specificity of endoglucanases, cellobiohydralases and related enzymes.

6.3.11.2   Bio-polishing

Bio-polishing is a finishing process for cellulosic fabrics and garments. It is the process of improving surface of the material. It is also an "Enzymatic singeing". Bio-polishing with acid cellulose effectively reduces fabric fuzz

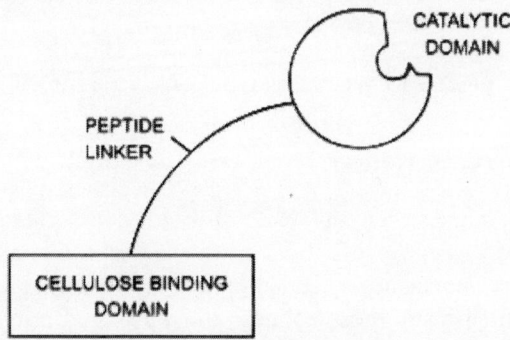

Figure 6.14. Structural parts of acid cellulose

and the pilling on denim materials. A cellulasic treatment gives the fabric a durable improvement of:

- Resistance to pilling
- A clear lint and fuzz free surface structure

- Bio-polished garment looks new even after repeated wear
- Bio-polishing permanently enhances fabric look without any chemical coating
- Bio-polishing improves drapebility and softness.

The fabric surface obtains a silky sheen similar to that resulting by traditional mercerisation.

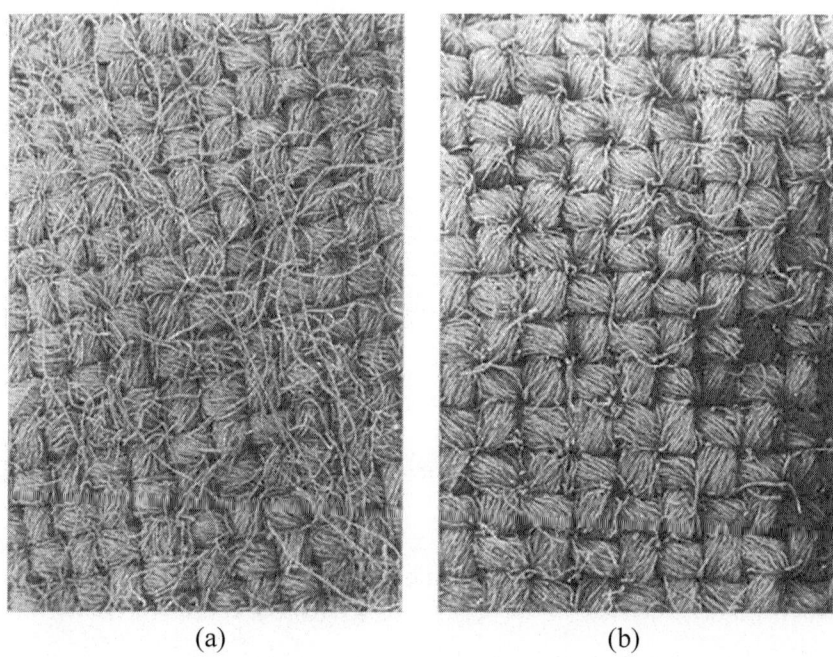

(a)   (b)

Figure 6.15. Effect of bio -polishing (a) Before treatment (b) After treatment

6.3.12   Enzyme fading

The popular washed-down or worn look of denim jeans has usually been obtained by the use of pumice stones that are soaked in sodium hypochlorite or potassium permanganate (oxidizing agents). When jeans are tumbled with these stones, the dye is eliminated when the stones are rubbed against the fabric. Pumice stonewashing has caused several problems, including rapid wear and tear of washing machines, a large number of second-class garments, unsafe working conditions, environmental pollution and the need for manual removal of pumice from pockets and folds of garments. In the mid-1980s bio -stoning with cellulases appeared as a perfect alternative. The action of cellulases and mechanical agitation, simultaneously or sequentially, abrades the fibre surface, releasing dye and dye aggregates from yarns and resulting in contrast in the blue colour. Commercially used acid cellulases mainly originate from *T. reesei* and their low-price is one of the reasons why they are so frequently used. As these cellulases are applied under acidic conditions, cellulose degradation may occur. Neutral cellulases have been tested recently to minimise fibre degradation, obtaining the desired colour fading with a low-degree of hydrolysis. One of the best known applications of enzymes today is the bio -stoning, an alternative to stone washing on denim and other garments to give fading or vintage look. Cellulace enzymes can be used in place of pumice stones and achieve less damage to fabric, machinery and environment. Even though this process gives a more even fading, it had many disadvantages.

Following are the disadvantages:

1. Process control is difficult because of the size and variation in quality of stones, leading to inconsistent quality of the washed garments in terms of quality of fading, strength, smoothness, etc.
2. Sometimes stones become powder during the processing making the garment greyish in colour and also harsh.
3. Stonewashing gives harsher feel than enzyme wash.
4. Stones may damage the machine parts.
5. Stone sometimes causes damage and fading to the accessories of the garments.

Due to the many disadvantages of this process has been mostly replaced by treatment with cellulose enzymes or sometimes in combination with pumice stones. There are two types cellulase enzymes used for this purpose– acid and neutral.

Cellulase enzymes can be projected as having three regions–a globular head which is the core of the catalytic action and a tail which is a cellulose binding domain (CBD) and both head and tail are joined by a small portion called a hinge. However, the effectiveness of the catalytic core (to hydrolyse glycosidic linkages) depends greatly on the adsorption by the CBD tail. When tightly bound to the substrate, certain endogluconases disturb the crystalline structure and induces defibrillation, a result of the mechano-chemical effect by the binding domains

A cellulase enzyme has various activities, which are named as endoglucanase EG I, EG II and CBH II (exogluconase or cellobio hydrolysate), EG III and EG IV. These various activities have varying effect on the crystalline and amorphous regions of cellulose. Regular whole acid cellulases have a higher CMCase (Carboxy methyl cellulase activity) activity hence attack the amorphous regions producing better cutting, good puckering effects but with higher back staining.

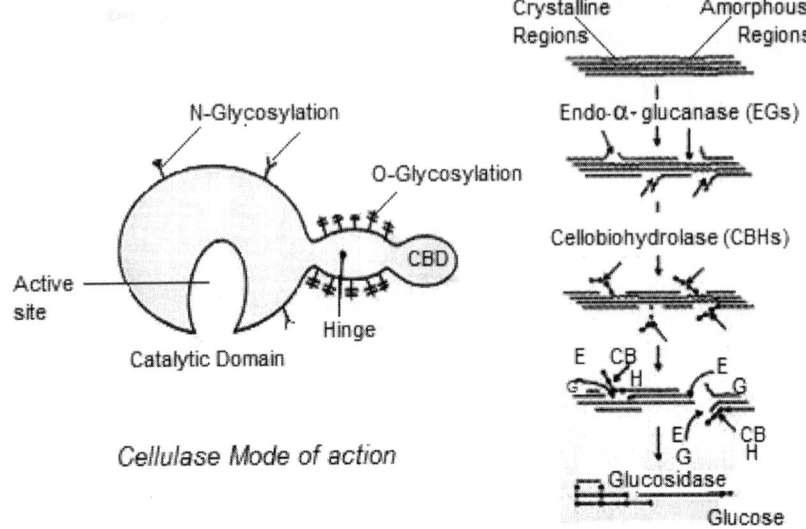

Figure 6.16.

In case of engineered acid cellulases the CBD is reduced considerably and hence these products do not get attached to the cellulose substrate easily and hence the cutting is not very aggressive as in the case of the whole acid cellulases. Also these products are engineered to produce peak activity at near neutral pH.

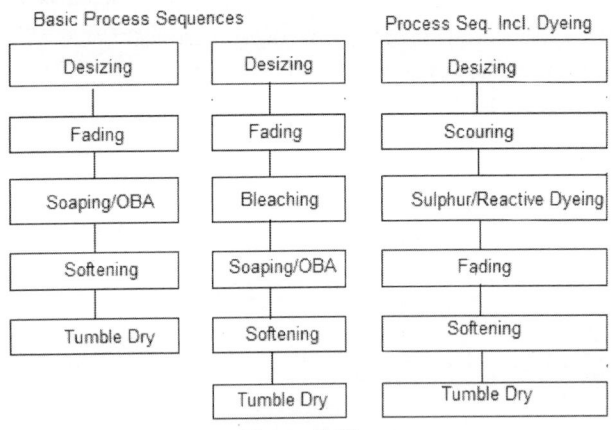

Figure 6.17.

The neutral enzyme does not have a CBD and hence does not attach itself to the cellulose surface. This is why we get a more surface cutting and not "cutting in" activity. This leads to lower back staining, a more even effect on the surface of the fabric, which produces a rich and more pleasing look.

Cellulase–Mode of action

Cellulase enzymes are natural protein which accelerate the biodegradation of cellulose by hydrolysing the β-1,4-glucoside bond of the cellulose molecule.

Effects of cellulase enzymes on Denim

1. In fading the desired effect is wash down look on the garment making the seams, hems, and pockets more prominent.

2. Other than look colour contrast commonly known, as salt pepper effect is also one of the effects produced by fading. Faded garment with acid cellulase enzymes gives less colour contrast as compared to garment washed with neutral cellulase enzymes.

Recipes and process

Recipe

Quantity	Unit	Additions
3–6	g/kg (owf)	Cellulase enzyme
1–2	g/l	Wetting agent

Run at 55°C and pH 4.5–5.0 for 60 min.

6.3.13   Solvent scouring

Certain organic solvents will readily dissolve oils, fats and waxes and these solvents can be used to purify textiles. Removal of impurities by dissolution is called extraction. There are commercial processes where textiles are cleaned with organic solvents. Fabrics processed this way are said to be "Dry Cleaned". Although not widely used as a fabric preparation step. It is an important way of removing certain difficult to remove impurities, where a small amount of residuals can cause downstream problems. Garment dry cleaning is more prevalent.

For fabrics that do not contain any size, solvent scouring is an effective way of removing fibre producer's finishes, coning and knitting oils. Solvent extractions are particularly useful in the laboratory by determining the amount of processing oils added to man-made fibres and the residual amounts of oils and waxes left by aqueous scouring. Properly controlled, fabrics can be produced with very little residual matter. Above figure shows a schematic diagram of a continuous, solvent scouring range. The entire ranges enclosed so the vapours are contained and not allowed to escape into the atmosphere. Recovery units are installed on the range to ensure that none of the solvent is allowed to vent to the environment. Usually carbon adsorption towers are used for this. Also a solvent distillation unit is needed to recover the pure solvent and separate the removed contaminants.

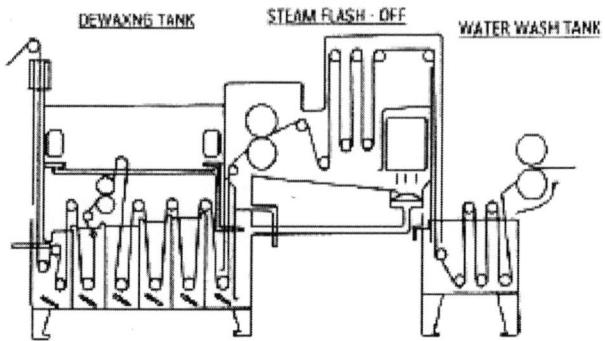

DEWAXNG TANK   STEAM FLASH - OFF   WATER WASH TANK

Figure 6.18. Schematic diagram of a Solvent extraction unit

The term solvent scouring is also used to describe processes where amounts of organic solvents are added to aqueous scouring formulations to assist in the removal of oils and waxes.

6.3.13.1   Advantages and disadvantages of solvent scouring

Solvents dissolve almost all oils and waxes. They have low-liquid surface tensions and quickly and easily wet out and penetrate fabrics with waxes and oils. They are much easier to evaporate than water, requiring less time and energy. On the negative side, hydrocarbon solvents are flammable and present explosion hazards. Most chlorinated solvents are proven or suspect carcinogens and some are known to contribute to atmospheric ozone depletion. Chlorinated hydrocarbons thermally decompose to form phosgene and hydrochloric acids. These decomposition by-products are corrosive to metals and also damage cellulosic fibres. Solvents are expensive so they must be recovered and purified by distillation requiring special equipment. The distillation residue becomes a solid waste disposal problem. Solvents do not aid in the removal of motes, metal ions, starch and other solvent insoluble impurities.

Table 6.7. Properties of solvents used in solvent scouring

	B.P.°C	Latent heat of evaporation (Cal/g)	Specific heat (Cal/ g/°C)	Surface tension (Dyenes.cm)
Water	100	545.1	1	72
Trichloroethylene	87	57.3	22	32
1,1,1 Trichloroethane	74	58.5	0.25	26.4

| Perchloro ethylene | 121.2 | 50.1 | 0.21 | 32.3 |
| Trichlorofluoroethane | 47.6 | 35.1 | 0.21 | 17.3 |

Listed above are some of the more common solvents used commercially. These are among the safest as they are generally non-flammable. However, they must the handled with care because the chlorinated ones are on the suspect carcinogen list of regulated chemicals.

6.4   Mineral earth–Technology for preparation of cellulosics

Clay is absolutely required for the fertility of the earth. All organic and inorganic chemicals needed for the growth of plants are adsorbed on the clay-particles in the soil. Clay has been used over many years in various applications like bricks, terracotta, ceramics (China), drainage pipes, wine and oil cleaning, face masks for cleaning, paper industry, etc. High-grade clay consists of inorganic, non-toxic elements like Al, Si, O, Mg, Ca with a thickness of 1 nm. They have a sheet-like flexible shape, with a strong negatively charged surface.

When a clay compound is put in water, all particles automatically separate due to the strong negative charge, and the total surface becomes available for adsorption of metal ions, hydrophobic waxes and oils, and hydrophilic impurities and polymers. The elimination of metal ions like Fe (III)–which causes catalytic disruption of hydrogen peroxide–guarantees good stabilisation of the bleach bath. The removal of heavy duty oil and even silicone softeners and fluorocarbon resins, opens new applications in textiles for clays.

Innovative products are based on selected, high-grade clays with extremely small particle size (up to ± 2 μm). These particular clays are being found in deposits of volcanic ashes, and which are formed during millions of years. They have a very specific crystal structure. This structure provides very particular properties to the clays. Most of the clays have a sheet-like structure with a more or less strong negative electrostatic charge, which is responsible for the separation of the particles as well as the adsorption of manifold types of organic matter and metal ions. Depending on the type of clay, a huge active surface from 190 m^2/g up to 800 m^2/g is available.

Compound product based on selected clays, which generate excellent stabilising effect for hydrogen peroxide is now available. Using single auxiliary which combines all required functions for bleaching. Iron contaminations (ions) are well-being eliminated (see photo). Excellent whites and level dyeings are being achieved. These chemical is a green product contributing negligible BOD or COD value as organic substances are

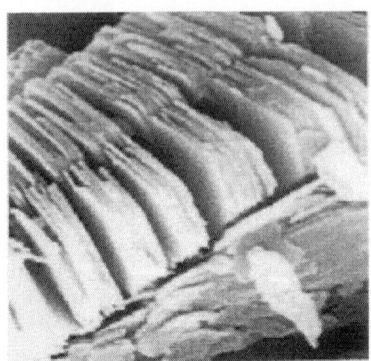

Figure 6.19. Mineral earth

adsorbed by the clay. They are one shot product combining all properties for bleaching. They are special clay based scouring and adsorption properties and very effective for sequestering of metal ions and cleaning of cotton. They also have excellent stabilising power in peroxide bath (no side reactions of hydrogen peroxide) and thus support very controlled bleaching conditions. It shows excellent lubricating properties on all cellulosic fabrics and synthetics. (No need to add separate lubricant.) This also helps to achieve higher degree of whiteness. The clay and does not contribute to TDS and low in foam formation. These can be used for pre-treatment of viscose or sensitive fabrics like bamboo, pre-treatment can be done at lower pH and without a pre-wash (demineralisation step).

6.4.1   Characteristics

Appearance at R.T	Grey dispersion
Ionic nature	Not applicable
pH of 2% solution	Slightly alkaline
Solubility	Can be dispersed well in water
Stability to hard water	Good
Stability to pH	Stable in normal conditions
Compatibility	Compatibility with non-ionic and anionic products. Not compatible with cationic products and lubricating polymers (polyacrylamides)
Storage stability	Must be stored at minimum +5°C To ambient temperature bring back to normal Working temperature before use.

6.4.1.1   Recipes and processes

Pre-bleach

Prebleach recipe

Quantity	Unit	Chemicals
1–1.5	ml/l	Clay product
2–3	g/l	Caustic soda 100%
2–3	g/l	Hydrogen peroxide 50%

30–45 min at 98°C.

Bleaching for white

Full bleach recipe

Quantity	Unit	Chemicals
1.5–2	ml/l	Clay product
3–4	g/l	Caustic soda 100%
6–8	g/l	Hydrogen peroxide 50%
0.6–0.8	%	OBA

40–60 min at 98°C.

Viscose bleaching

Recipe for viscose bleaching

Quantity	Unit	Chemicals
1–1.5	ml/l	Clay product
2–3	g/l	Sodium carbonate
4–6	g/l	Hydrogen peroxide 50%

30–45 min at 90–95°C.

6.5   Emerging technologies

There are many emerging technologies being tried in the textile field in which some are already being introduced in the bulk production or trials and some under investigation. These technologies are used mainly for cost reduction, better quality of process, flexibility, environmental conservation and sustainability.

Plasma Technology

Plasma can be described as a mixture of partially ionized gases, where atoms, radicals, ions and electrons can be found, sometimes referred to as the fourth state of matter. Plasma used in the textile industry consists of gases such as oxygen, air, nitrogen argon, helium or fluorine. There are several types of plasma that can be divided into two categories: non-thermal and thermal plasma. The temperature in thermal plasma is very high and is therefore not used for textile material, which often consists of heat sensitive material. Non thermal plasma, also known as cold plasma, is more suitable to use in the textile industry. Non-thermal plasma can further be divided into atmospheric and low pressure plasma. The most used and established plasma is corona plasma, which is a type of atmospheric plasma. Corona plasma is generated through low frequency and high voltage between two electrodes during atmospheric pressure. The process of low pressure plasma technology is easier to control, which make the results reproducible. Low pressure plasma contains high concentration of reactive species compared to atmospheric plasma. The disadvantages with low pressure plasma are longer process time and that only batch processes can be performed.

Plasma technology can be used for both natural and manufactured fibres. The technology is used for surface modification in order to increase the efficiency of several treatments and processes. Plasma technology is one of the most promising emerging technologies for surface modification in the textile industry. The physical and chemical changes on the fibres are obtained by activation, cleaning, oxidation, polymerisation, radical formation and creation of nano-structure caused by the plasma. The wide range of application areas for plasma technology includes pretreatment (desizing), change of wettability, pretreatment for printing and dyeing (increased affinity for dyestuff), antibacterial finishing, improvement of efficiency of wet finishing processes and influence of the physical properties.

Application of plasma in Preparation of fabrics

Plasma technology is used in many areas of textile processing. Generally plasma interacts with materials in different ways (1) formation and ions which will lead to the formations of ozonisation, NH3, NO2, etc. (2) in can initiate polymerisation and deposit thin film of polymers.(3) It can react on the surface of the material like fibres and break organic bonds with the evolution of gaseous products (e.g., hydrogen from hydrocarbons and forms carbon free radicals. These radicals can involve in chemical reactions. The latter reaction is widely used in textile processing

The plasma can cause cross-linking, solid-state polymerisation, etching action, radical formation and degradation. The required transformation for a particular process using plasma is controlled by the nature of gas, pressure and the power used for the discharge. For example oxygen containing plasma can increase the wettability of various synthetic fibres and substances by formation of polar surface of ketone, hydroxyl, ether, peroxide and carboxylic acid groups that are much more hydrophilic (wettable) than the untreated surface. Various transformations achieved by plasma is more applied in the finishing area which will be explained in detail including the creation of plasma etc. in the book on Finishing in the same series by the same author.

Plasma may be used for removing the contaminants, finishing and sizing agents from the fabric. Desizing of polyester fabric that used polyvinyl alcohol as the sizing agent can be removed by plasma treatment. Using nitrogen plasma wettability of cotton, silk, polyester etc. fabrics can be increased few times which can make the further processes more effective. Using air and oxygen plasma the scouring process can be eliminated in case of many fibres.

High energy radiation can be used for the bleaching of textiles. The bleaching of textiles can be done either by radiating the bleach bath before the entry of the fabric into the bleaching solution or the fabric may be exposed to radiation and stored and then bleaching is carried out. The bleach bath containing sodium chlorite (5 - 110 g/l) can be activated by high energy radiation, with pH adjusted to 9 to 11. The application baths also contain other usual additives such as optical brighteners, wetting agents and auxiliaries and bleaching can be done using continuous or batch methods at or below normal processing temperature. The free radicals are formed when textile materials are subjected to moderate amounts of radiation (1-2 Mrad), so that the fibre degradation can be kept to minimum.

It has been found that the plasma treated PE/C fabric dyed by one bath one dye class method (Reactive, Direct dyes). Corona discharge treatment is supposed to enhances hydrophilicity of cotton and PES fibers due to surface modification of the material and formation of C-O, C=O, and COOH groups. The hydrophilicity and dyeability is further improved, especially of PES fibre, by a second subsequent treatment with biopolymer chitosan which again depended on the concentration of chitosan treatment solutions. Satisfactory values of dye fastness and fixation degree of reactive dye can obtained.

A short corona plasma treatment of cotton fabric increase the wetting properties of cotton with the functionalisation of the surface with oxygenated moieties, without any significant alteration in surface topography of the fibres. Further cationising of the cotton fabrics using an epihalohydrin as cationising agent increases the exhaustion of the dyestuff as high as 90%, and produces a dramatic improvement (80% increase) in the colour intensity (K/Scorr) on both sides of the fabrics. Plasma treatment on one side only can give a two sided effect proving the changes are only on the surface and low penetration of the plasma on the fabric The improvement in colour intensity of the cationised cotton fabrics can be explained taking into account that the hydrolised reactive dye has high anionic character which can be bound to the cationic amine of the cationic agent on the cotton fabrics. Following figure shows how plasma treatment and cationisation increases the functional groups on the surface of cellulose and increases the dyeability

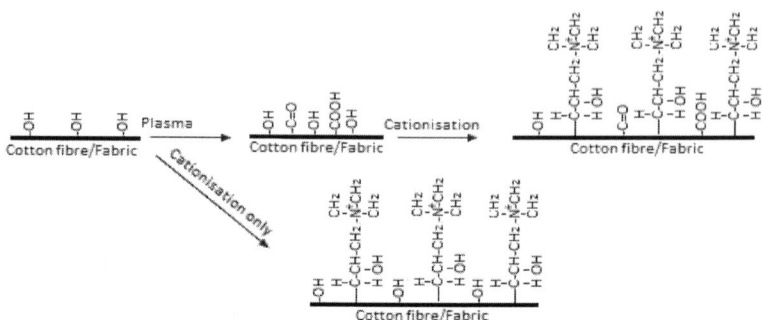

Surface functionalisation of cellulose by Plasma.Plasma+Cationisation and cationisation only

Ozone bleaching

Ozone bleaching is also chemically an oxidative bleaching. It is relevance is in the present importance of environmental conservation. Ozone (O3) is for technical purposes produced by an ozone generator. In the generator, air or oxygen pass through a strong electrical field which splits the oxygen molecule into two unstable atoms that can react with oxygen molecules and produce ozone. Ozone is a strong oxidising agent and has an oxidation potential of 2.07 electron-volts (eV), which can be compared to 1.77 eV for hydrogen peroxide. Ozone is thermodynamically unstable, which means that ozone will spontaneously convert back to oxygen. The fact that ozone is chemically unstable is an advantage compared to other bleaching agents, since no secondary derivatives remain in the bleached fabric. Ozone can be used for bleaching loose wool, silk, cotton and also yarn and fabrics made of these fibres.

Ultrasonic treatment

Ultrasonic waves are acoustic waves with frequencies between 20 KHz to 20 MHz. Ultrasonic waves have a frequency of 16 kilohertz and have been investigated as a solution for the textile industry to optimize cost, reduce the environmental impact and meet the quality requests. When ultrasonic energy is used in a liquid it cause two things: heating and cavitation. Cavitation is a phenomenon of formation and collapse of bubbles and ultrasound, can be used for improvement of transport of molecules to the surface of the fibre, which also increase the reaction rate. A major benefit with ultrasonic treatment is that the equipment can be installed within existing machines.

Ultrasonic energy has also been studied as an alternative for conventional methods to accelerate mass transfer in desizing, scouring, bleaching, mercerizing and dyeing of cotton fabrics. Several studies, have also investigated the use of ultrasonic energy in enzymatic pretreatment processes of textiles, increase the efficiency of textile wet processes etc. It has been known that ultrasonic waves in liquids are used to clean surfaces. They produce bubbles in liquids and when the bubbles burst substantial amounts of energy is released. When compared with hand washing, ultrasonic agitation has many advantages. It has negligible effects on the strength and colour of fabrics and causes less fibre migration. On carrying dyeing in ultrasonic bath of cotton fabric pretreated with enzymes show better and faster dye uptake after enzyme pretreatment on cotton fabric and results of dyeing are better than those obtained using metal mordanted fabric.

Nanotechnology

Nanotechnology deals with materials having at least one dimension less than 100 nm. It includes nano-particles, nanorod, nanowires, thin films and bulk materials made of nano scale structures. Application of nanotechnology on textile materials could lead to the addition of several functional properties. Deposition of silver nano-particles can be used to create shiny metallic yellow to dark pink colour while simultaneously impartingantibacterial properties to the fabric. Gold nano-particles allow the use of molecular ligands so that the presence of biological compound surroundings is rapidly detected. Metal oxidenano-particles such as TiO_2, Al_2O_3, MgO and ZnO possess photo-catalytic and antibacterial activity and UV absorption properties. More often these nano-particles can be impregnated. These are explained in detail in the volume on Finishing in the same series of books by the same author.

Radiation induced surface grafting

Many new technologies have been introduced in different textile processing areas. Surface grafting is a relatively a new technique which is used to alter the surface morphology and chemical composition of textile substrates and, thus imparts new or improved functional properties. The surface grafting can be achieved by a) light (generally UV light)induced grafting, b) plasma induced grafting and c) irradiation (γ-rays, electron beam)induced grafting. These three methods share the same objectives i.e. creating free radicals onto the polymeric substrates and then these radical sites are used as initiators for copolymerization reactions with vinyl monomers present in the grafting solution. Variousparameters such as concentration of the monomer, time of treatment, radiation dose, type and concentration of catalyst used, if any, could greatly affect the grafting efficiency and need to be optimized.

Radiation induced grafting involves the use of high energy radiation to create free radicals onto the textile substrates. It can be carried out in two ways: in-situ grafting and postexposure grafting. Compared with chemical grafting, radiation-induced grafting has many advantages such as no chemical initiators are used and grafting yield can be controlled by controlling dose and the time of irradiation. Plasma induced grafting has been explained in the para on plasma technology.

Light induced grafting basically uses, UV light energy for creation of free radicals onto the textile surfaces. UV radiation induced surface grafting of textiles is another technique that has attracted attention because of its simplicity. Radicals thus generated onto the surface are used to initiate copolymerization reactions with various monomers. UV radiations are not as penetrating as high-energy gamma (γ) radiations, free radicals thus produced mainly close to the surface rather than uniformly distributed throughout the fibres. Radiation induced grafting using high energy sources such as UV radiation or electron beams has been explored as an alternative to chemical grafting. However, if the conditions are not controlled, radiation grafting may result in deterioration of physical properties of the substrate and may lead to photo-degradation of the substrate. The parameters affecting the efficiency of grafting, such as time and temperature of the reaction as well as concentrations of the initiators, were optimized in this study to maximize the amount of grafted monomer on the PP fabric surface

References

1. Basacryl Dyes, Acryl Dyes on Acrylic Fibres – Manual for Dyeing and Finishing of Acrylic Fibres, Alone and in Blends with Other Fibres, Badische Anilin- & Soda-Fabrik AG: 1972.

2. Mike Flatau, (2007). Trends in Continuous Pretreatment, Proceedings of Texsummit 20077 - Contemporary Processing of Textiles and Apparels: Challenge sand Opportunities.

3. BASF, Technical Information, TI/T 7036 e March 2002, Kierlon JET-B Conc, 2002.

4. BASF, Technical Information, TI/T 7036 e November 2003, Performance Chemicals for Textiles-Luzyme PK Conc, BASF India Ltd. 2003.

5. BASF, Technical Information TI/T 7156 e October 2003: Ultraphor TC Liquid, 2003.

6. BASF, Technical Information, TI/T 7114 e September 2000 A: Colorants and Finishing Products, UltraphorRN Liquid, 2000.

7. BASF, Technical Information, TI/T 356 e May 2000 (RV): BASF Auxiliaries for Pretreatment.

8. BASF, Technical Information, TI/T 341 e September 1999 (RV): Colorants and Finishing Products - BASF RedEx BLEACH process, 1999.

9. Boehme, (2003). BTR·TC / wh –ho /10.2003 - Products for the Textile Industry.

10. Edward Menezes, (2008), Sequestering Agents: Asian Dyer. April, 43–46.

11. Novo Nordisk, BI 194a, GB BioPrep - Product sheet - Enzyme Business.

12. Alat, D.V., and Dr. Saraf, N.M., (2002), Bioscouring: An Environment-Friendly Process, *Chemical World*, September 16, pp 20.

13. Edward Menezes, (2007), Enzymatic Approach to Textile Processing. *Proceedings Contemporary Processing of Textiles and Apparels: Challenges and Opportunities - Tex Summit.* pp 35–40.

14. Hempel, W.H, (1991), The surface modification of woven and knitted cellulose fibre fabrics by enzymatic degradation, *ITB Dyeing/Printing/Finishing.* 3, pp 8–14.

15. Enzyme for Biowashing. *Rossera,* – March–April 2002.

16. Clarient, Technical Literature, Continuous Bleaching.

17. Dr. Th. Bohme K.G., Continuous Process - Oxidative Desizing Bleaching (Pad Batch) +Bleaching (Pad Steam), Continuous Process Demineralisation/Desizing (Pad Batch) + Bleaching (Pad Steam); Laborder: 2005020402-2005-10-07.

18. Robert G.B., (Ed): *Review Ciba – Geigy Ltd.*, Dyestuffs and Chemicals Division CH-4002 Basle, Switzerland, 1973/1.

19. Sadov, F., Korchagin, M., Matetsky, A., (1973), *Chemical Technology of Fibrous Materials*; MIR Publishers, Moscow.

20. CHT Technical Literature: *Detection of sizes Recipes for desizing.*

21. Thomas Hipp, (1996), High temperature Bleaching – An out dated procedure or an attractive process variation. *Lecture at Sympotex.* South Africa, Durban, pp 1–12.

22. Behenke, H., (1996), The demineralization process, *Colourage*, September, pp 27–42.

23. Sandoz (Pakistan) Ltd, Chemicals Division – Technical Literature - *Processing of Hosiery.*

24. Sandoz, Chemicals from SANDOZ for textile pretreatment – Bleaching.

25. Sandoz, Chemicals from SANDOZ for textile pretreatment – Mercerising and Causticising.

26. Sandoz, Chemicals from SANDOZ for textile pretreatment – Scour boiling/alkaline stage.

27. Sandoz, Chemicals from SANDOZ for textile pretreatment – Scouring and washing - Removal of water soluble starches.

28. Sandoz, Chemicals from SANDOZ for Textile Pretreatment –Desizing of fabrics of cellulosic fibres and their blends with synthetic fibres.

29. Sarex; Cotton Preparation Manual.

30. Cotton Incorporated ISP 1003; Enzyme Technology for Cotton Products, 2002.

31. Prof. Dr. Rer. Nat. (2000), Hans-Karl Rouette: Encyclopedia of Textile Finishing; Springer.

32. Anupama, M. and Anita, R., (2007), Enzymes and Textile Sector; Asian Dyer. December. pp 31–38.

33. CIBA - Ciba pretreatment on Benninger woven and knitgood ranges Continuous success - Ciba Specialty Chemicals Inc. 2003.

34. Strohle, J.,Open width Processing of Knit Goods - International Dyer, pp 13–14.

35. Gunter Euscher, (1982), Medium Knit Mercerizing, Textile Asia - September.

36. Gunter Euscher and S Jayachandran, (1997), Knit Goods Processing, The Indian Textile Journal. April, pp 17–21.

37. Gordon, R., Nearchimica, S.P.A., (2006), Textile Auxiliaries Preparation, The dyeing and Finishing of cotton knit goods. June.

38. Rakesh, G., and Prabhu, C.N., (2007), Mercerization of Knit Fabrics, Colourage, October, pp 114–116.

39. Strahle, J., and Schramek G., (2007), Open-width Mercerizing of Knitwear, ATA Journal. October/November, pp 54–55.

40. Carmine, M., and Paola, Z., (2002), Knitting, Fondazione, ACIMIT, January.

41. Du Pont Technical Information, (1976) Fibres – Physical and Chemical properties of Lycra – as Related to end Use requirements. December.

42. Strohle, J., and Gehrlein, H., Benninger A.G. Tension-Controlled Pre-Treatment of Cotton Fabrics Containing Elastane.

43. Rossari Bio-tech: Continuous Bleaching Range Package.

44. Huntsman, (2007), Pretreatment of Viscose (CV) including polynosic (CP) and Modal (CM); Textile Effects.

45. Karmakar, R.R., (1999), Textile Science and Technology 12- Chemical Technology in the Pretreatment Process of Textiles, Elsevier Science B.V.

46. Colorband Dyestuff (P) Ltd, & Farbotex S.P.A, – Clay in Textile Wet Processing.

47. Sandoz - Technical Information; Alkaline pretreatment by the cold pad batch process, Sandoz Ltd, Basle, Switzerland 31.1 1975.

48. Flensberg, H., Hammers, I. (1987), Pretreatment and dyeing of Silk – Lecture presented at the symposium , Naturseide Forschung and Praxis, von der Textile chemie biz zur Fertingungon.

49. Huntsman Textile effect – Pretreatment: Pretreatment of Linen/Flax fibers Huntsman, 2007.

50. Ciba – Textile effects: Introduction to Pretreatment, Presentation, Ciba.

51. Cavaco-Paulo. A., and Gübitz, G.M. (ed), (2003), Textile processing with enzymes - Woodhead Publishing Limited in association with The Textile Institute.

52. Dr. Padhye, R.N., and Ms. Sakha, M., (1992), Degumming of Silk at Low Temperature – Colour Chronicle, Sandoz Limited, India, Jan-March.

53. Iyeriyer, N.D., (2004), Silk - The queen of textile fibres, Colourage, June, P 75.

54. Dr. Mahapatra, N.N., (2004), Degumming of Silk in Industries. Colourage, October pp 53–54.

55. Mahall, K. (1985), Silk and its Treatment, Textile Asia, October. pp 95–101.

56. Sridharan, V., (1981), Wild Silk Processing, Colour Chronicle, Sandoz Limited, India, January-March.

57. Sandoz, Silk and Colour, Brochure No. 9307.00.88.

58. Wescher, H., and Zeller, R. (1950), The Silk and Velvet Industries of Crefeld, Ciba Review, December 83, pp 3002–3026.

59. Iyer, N.D., (2005), Continuous processing of silk fabrics, Colourage April, pp 49–50.

60. Iyer, N.D., (2005), Colouring Test – Colourage, February pp 85–86.

61. TEGEWA: Textil Praxis, 36 (1981) 1331–1332, 1349–1350.

62. Brown, Olson, Keegan, American Dyetuff Reporter (1967) 703–707.

63. Losonczi, A, Csiszar, E, Szakacs, G, Kaarela, O. (2004), Bleachability and Dyeing Properties of Biopretreated and Conventionally Scoured Cotton Fabrics. Textile Research Journal, June.

64. Bhattacharya, S.D., and Shah, J.N., (2004). Enzymatic Treatments of Flax Fabric. Textile Research Journal, July.

65. BASF – Manual – Dyeing and Finishing of Polyester Fibres and Their Blends with Other Fibres.

66. Dr. Charles Tomasino, (1992), Chemistry & Technology of Fabric Preparation & Finishing; Department of Textile Engineering, Chemistry and Science, College of Textiles, North Carolina State University, 1992.

67. Michael, S. Showell (ed), (2006), Handbook of Detergents Part D: Formulation CRC Press.

68. Chakraborty, J.N., (2010), Fundamentals and practices in colouration of textiles: Woodhead Publishing India Pvt. Ltd.

69. Anna-Maria Janßen-Tapken, and Mike Flatau, Handbook for Pretreatment; CHT R. BEITLICH GMBH.

70. Clark, M. (ed), (2011), Handbook of textile and industrial dyeing Volume 1: Principles, processes and types of dyes. Woodhead Publishing Limited.

71. Sandoz (India), Pre-treatment of Linen, Colour Chronicle; October 2005.

72. Edward Menezes, (2007), Wet processing of polyester microfibres, colourage, October.

73. Fritsch, J., and Praveen, A., (2007), Innovative Techniques in Man Made Cellulosic Wet Processing - Cellulosics: Coloration, Innovations & Regulations; Colour Trends.

74. Pietro, B., Ferruccio, B., Ester, F., Giuseppe, R., and Sergio, V., (2001), Reference Book of Textile Technologies; ACIMIT Foundation; November.

75. Gries, V., and Wulfhorst, (2015), Textile Technology: An Introduction, Hanser Publishers, Munich.

76. Bhuvalesh, C., Goswami, R.D., Anandjiwala, and David, M.H., (2004), Textile Sizing, Marcels Ekkerin, C. Dekker, New York, Basel.

77. Engr Shah Alimuzzaman Bela, Understanding for a textile Merchandiser; BM N³ foundation Dhaka, Bangladesh.

78. Near Chimica, Viscose - Recommendations for Pre-treatment, Dyeing and Finishing, Near Chimica 2006.

79. Arindam B., (Ed), (2015), Advances in Silk Science and Technology, Woodhead Publishing Limited, Cambridge CB21 6AH, UK in association with The Textile Institute.

80. Nierstrasz, V.A., and Cavaco-Paulo, A. (Ed), (2010), Advances in Textile Biotechnology, Woodhead Publishing Limited, Cambridge CB21 6AH, UK, in association with The Textile Institute.

81. Gulrajani, M.L., (ed), (2013), Advances in the Dyeing and Finishing of Technical Textiles, Woodhead Publishing Limited, Cambridge CB21 6AH, UK in association with The Textile Institute.

82. Johnson, N.A.G., and Russell, I.M., (ed) (2009), Advances in wool Technology, Woodhead Publishing Limited, Cambridge CB21 6AH, UK in association with The Textile Institute.

83. Robert, R.F., (2005), Bast and other Plant Fibres, Woodhead Publishing Limited, Cambridge CB21 6AH, UK in association with The Textile Institute.

84. Murugesh, B.K., (2013), Silk: Processing, Properties, and Applications, Woodhead Publishing Limited, Cambridge CB21 6AH, UK in association with The Textile Institute.

85. Robert, R.F., (2001), Silk, Mohair, Cashmere and other Luxury Fibres, Woodhead Publishing Limited, Cambridge CB21 6AH, UK in association with The Textile Institute.

86. Simpson, W.S., and Crawshaw, G.H.. (ed), (2002), Wool: Science and technology; Woodhead Publishing Limited, Cambridge CB21 6AH, UK in association with The Textile Institute.

87. Sandoz, Wool - A Sandoz Manual, Sandoz, Basle.

88. CIBA, Wool Dyeing, CIBA Limited, Basle, Switzerland, 1955.

Index

A

Absorbency 19, 104, 116, 119, 124, 131, 142, 145, 155, 157, 161, 167, 173, 178, 182, 205, 243, 332, 412, 416, 418, 419, 420, 431
Acetate fibres 368
Acid hydrolysis 68
Acrylic fibres 336, 345, 346, 347, 348, 350, 368
Action of acids 74, 278
  on wool 281
Activators 46, 167, 191, 414, 415
Air pollution 178, 304
Alkali cellulose 25, 78, 85
Amino acid 293, 300, 302, 430
Ammonia mercerisation 77, 112
  advantages of 115
  disadvantages of 116
Amorphous region 245, 413, 435
Amphoteric 131, 132, 136, 293, 305, 357
Amphoteric group 126
Amphoteric surfactant 136, 137
Amylase 24, 43, 44, 45, 46, 47, 248, 410, 411, 412, 414, 417, 419
Amylopectin 22, 23, 24, 45, 46
Ash content 6, 7, 8, 58, 70, 140, 169, 182
AOX 188, 429
Available chlorine 163, 164, 165, 166, 168, 231, 232, 233, 235, 238, 239, 240, 242
Available chlorine, effect of concentration of 165

B

Barium activity number 109, 243
Bast fibres 166, 167, 176, 218, 220, 274, 431
Bio-degradation 437

Bio-polishing 417, 418, 420, 431, 433, 434
Bio-technology 407
Bleaching agents 161, 173, 181, 182, 207, 444
Bleaching of coloured woven goods 166
Bleaching powder 161, 163, 164, 165, 166
Brushing 14, 280, 316
BOD 151, 152, 409, 439
Boiling off 113, 146, 147, 153, 154, 155, 212, 213, 270, 408, 429
Builders 140
Burrs 418

C

Calcium hypochlorite 166, 232, 238
  formation of 238
Causticising 257
Causticisation 122, 254
Catalytic damage 123, 144, 145, 146, 181, 212
Cationic surfactants 26, 134, 136
Cellulase treatment 418
Cellulose 4, 5, 21, 25, 36, 178, 264, 323, 364, 389, 416
Cellulose I 78, 81, 83, 85, 88,
Cellulose II 88
  chemical structure 25
Chain mercerizing 96
Chain mercerising machine 96, 97, 98
  Disadvantages of 98
Chelating agents 8, 137, 138, 223, 298, 419, 422
Chlorite bleach 271, 320
Chlorite bleaching 167, 169, 173, 191, 320
CMC 21, 25-27, 29, 36, 37, 38, 40, 41, 43, 47, 48, 53, 60, 127, 128, 129, 130, 323, 364, 435